# Forensic Engineering

## The Art and Craft of a Failure Detective

# Forensic Engineering
## The Art and Craft of a Failure Detective

Colin R. Gagg

CRC Press
Taylor & Francis Group
Boca Raton  London  New York

CRC Press is an imprint of the
Taylor & Francis Group, an **informa** business

CRC Press
Taylor & Francis Group
6000 Broken Sound Parkway NW, Suite 300
Boca Raton, FL 33487-2742

© 2020 by Taylor & Francis Group, LLC
CRC Press is an imprint of Taylor & Francis Group, an Informa business

No claim to original U.S. Government works

International Standard Book Number-13: 978-0-367-25168-0 (Hardback)

**Visit the Taylor & Francis Web site at**
**http://www.taylorandfrancis.com**

**and the CRC Press Web site at**
**http://www.crcpress.com**

*Dedicated to Ken Reynolds (1932–2010):*
*colleague, mentor and great friend.*

# Contents

## 2   Initial Aspects of Forensic Failure Investigation          51

## 3   A Framework or Methodology for Forensic Investigation       79

## 10    Component Failure in Road Traffic Incidents and Accidents                                            317

## 11    Fraudulent Insurance Claims                            343

**Proviso: for space restrictions alone, all case study input throughout this book is heavily abridged. Supporting details have been omitted except where they directly relate to the findings.**

The content of this book will not present a rigid guideline for the investigator to faithfully follow, simply as no single set of rules could encompass every possible failure scenario. Consequently, this book will develop a logical generic approach to failure analysis that can be adopted or adapted to suit the needs of most failure investigation. Therefore, it is hoped that every book purchased will become ragged and creased from constant reference and use, rather than just sitting in decorative inactivity on a bookshelf.

# Preface

Forensic methods have improved dramatically in recent times, increasing the chances of catching the lawbreaker, enhancing the understanding of premature demise and resolving disputes. It is common knowledge that forensic science has enabled many old, cold cases to be solved, particularly unsolved murders committed years ago, provided the evidence had been preserved at the time, and is now available for modern analysis. However, there has been similar (although less well-known) progress in the field of forensic engineering. Forensic engineering is the subject which deals with accidents, disasters and product failure of all kinds. The application of modern analytical techniques has, likewise, shed light on tragedies from different ages of technology – the Tay and Dee Bridge catastrophes of Victorian Britain, for example. The re-examination of evidence from past railway accidents is a further arena of investigation that has shown to benefit from the application of modern investigative methods and techniques. A recent examination of historic evidence revealed the previously unknown nature of a failure which derailed an entire train, causing 34 deaths among the passengers. The failure in question was found to have been initiated by the mechanism of metal fatigue. Metal fatigue was subsequently found to have been a significant mode of failure in many Victorian disasters. Yet, fatigue as a failure mechanism went unrecognised by engineers and designers of the time, with failures continuing without respite. However, as a word of caution, despite the advance of understanding in the 21st century, the problem of fatigue continues down to the present, and is found across all engineering fields.

Current forensic investigation will cast the engineer into the role of the detective, using a range of contemporary analytical methods for investigation. He/she will apply those techniques in ways that avoid flawed reasoning, while delivering convincing reconstruction scenarios. The aim of a forensic investigation is to identify the 'root cause' of failure, and thus provide valuable information that may prevent similar failure in the future. As everything physical is made from a material of some type or form, the professional working in a forensic failure analysis arena can expect instruction over an incredibly wide breadth of application. Their brief may range from investigation of an air accident, through a railway disaster, to a domestic product failure.

The investigator is not simply confined to the laboratory or the field. The unwanted failure of a complete structure or structural component can lead to injury, loss of life, loss of money, loss of company reputation or an environmental disaster. As such, the forensic engineer may frequently find himself/herself involved in the legal process, acting as an expert witness in both civil and criminal cases. Furthermore, he/she may be called to act as a single joint expert, operating as an arbiter in disputes where all parties agree to abide by his/her final decision. It is these particular legal aspects that differentiate a forensic analysis from a failure analysis. Failure investigation '*per se*' tends to be restricted in that the centre of focus is on material selection, design, product use/abuse, manufacture and failure mechanics within the part itself. Subsequent reporting will therefore present a somewhat narrow outlook, not fully exploring causes of failure or identifying potential ramifications that may ensue. Forensic engineering is therefore the application of the art and science of engineering, undertaken within the science or philosophy of law.

Requiring the services of legally qualified professional engineers, forensic engineering usually includes investigation of the physical causes of accidents and other sources of claims or litigation. The forensic engineering practitioner will be responsible for the preparation of engineering reports, testimony at hearings and trials in administrative or judicial proceedings and submitting advisory opinions to assist the resolution of disputes affecting life or property.

There exists a growing arena of forensic investigation that would appear to fall outside the conventional notion of failure analysis. The arena in question is that of intellectual property rights (IP). As businesses increasingly seek competitive advantage, reliance on intellectual property protections, such as copyrights, patents, design rights and trade secrets, is increasing. However, companies whose intellectual property is compromised often find subsequent damage significant, both in terms of lost sales and market share. Litigation involving infringement and misappropriation of intellectual property raises a unique set of issues. It is in this context that the forensic engineer can facilitate clients, companies and counsel in unscrambling each aspect of any IP dispute throughout each stage of the litigation process.

## Summary

From the brief preceding introduction into the art and science of forensic engineering, it is safe to suggest that the reader will gain an understanding of the breadth and depth of knowledge and experience demanded of the practitioner. This book will reference the work of many scientists, engineers and legal practitioners working in the arena. Their combined body of work represents a backbone of forensic engineering analysis. Furthermore, I hope to stand amongst such eminent colleagues and raise the profile of forensic failure investigation, analysis and dissemination of knowledge. The content of this book will not concentrate on an in-depth presentation of a particular academic theory or subject-specific science; such knowledge is contained in perfectly adequate textbooks. The regurgitation of such information is therefore not considered to be of necessity for the purposes of this publication. However, the exception to this rule is presented in Chapter 5, where basic sources of stress and service failure mechanisms of engineered components and structures are reviewed. This exception is considered a necessity, as the way in which engineered products fail in service can often be directly related to operational conditions at play, both prior to, and at the instant of, demise. Therefore an understanding of the source or sources of service stress is a paramount pre-requisite for the failure investigator. Furthermore, there is an additional number of scientific techniques, basic investigative methods and background concepts or knowledge that are frequently overlooked by textbooks and in the classroom. It is these understated, often ignored and somewhat intangible assets that are essential tools for the investigator and which, in the majority of cases, are best learned by practice. The art and craft of forensic engineering is therefore an applied activity, rather than being a purely theoretical pursuit. As such, the bulk of this book will be dedicated to presenting the *applied* engineering background knowledge required of the investigator, or failure detective, along with a range of practical investigative concepts. Moreover, and perhaps more importantly, a review of those concepts in practice will be brought clearly into focus. Knowledge and experience gained from practice will be referenced throughout. Furthermore, in an attempt to both reinforce the subject of forensic engineering, and to bring it to life, the use of appropriate historic and current case study analysis will be found throughout the text. Finally, legal aspects and professional ethics demanded of the practitioner will be highlighted as and where appropriate.

# Acknowledgements

First and foremost, my thoughts turn to Mr Ken Reynolds, who was both a mentor and friend. Ken first introduced me to the world of forensic engineering some 30 years ago. In the intervening years, we collaborated on numerous case files, ranging from murder investigation to simple ladder accidents. It was his foresight and faith in my ability that has given me one of the loves of my life, that of forensic engineering investigation. For this, I am eternally grateful.

Particular thanks must go to the numerous manufacturers, insurance companies, loss adjusters, barristers and lawyers. Instruction received from these professional agents has been varied, stimulating and at times a challenge. Many of the case studies used to illustrate this book have derived directly from instruction received from them. My thanks are extended to each and every one for the confidence they expressed in my ability to investigate their particular failure occurrence. I would also like to extend my appreciation to fellow experts with whom I have collaborated over the passing years, all of whom have been open to both argument and (some very lively) discussion.

Having worked side-by-side over many years with Mr Paul Fagan, an internationally recognised fire investigation expert, his quiet demeanour coupled with a dogged determination to uncover the truth in human affairs has been an inspiration. Many thanks, Paul, I have enjoyed our professional journey together, and you will no doubt recognise a number of the case studies presented within the text body of this book.

Dr Peter Lewis, of The Open University, is both a work colleague and friend. We have collaborated on numerous research papers, two textbooks and as contributing authors for a post-graduate course on forensic engineering, T839 Forensic Engineering, presented as part of the post-graduate engineering offering at The Open University, UK. Peter has been a source of continued encouragement, patiently persuading me to continue the publication of my investigations, as a contribution to the growing body of knowledge. Again, my unreserved thanks are offered to Peter.

Thanks must be extended to the Royal Academy of Engineering (UK) and The Open University for their financial support for continued participation at the International Conference of Engineering Failure Analysis (ICEFA) series; The Open University must also be thanked for permission to use case study material that I first developed for the post-graduate course, T839 Forensic Engineering.

Special thanks must go to both Dr D.R.H. (Dai) Jones of Cambridge University, and Dr Richard Clegg, Editor-in-Chief, *Engineering Failure Analysis* for their unfailing encouragement to publish engineering failure case studies in the journal *Engineering Failure Analysis*, and for their continued support, and confidence in me.

Final thanks must be reserved for my family and friends. Their continued support, patience and understanding over the past years have been unshakable; I can clearly remember hearing their excitement (or was that groans from my wife?) when I informed them of my intention to write another book.

When undertaking practice in any chosen vocation, development as a professional will continue throughout the individual's working lifetime. Therefore, the philosophy for my growth, ongoing training and experience as a forensic engineer should now be clear: it will continue year on year. It follows logically that as a practicing professional, I am as much a work in progress as is the subject area. Therefore, my writing of this book will be open-ended, giving both confidence and credibility to the quotation:

> This book has a beginning but not an ending; it is, like the writer, a 'work in progress'.

**Dr Colin R Gagg**

# Author

Dr Colin Gagg earned an honours degree in engineering technology (BA Hons), a master's degree in management and technology of manufacturing (MSc) and a doctorate in forensic engineering (PhD). He is a Chartered Engineer (CEng) and Fellow of the Institution of Mechanical Engineering (FIMechE). His practical experience includes 2 years at the Structures Laboratories of the Imperial College of Science, Technology and Medicine, 4 years at the Engineering Department of the University of Toronto and some 30 years at The Open University. He held the post of research projects officer at The Open University and, for 15 years, was a member of the Forensic Engineering and Materials Group. Having now retired from academia, he continues to pursue the field of forensic engineering by acting as an independent consultant.

Dr Gagg's prior research interests focused on advanced metal processing techniques such as Ospray processing, thixo-casting and melt spinning, from which a long-standing avenue of interest developed, that of tailoring materials for the use of engineering in medicine. His research interests have focused on two distinct arenas: solder studies and forensic engineering. Solder assessment focuses on creep, creep rupture and monotonic characteristics of lead and lead-free solder systems. A principal aim of the work is to ensure that stress analysis and modeling for life prediction employ appropriate and reliable data. In addition to research publications, results generated have been fed directly into mainstream course production and into a specifically tailored course for the Taiwanese market.

His second arena of interest is forensic engineering, an area of research and practice that was fed directly into the teaching stream as case study input. He was a contributing author for a postgraduate forensic engineering course (T839) at The Open University. He maintained his enthusiasm for teaching the subject by tutoring 20 to 30 postgraduate students per year over a 10-year period. During that time frame, he served as a member of the examination board for the course, and was also an examiner for MSc dissertations in the manufacturing program at The Open University.

Dr Gagg has published 40 academic papers in peer reviewed learned journals, with the body of work being credited with 515 citations, an RG score of 19.6 and an 'h' index ranking of 13 – thereby contributing to the knowledge base. He has co-authored two prior textbooks: *Forensic Engineering – Case Studies*, published by CRC Press, Boca Raton, FL, ISBN: 0-8493-1182-9, 2003 and *Forensic Polymer Engineering: Why Polymer Products Fail in Service*, published by Woodhead, Oxford, UK, ISBN 1-84569-185-7, 2010.

Dr Gagg co-chaired both the 4th International Conference on Engineering Failure Analysis (ICEFA-IV), held at Churchill College, Cambridge, UK, in 2010, and the 5th International Conference on Engineering Failure Analysis (ICEFA-V) held at The Hague, the Netherlands, 2012. He has served as a member of the Editorial Advisory Board for the journals *Engineering Failure Analysis* and *International Journal of Forensic Engineering*.

Dr Gagg has accepted instruction in over 880 independent forensic investigations, in addition to a further 200 joint instructions dealing with the failure of both metallic and non-metallic products. Clients include major insurance companies, solicitors and barristers, the UK police force, ambulance authorities, the health service, major motor manufacturers and assessors, major retailers and private individuals. He has acted as a single joint expert in product failure difficulties, personal injury disputes and (equipment) failure during surgical procedures. He has appeared as an expert witness in court proceedings related to both personal and fatal injury.

# Failure Analysis or Forensic Engineering?*

1

## 1.0 Synopsis

Lessons derived from all forms of structural and equipment failure have been learned since the dawn of modern civilisation. Techniques and tools for investigating failure have been honed over time, by both scientists and practising engineers. However, until quite recently, failure investigation was restricted in that focus centre on material selection, design, product use/abuse, manufacture and failure mechanics within the part itself. Subsequent reporting therefore presented a somewhat narrow outlook, not fully exploring causes of failure or identifying potential problems that could follow. In more recent times the analytical discipline of forensic engineering has evolved, demanding a systematic approach to failure that will include any and all managerial and legal aspects surrounding an incident. The process will encompass and emphasise the whole system, rather than simply concentrating on the failure of a specific part. Service failure of any part or system will be accompanied by a clear potential for personal injury, monetary loss and/or damage to property. As such, forensic investigation and analysis are performed under an overarching umbrella of the legal system, with the consequences of failure being judged under the law of product liability. As a subject, forensic engineering is most commonly applied in civil law cases. However, its profile is also being raised in the arena of criminal law. In general, the objective of any forensic engineering investigation is to determine cause or causes of failure with a view to improve the performance or life of a component and/ or to assist a court in establishing the facts of an event. The subject will also embrace the investigation of intellectual property claims, particularly patent disputes. Case study analysis has been identified as a prime route for promulgating failures in an attempt to limit (if not eliminate) further re-occurrence of identical failures. Furthermore, case study analysis has been accredited as a vehicle to inform a new generation of engineers of the downsides to new product development, and the serious consequences of product failure in the marketplace. Direct practical experience of failed products shows that

---

* Certain brief passages in this chapter are incorporated, with permission, and used verbatim from the Elsevier book – *Forensic Polymer Engineering: Why Polymer Products Fail in Service, Second Edition* (2010) – co-authored by the author and cited within the References section at the end of this chapter.

there is still much room for progress in preventative engineering, and raising awareness among product designers and manufacturing personnel of the need to examine and analyse failures utilising appropriate tools which have been developed to a sophisticated level over the past decade.

## 1.1 Historic Failure Analysis

From classical temples to 21st-century corporate towers, engineers have learned more about safe and reliable design from past *failures* than from any perceived success.[1] Failure engineering is the investigation of any unacceptable difference between expected and observed performance.[2] As a subject, it is a recognised and researched arena in its own right, being of interest to both the academic and engineering communities.[3,4] Prevention of failure is a key underlying principle for all engineering disciplines, with the failure analyst normally taking a lead role in any subsequent investigation. Some of the best-known examples of failure analysis have been associated with high-profile disasters.[5,6] However, when *any* structure, product or process fails prematurely, the question will be raised as to 'why', particularly if failure causes injury or death. From the far reaches of time, there has been a history of penalties for 'failure', with blame for any mistake or loss being firmly apportioned. Historic law includes:

1. The King of Babylonia's 'Code of Hammurabi' in 2200 BC (Figure 1.1):

The code contains a section covering personal injury, indicating that penalties were imposed for injuries sustained through unsuccessful operations by doctors, and for damages caused by neglect in various trades. Rates are

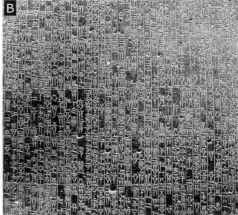

**Figure 1.1** (A) The Code of Hammurabi, 2200 BC. (B) Closer view of inscription. (Courtesy of Wikimedia Commons, licensed under Creative Commons.)

fixed in the code for various forms of service in most branches of trade and commerce,[7] e.g. for builders whose structures fall down – if the owner of the building dies, the builder shall be put to death. If the owner's son is killed, the builder's son shall be put to death.

2. The 15th-Century Common Law in England:

If a carpenter undertakes to build a house and does it ill (not well), an action will lie against him.[8]

3. The Napoleonic Code of 1804:

Stated that if there is a loss in serviceability in a constructed project within ten years of its completion because of a foundation failure or from poor workmanship, the contractor and architect will be sent to prison.[9]

With any law or decree surrounding failure, care must be exercised to ensure that a culture of 'blame' does not obscure the main priority of failure investigation.[10,11]

July 2003, six railway engineers – among them chartered professionals – stood in the dock at Hertfordshire Magistrates Court (UK) to face manslaughter charges levelled against them for their role in the Hatfield rail crash of October 2000. That they were there simply highlights the absurdity of society's ever-increasing desire 'to see someone pay'. It highlights the growing trend by the public and the media to look for someone to blame and make an example of them.[12]

When analysing a failure, there is an implicit need to understand how a component or system works in its present configuration and the most likely way it will fail. This is a key requirement if reliability is to be improved.[13] However, true understanding of any failure can be completely biased by a 'blame culture' search for the cause or causes of failure. Human nature being what it is, the individual will often look for someone else to blame for misfortunes of their own making, thus avoiding the embarrassment of having to admit to a mistake. If this situation prevails, it will lead to poor investigation of an incident or accident, with only the basic cause being identified and not the underlying cause or causes.[14]

The European Community, at an International Civil Aviation Organisation meeting,[15] in describing 'The need for a Just Culture', reported:

'Lack of full and open reporting continues to pose a considerable barrier to further safety progress in many areas. Major impediments are a fear of prosecution and a lack of appropriate confidentiality. The effectiveness of reporting is totally

**Figure 1.2** Boeing 307 NC19903 crashed into Elliott Bay near Seattle, 28 March 2002. The U.S. Coast Guard responded to rescue the victims. Probable cause: 'Loss of all engine power due to fuel exhaustion that resulted from the flight crew's failure to accurately determine onboard fuel during the pre-flight inspection. A factor contributing to the accident was a lack of adequate crew communication regarding the fuel status.' (Credit: United States Coast Guard, PA2 Sarah Foster-Snell, public domain.)

> dependent on a conducive reporting environment – a Just Culture – defined as a culture in which front line operators are not punished for actions or decisions that are commensurate with their experience and training, but also a culture in which violations and wilful destructive acts by front line operators or others are not tolerated (Figure 1.2).'

Blame culture reporting, of any failure in question, will not elucidate the true cause or causes of demise. It follows that any proposed remedial action determined by biased investigation or reporting may lead to the implementation of fruitless or inefficient corrective measures.

## 1.2 Conventional Failure Analysis

'Eventually, all systems fail.'

Although at first sight this appears to be an unassuming sentence, it encapsulates the reason why extensive effort may be expended to determine the life span of an engineered system. It is a simple fact that all engineered systems will fail at some point. What is often unknown is exactly *when* the system will fail. For safety-critical systems, people will be ill at ease in using such systems without an estimate of their time to failure. Hence, manufacturers specify the life of their product or system in terms such as hours, or operating cycles.[16] When these lives are reached, action is required ranging from repair or refurbishment to total replacement.

One of the first concerns of the design engineer is to bring a product to market having designed out or mitigated any potential failure modes prior to production release. A prime objective will be to ascertain any harsh operating conditions that a product will be exposed to when in service.[17] There are systematic approaches to the design process – such as 'design for reliability' – that are sharply focused on reliability and firmly based on the physics and chemistry of failure.[18] Design approaches of this nature impart an understanding of how, why and when to use the wide variety of reliability engineering tools available and offer fundamental insight into the total design cycle. The product or structure can then be designed to minimise potential hazards associated with any worst-case scenario. However, to establish a worst-case state requires a complete understanding of the product, its loading and its service environment.[19,20] Unfortunately, obtaining a complete understanding of any product service environment can be an extraordinarily difficult exercise. For example, unexpected temperature excursions may be encountered; unanticipated vibrations develop; and the product is used, abused or serviced in unintended ways.[20–22] With any unanticipated factor, mechanical failure may well be a resultant outcome, due to a diverse range of (unforeseen) reasons. More often than not, such failures frequently involve loss of life, personal injury and/or significant financial losses.[23–25] Rather than being the result of an accident or act of God, failure can clearly be defined as the results of human error.[26,27] Such human error may originate from oversight, carelessness, ignorance and/or greed.[28] Take for example:

## Loss of the Space Shuttle *Challenger* – January 1986

Loss of the Space Shuttle *Challenger* was inevitable once the O-rings on one of the booster rockets failed. The booster propellant exploded though the O-ring gap, engulfing the entire structure shortly after take-off. The Viton rubber O-ring (a fluorinated copolymer) was very inflexible at the near-freezing temperatures at lift-off, and could not correctly and efficiently seal the booster sections. The Viton material limitation was well-known, but managerial staff chose to ignore the warnings.

## Mihama, Japan – 9 August 2004

A steam eruption occurred from failed piping system in a pressurised water reactor (PWR). Here, six people were killed and five injured. An orifice plate was inadvertently inserted to the pipe network to measure the coolant flow rate. This caused a localised metal loss (thinning) in the cross-sectional area of the piping, year-on-year. Accordingly, stress levels became higher, plastic collapse levels were reached and the pipe segment ruptured. Unfortunately, the pipe thickness was not checked for 27 years due to 'oversights'. The rupture is expected to cost Kansai Electric more than 10 billion yen (approximately $71 million). Safety upgrades are due to be completed by March 2020 and will cost about 165 billion yen ($1.51 billion).

## San Bruno, CA – Pipeline Explosion 2010

On 9 September 2010, a 30-inch diameter steel natural gas transmission pipeline ruptured in a residential area in San Bruno, California. The rupture caused an estimated 47.6 million standard cubic feet of natural gas to be released. The discharged gas ignited, resulting in a fire that destroyed 38 homes and damaged 70 others. Eight people were killed, numerous individuals were injured, and many more were evacuated from the area. The National Transportation Safety Board said that a segment of pipe 28 feet (8.5 m) long blew out onto the street, landing about 100 feet (30 m) away. The blast crater formed was 167 feet (51 m) long and 26 feet (7.9 m) wide. Inspection of the severed pipe chunk (Figure 1.3A) revealed that it was fabricated from several smaller sections that had been welded together, with a seam running its length. In January 2011, federal investigators reported the discovery of numerous defective welds in the pipeline. In addition to pipe thickness variation, it was found that longitudinal seam welds had been fabricated using a double submerged arc welded (DSAW) process that had deposited weld metal along the outer portion of the seam, but left an un-welded region along the inner portion of the seam. (Figure 1.3B). The process had, in effect, 'manufactured-in' a latent crack defect. An exemplar fabrication weld is shown in Figure 1.3C, where it can be seen that the outer and inner weld passes overlapped, resulting in full penetration in the joint, with each weld exhibiting a raised weld reinforcement with a smooth surface. As the gas company increased the pressure in the pipes to meet growing energy demand, the defective welds were further weakened until their failure.[29] As the pipeline was installed in 1956, modern testing methods such as X-rays were not able to detect the problem. The NTSB held a three-day public hearing in March 2011, to gather additional facts for the on-going investigation of the pipeline rupture and explosion. The U.S. government passed a bill in December 2011 that doubled the maximum fine for pipeline safety violations, but ignored several key recommendations arising from past investigations of deadly natural gas explosions and high-profile oil spills. The final outcome of this disaster investigation is still undecided at the time of writing. However, on 13 March 2012, Pacific Gas & Electric agreed to pay the Californian city of San Bruno $70 million as compensation for the gas pipeline explosion. In April 2015, the Public Utilities Commission fined PG&E $1.6 billion. On 21 January 2017,

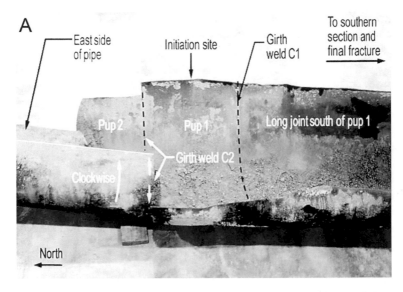

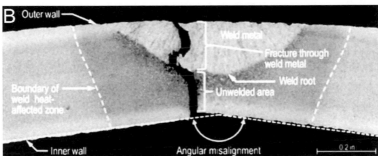

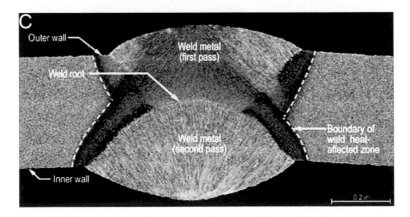

**Figure 1.3** (A) A section of the San Bruno pipeline with white arrows showing the direction of crack propagation. (B) Cross section of the double submerged arc welded (DSAW) longitudinal seam. (C) A typical cross section of a longitudinal seam in an exemplar DSAW steel pipe fabrication. (From "Pacific Gas and Electric Company Natural Gas Transmission Pipeline Rupture and Fire", NTSB Report PAR-11-01.)

PG&E was fined $3 million and ordered to perform 10,000 hours of community service for criminal actions of violating the Natural Gas Pipeline Safety Act and for obstruction of justice.

### Genoa, Italy – Morandi Bridge Collapse 2018

The Morandi Bridge, in Italy, carried the A10 toll motorway, linking northern Italy to France. On 14 August 2018, a large section of the bridge collapsed during heavy rainfall. Dozens of vehicles to plunged 45 m to the ground below, with the tragedy culminating in the loss of 43 lives. To date (December 2018), as investigations continue into the reasons for the collapse, it appears that initial structural issues, reduced maintenance spending, extreme weather conditions and traffic use far beyond expectations will all have played a part in the bridge demise. Without doubt, one outcome of the failure will be renewed inspection of bridges, both in Italy and further afield. Italian road spending will, in all probability, increase in the short term. However, without doubt, legal action will follow in the event of any proven negligence by any party involved.

With any failure investigation, there can be a tendency or temptation to focus on the component or fracture in isolation. There may be little or no account taken of all circumstances at play, or of unfolding events surrounding failure.[30,31] Even in an authoritative text on failure analysis,[32] a full account of encompassing events, circumstances and potential repercussions is not fully considered. Failure cases are presented that are too narrow in scope, and simply consider the failed component. After having taken such a 'focused' view of failure, it follows that the notion of limited outcomes must also hold true when considering any mode of failure at play. All too frequently, an investigation will focus on the mode of failure in isolation; 'it failed by fatigue' will often be the cry.[33,34] There may be little or no effort to review or determine additional structural, environmental or operating parameters that may have been at play. The presence of other physical events may well conspire to sensitise the system to the onset of failure by fatigue.[35,36] For example, additional mechanisms at play were contributary to the unexpected demise of yet another natural gas pipeline, that failed with catastrophic results:

On 30 July 2004, the rupture of a major underground high-pressure natural gas pipeline in Ghislenghien, Belgium resulted in 24 deaths and over 120 injuries[37]. It is postulated that the failure initiated with a through wall defect. Heat transfer effects of the fluid escaping through the defect would induce significant cooling of the surrounding pipe-wall, with the localised temperature dropping below the ductile–brittle transition temperature of the pipe material. The resulting significant reduction in the fracture toughness coupled with the accompanying thermal and pressure stresses then resulted in the transition of the initial defect into a running fracture and hence catastrophic pipeline failure. This failure highlights the importance of taking into account any expansion-induced cooling effects as a credible failure scenario when undertaking safety assessment of pressurised pipelines.

## 1.3 Product Defects

Both design and manufacturing engineers generally operate rigorous 'quality control' procedures at every stage of design and manufacture. It is therefore unusual for faulty products to enter service. However, on occasion, faulty goods do manage to avoid the most rigorous of checks and enter service with an inherent deficiency or defect.[38] A defect is simply an imperfection that renders a product unsafe for its intended use and is usually introduced at either the design stage or at the point of manufacture.[39] A design defect exists when a whole class of products are inadequately made, and often poses unreasonable risks to consumers. A car manufacturer's design of a vehicle with the fuel tank positioned so that it explodes in low-speed collisions is defective, for example.[40] When the design is defective, even products perfectly manufactured will be defective. On the other hand, a production or manufacturing defect will arise when a sound design solution is not followed and, subsequently, the product is improperly manufactured. Such 'manufactured-in' defects may be patently obvious, or of a latent nature, i.e.,

- 'Patent defects' are defects that are plainly visible or which can be discovered by reasonable inspection or customary tests – hence the phrase 'patently obvious'. Visible surface cracking or blow-holes are two examples of patent defects.
- 'Latent (or hidden) defects' are defects which are not plainly visible and which cannot be determined by reasonable inspection or customary tests, and which are unknown when the item is accepted. Internal voids or sub-surface cracking provide examples of a latent defect.[38]

An example of a latent defect that led to serious loss of life is clearly demonstrated by the San Juan, Puerto Rico gas explosion of 21 November 1996 (Figure 1.4). Here, 33 people lost their lives, with more than 80 others injured. To date a further six people are still unaccounted for, presumed crushed under the rubble. It transpired that intensified bending stresses led directly to the formation of brittle-like cracks initiating at the internal wall of the gas pipe. Clearly, a latent (hidden) defect directly led to catastrophic pipeline failure.

**Figure 1.4** Interior of polyethylene pipe from San Juan pipeline accident showing brittle-like crack with no visible deformation. (From "San Juan Gas Company, Inc./Enron Corp. Propane Gas Explosion", NTSB Report PAR-97-01.)

Consideration must also be given to 'consequential (or secondary) damage' that may be an indirect result of either a patent or latent defect. An example would be secondary (or after-the-event) damage to surrounding equipment or structure, inflicted by a flailing drive shaft. In this case, initial shaft failure was the traumatic failure event, with subsequent flailing damage – inflicted by the now liberated shaft – being secondary.

## 1.4  Causal Analysis

At the outset of any failure study, the investigator will intuitively gather data from a number of sources, such as direct observation, physical artefacts, documentation, interviews, etc. By using common-sense physical reasoning, he/she will build a range of possible failure scenarios (situations) and, wherever possible, corroborative evidence will be sought. A database will be constructed, so that all evidence may be available for subsequent review. Prior to analysis, the data may be manipulated in a number of ways.[41] Evidence may be placed in a matrix of categories, and graphical displays such as flow charts may be used. For example, accident investigators place witness evidence in a matrix, so that apparent inconsistencies may be elucidated, and flow charts have been advocated in accident investigation.[42]

Analytical techniques available to the failure investigator can be roughly divided into three types: those of qualitative analysis, semi-quantitative analysis and quantitative (or physical) analysis.[43]

## 1.4.1 Qualitative Analysis

Quantitative analysis techniques are more abstract, usually beginning with the investigator conceptualising probable trains of events or failure scenarios. These abstract 'data' are often stored within the mind of the investigator, but can be assembled pictorially for ease of understanding by others. Types of conceptual representation are

- *Affinity diagrams*: a tool used to organise thoughts and data. It allows large numbers of ideas to be sorted into groups for review and analysis. It has advantages when investigating complex failures or disasters.
- *Flow charts*: a pictorial representation describing a failure process under investigation. It is an ideal aid to the understanding of the interaction between various component parts of a system. Flow charts tend to provide a common language or reference point when dealing with a process.
- *Check sheets*: are simply a structured form, table or list, for identifying debris or data. It can be a vital aid when classifying accident or disaster debris for further analysis. It is a generic tool that can be adapted for a wide variety of analytical purposes.
- *Mind maps*: diagrams are used to represent words, ideas, tasks or other items linked to and arranged around a central failure. They are used to generate, visualise, structure and classify possible failure scenarios.

There is no single, universal 'right' or 'best' qualitative representation. From the above list, there exists a range of choices, each with its own advantages and disadvantages. What each and every one of them has in common is that they provide notations for describing and displaying individual thought processes and reasoning around a failure event.

## 1.4.2 Semi-Quantitative Analysis

Semi-quantitative analysis (SQA) is employed to identify investigative priorities and identify failure or operational features that require further study and analysis. Techniques can include

- Fault tree analysis (FTA) is one of the most widely used methods in failure or system reliability analysis. It is a deductive procedure for determining the various combinations of both hardware *and* software failure, and human errors that could result in the occurrence of failure (the top event). A deductive analysis will begin with a general (failure) condition, and then attempt to determine specific causes of failure. This is a 'top-down' approach to failure analysis.

- Cause-and-effect diagrams (CED) – also known as Fishbone diagrams or Ishikawa diagrams – are a tool for breaking down potential causes of failure into more detailed categories so they can be organised and related into factors that help identify the root cause of failure. In addition, a cause and effect diagram is considered one of the seven basic tools of quality management.
- Failure modes and effects analysis (FMEA) is a method for reliability analysis used to identify potential failure modes that would exert significant influence on the demise of a system. In essence, FMEA is the deliberation of the 'criticality' and 'probability of occurrence' of potential failure modes. It is now widely used in manufacturing industries throughout various phases of the product life cycle.
- Bathtub (reliability) graphing. The bathtub curve describes the failure rate of a product, plotted against that product lifetime. The trend will show an early, high mortality (infant mortality), followed by a stable mid-life region of random failures and finally a rise in failure numbers as critical components wear out.
- Pareto bar charting. Analysis used to differentiate between the vital few and the trivial many. Also known as the 80–20 rule, for many events, roughly 80% of the effects come from 20% of the causes.

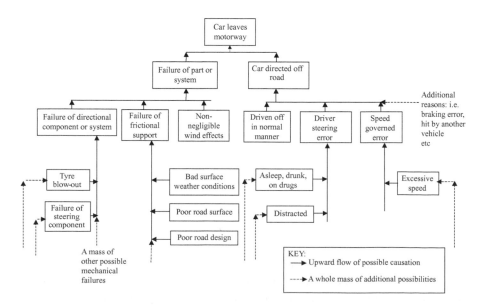

**Figure 1.5** A simple illustration of a fault tree analysis (FTA) of reasons why a car left a motorway.

Semi-quantitative methods of event analysis are simply aids for evaluation of the engineered system or failure under study, where numerical values are tentatively incorporated.[44] There are many qualitative methods and statistical tools available to perform failure analyses over and above those listed here.[45] What is of importance is that the investigator can choose (or mix and match) the most appropriate qualitative and semi-quantitative analysis method that best suits him or her, and that is relevant to the issue at hand. An example of FTA event analysis is shown in Figure 1.5.

### 1.4.3 Computer-Aided Causal Analysis

Both qualitative and semi-quantitative techniques are methodologies designed to identify potential failure modes for a product or process, to assess the risk associated with each mode and to rank the issues in terms of the most serious concerns. The development of any technique requires time, expertise and often (as with large complex failures) the involvement of multiple team members. However, there are a number of commercially available computer-aided analysis packages that will help to lower the (analytical) workload. Such an approach will not only provide quick and reliable results but will also facilitate validation and repeatability of the analysis.

As an example, one particular off-the-shelf package contains three modules:

- *Fault tree analysis*: facilitating both the construction and analysis of fault tree diagrams
- *Event tree analysis*: allowing analysis of possible outcomes of an event occurring
- *Markov analysis*: enabling the construction of Markov models for failure analysis of components with large interdependencies

After possible site inspection and gathering relevant information, there is a range of available methods (or tools) for analysing failure debris. Use or implementation of such techniques will enable the investigator to glean information from a failed artefact. Available tools may range from a simple eyeglass and macro-photography on the one hand, through to non-destructive testing, hardness testing, mechanical testing, microscopy (both optical and electron), chemical, thermal and microanalysis on the other. Initially, any analysis should be approached using non-destructive methods, leaving any destructive methods as a last resort, and then, only with the specific consent of all parties involved. The aim of any failure investigation is to determine the 'root cause' or causes of failure, with each investigation being unique in its own right. An investigator is therefore required to have an implicit understanding of available analytical 'tools', along with an implicit understanding of any associated limitations and compromise associated with their use. Such an understanding will allow more informed decisions regarding the most appropriate technique(s) and tools to use during the course of an investigation. A comprehensive review of limitations and compromise associated with each of the more popular investigative tools will be presented in Chapter 4.

### 1.4.5  Case Study: Failure of a 4.5-inch Kitchen Knife

A straightforward case concerned the failure of a 4.5-inch blade utility kitchen knife (Figure 1.6A). At the instant of failure, the knife was being used to cut a pineapple, when the blade snapped. As a result of the unexpected failure, the user sustained severe injury to the hand/wrist area.

Simple measurement determined that the blade section thickness was equal to or better than the specified design value of 1.8 mm, with measured cutting surface hardness being 58–60 HRc – this value being higher than the specified values of 52–53 HRc. Visual observation under low-power magnification revealed evidence of primary processing flaws such as porosity (Figure 1.6B), surface cracking (Figure 1.6C), inclusions, laps and seams (Figure 1.6D), along with blow-holes (Figure 1.6E). Furthermore, the process of final grinding/polishing appeared to be exceptionally coarse. This level of abrasive grinding would constitute a secondary processing or finishing flaw. Certainly,

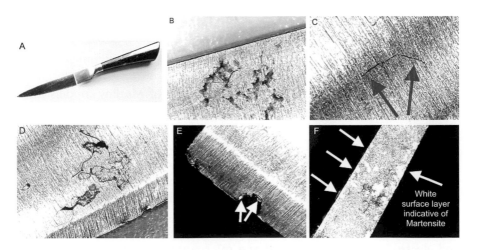

**Figure 1.6** (A) The subject kitchen knife. (B) Primary materials and processing defects include porosity, (C) surface cracking, (D) inclusions and laps, (E) blow holes (F) martensitic surface layer.

heat generation and inadequate cooling from this grinding process had introduced a (brittle) surface martensitic layer (Figure 1.6E), thus providing an explanation for the elevated hardness values found at the heavily ground cutting surface.

It was evident that the failure initiated as a result a number of defects, both inherent within the material of manufacture (primary defects) and as a result of a secondary finishing flaw. Although the defects were latent in nature (in that their presence would be difficult to detect with simple visual observation alone), it would be expected that the issues noted should have been readily detected during the course of any robust quality control procedure on the factory floor. It could be clearly stated that the noted defects were 'manufactured in' at the initial fabrication stage, and therefore present at the time of purchase by the injured party. Therefore, it was considered prudent to advise the manufacturer of the need to undertake a review of their in-house quality control procedures, as a matter of urgency.

When considering the action of cutting a pineapple, there would most certainly be a degree of bending introduced into the product blade (attempting to 'lever' the cut open, for example). It could therefore be argued that the product blade had not been designed to withstand bending whilst in use – therefore failure was as a result of abuse by the user. However, it could equally be argued that any such action was 'foreseeable' and, as such, should be recognised in the design solution. Equally, it could also be argued that, had the product blade not been in a compromised condition by virtue of inherent processing defects, the product failure in question may well not have occurred.

Whatever analytical technique or combination of techniques used by the failure investigator, continual iteration and evaluation of evidence is essential. An investigation is not undertaken simply to provide scientific observation for further debate at a later date. The aim is a root cause finding and recommended corrective actions to prevent future failures, with the entire *perspective* of failure being critical. As any investigation progresses, it will demand continual evaluation of detail against the big picture and a sustained assessment of unfolding implications.

## 1.5 Design Calculations and Modelling

The process of failure analysis often requires review and analysis of initial design calculations and modelling.[46] At the outset of any design process, the use of first principle formulae and/or empirical equations are used to describe or predict the response of a system, component, material, etc. to applied stimuli. Mathematical modelling typically consists of three types of equations: balance equations, constitutive equations and constraint equations.[47] Analysis has traditionally assumed that basic properties such as elastic modulus (E) possessed a single value. However, be aware – reported values of Young's modulus for two common lead-bearing solders, for example, range between 5.7–40 GPa (Sn-37Pb) and 16–38.6 GPa (Sn-40Pb).[48] Furthermore, an assumption has to be made as to whether a material behaves perfectly elastically which may overestimate calculated stresses (subsequently resulting in a non-conservative prediction of life), or that it exhibits an ideal elastic–plastic response (which may have the opposite consequences). The choice of which assumption to incorporate in a numerical analysis will have a profound effect on output results. Therefore, when considering published values of materials properties in design calculations, it is important to consider the influence of parameters such as temperature, grain size, strain rate, strain hardening, etc. As an example, a popular empirical description of strain hardening is the Hollomon expression.

$$\sigma_t = k\varepsilon_t{}^n \qquad (1.1)$$

where

'n' is a strain-hardening exponent, and 'k' is a strength co-efficient. Table 1.1 lists the values of strain-hardening exponent 'n' and strength co-efficient 'k' for eutectic solder at three different temperatures and strain rates.

The substantial influence of temperature and strain rate on values of the exponent and co-efficient is clearly seen.[49] Consequently, the assumptions regarding material behaviour made in many stress analysis approaches are unlikely to produce accurate values of stresses and strains which, in turn, is a serious impediment to reliable life prediction. Much closer attention needs to be paid to material properties associated with conditions likely to be encountered in service.

## 1.5.1 Miner's Law for Life-Time Cumulative Damage Prediction

As an additional example, a fundamental method of life-time cumulative damage prediction was formulated by Miner. The damage from each cycle at

**Table 1.1 Strain Hardening Exponent ($n$) and Strength Coefficient ($k$) in the Expression $\sigma = ke^n$**

| Alloy | Temp. (°C) | Strain Hardening Exponent ($n$) | | | Constant ($k$) | | |
|---|---|---|---|---|---|---|---|
| | | Strain Rate (s$^{-1}$) | | | Strain Rate (s$^{-1}$) | | |
| | | $10^{-1}$ | $10^{-2}$ | $10^{-3}$ | $10^{-1}$ | $10^{-2}$ | $10^{-3}$ |
| Sn-37Pb | −10 | 0.93 | 0.43 | 0.32 | 208 | 96.7 | 92.5 |
| | 22 | 0.30 | 0.14 | 0.08 | 91.2 | 71.5 | 57 |
| | 75 | 0.21 | 0.12 | 0.048 | 40.5 | 29.8 | 15 |

*Source:* After Plumbridge and Gagg[49].

a certain stress level is defined as the reciprocal value of the number of cycles to fracture $(1/N_i)$. Fracture occurs when the sum of cycles at each level $(n_i)$ related to the number of cycles to failure $(N_i)$ is equal to unity. The mathematical (Miner's Law) expression is:

$$\sum \frac{n_i}{N_i} = 1 \qquad\qquad (1.2)$$

One limiting assumption here is that each loading cycle will drive the crack front forward. In numerous situations, this simply is not the case. For example, accumulated plastic deformations from thermal stress could weaken a material sufficiently so that fewer vibration cycles need be applied before the material fails. Assuming that damage mechanisms are not interactive can therefore lead to an overly optimistic fatigue life estimate. Conversely, plastic deformations from thermal stresses could create compressive residual stresses, thus retarding the rate of vibration fatigue damage accumulation. In this case, using Miner's hypothesis would lead to an overly pessimistic fatigue life estimate. Furthermore, actually quantifying service load cycles is not an easy matter, for example, load cycles experienced by a broken road wheel axle on a car, where road condition such as surface roughness and speed bumps, etc. would come into play. As with many other materials law, Miner's law should be regarded as a useful *approximation*, one that can be readily applied, but with accuracy limited to use in certain circumstances. However, damage accumulation in fatigue is usually a complicated mixture of several different mechanisms, such as corrosion, that influence failure. Therefore the assumption of linear damage accumulation, inherent in Miner's law, should be viewed sceptically. It generally ignores such effects, often failing to encapsulate the essence of the fatigue process. This shortcoming may not be appreciated, or simply ignored during the course of failure assessment.

## 1.6 Summation of Conventional Failure Analysis

Identification of the cause or causes of physical failure, of a component or system, is only the first element of a total investigation process. However, it is often this first step that will set the direction of investigation. A vital part of any enquiry is to determine a root cause of what went wrong, or what created the conditions for physical failure to occur. Progress in an investigation is often dependent on the attitude of the customer/client and expertise of the failure analyst – the human factors. In many cases of failure, investigation is simply terminated when the physical cause has been determined. Conclusions are therefore drawn with no attempt to establish the root cause of failure. From simply reviewing published literature, root cause determination is undertaken in only a handful of investigations, with the majority not going beyond the determination of a primary or subsequent cause of failure. The reason for early (premature) closure of an investigation is often due to a lack of initiative and/or information from the instructing party. However, the onus is on the failure analyst to pursue the root cause as, without this, an investigation will not be comprehensive or complete.[50-52]

In summation, failure analysis usually implies the discovery of how a specific part or component failed. Focus will centre on material selection, design, product use/abuse, manufacture and failure mechanics within the part itself.

## 1.7 Forensic Engineering

The last few decades have seen an increased focus on investigating failure causation, be it performance-related, or related to fires, explosions or accidents. Forensic engineering is a relatively new discipline in engineering that takes a systematic approach to failure, having an additional focus on managerial and legal aspects of failure. The investigative process will encompass and emphasise the whole system failure, rather than concentrating on the failure of a specific part.[53] The role of the forensic engineer therefore is to analyse, evaluate and interpret failures in engineered products or processes, and to identify new rules and practices that will avoid future failures of those systems. Furthermore, the principle aim of any forensic investigation is often to assist a court in establishing the facts of an event.[54] The subject can and does embrace investigation of intellectual property claims, particularly patent disputes.[55]

### 1.7.1 Case Study: Collapse of the Rana Plaza Factory

On 24 April 2013, the Rana Plaza factory building in Dhaka, Bangladesh collapsed, killing 1,127 workers in what is considered one of the deadliest ever structural failure accidents (Figure 1.7). A government-backed forensic report determined that the building was constructed with substandard materials and with blatant disregard for building codes.

The direct reasons for the building shortcomings were as follows:

- Building constructed on a filled-in pond, which compromised structural integrity
- Conversion from commercial use to industrial use
- Addition of three more floors in direct contravention of the original permit
- The use of sub-standard construction material, resulting in structural overload of the building (aggravated by vibrations due to large generators)

**Figure 1.7** The Rana Plaza factory building collapse of 24 April 2013 in Dhaka, Bangladesh. (Courtesy of Wikimedia Commons, licensed under Creative Commons.)

It also blamed the local mayor for wrongly granting construction approvals and recommended charges against the building's owner, and the owners of the five garment factories in the building. On 14 June 2016, the plaza owner and 17 others were indicted for violating building codes in the construction of Rana Plaza. The defendants filed appeals with the High Court of Bangladesh in August 2016, the outcome of which was the postponement of the trial.

Being of relatively recent origin as a subject in its own right, forensic engineering is the 'behind-the-scenes' detective process of deducing why and how components, devices or systems fail, and the subsequent ramifications of such failure.[55,56] The forensic engineering practitioner will train to become a 'failure detective', being

- Aware of the normal or expected location of a fracture or failure in any part or component in a load path (the weakest-link principle).
- Familiar with the idea that deviation from the normal or expected location must have been caused by additional (unknown) factors that have to be revealed.
- Aware that components or structures often fail in a much more complex version of the 'weakest-link' principle. However, it is a solid starting point for an analysis of the situation.

- Aware that investigation may also include examination of the physical causes of accidents, other sources of claims and litigation, preparation of engineering reports, testimony at hearings and trials in administrative or judicial proceedings and the rendition of advisory opinions to assist the resolution of disputes affecting life or property.

When analysing a failure, an understanding of how the system works in its present configuration is a key requirement if reliability is to be improved.[57] This aspect is often overlooked in the search for the cause of failure, consequently leading to ineffective or incorrect remedial measures.

## 1.8 Range of Competence

As a field of activity, forensic engineering draws on many different disciplines, requiring competence across the fields of engineering mechanics, materials engineering, engineering design, structural integrity, chemistry and electrochemistry, physics and mathematics, manufacturing processes, stress analysis, design analysis, fracture mechanics, along with a working knowledge of the legal system in the UK.

On the one hand, it can be clearly seen and demonstrated that forensic engineering is an inter-/multi-discipline subject area.[57,58] On the other hand, it is nearly impossible for any one person to be an expert across all the above fields. It is therefore of great importance that the practitioner is aware of, and can accept, their personal limitations, and thus recognise when to seek other relevant expert help.

In addition to discipline-specific expertise, the forensic engineer is required to have a clear understanding of legal practices required to demonstrate his expertise and findings in courts of law. Although the subject is most commonly applied in civil law cases, there is a continually growing demand for forensic engineering use in criminal cases. However, the rationale behind any forensic engineering investigation is the enhancement in both performance and/or service life of a component, by ascertaining the cause or causes of any accident or failure. Therefore, the over-riding priority of any forensic investigation is to provide impartial scientific evidence and opinion, with a view to ultimate use in the courts of law.[59,60]

Defence of findings in courts of law would suggest that competence and investigative skills are not the only pre-requisite for the forensic investigator – he/she is also required to hone a range of skills in presenting evidence in the courtroom witness box. A lack of experience in presenting scientific findings may lead to ultimate collapse of a valid case, as expounded by a quote from *Death of an Expert Witness*.[61]

> "The defence doesn't always accept the scientific finding. That's the difficult part of the job, not the analysis but standing alone in the witness box to defend it under cross examination. If a man's no good in the box, then all the careful work he does here goes for nothing."

It must be stressed that evidence given by experts is a duty directly to the court (a point made by the 'Woolf' reforms in the UK), rather than to the party who hires him or her, and pays for their time. Clearly, problems may arise from 'confusion' that some experts may have over their duty to the court, and not their paymaster, raising the spectre of the hired gun. If such experts show bias towards their client, they can expect a tough time in cross-examination on the witness stand.

The forensic engineer can appear in both criminal or civil cases and (as an expert witness) can provide opinion to both courts and/or lawyers. This concession is normally denied to a witness of fact (eyewitness), who can only offer evidence of their direct involvement in an incident, event or crime. An eyewitness can certainly provide testimony regarding an incident or accident he/she has personally observed, but that person is not qualified to express an opinion as to *why* that event may have occurred.

### 1.8.1  A Generic Failure Analysis or a Forensic Investigation?

To summarise the salient differences between a conventional failure analysis and a forensic investigation, the failure analysis will focus on how a specific part or component failed, whereas a forensic investigation will be conducted under an additional umbrella of managerial and legal ramifications of the failure. So, any forensic investigation will encompass and emphasise the *whole* system failure, rather than simply concentrating on the breakdown of a specific part or product.

## 1.9  Initial Forensic Approach to Failure

When considering the failure of engineered systems and products, it is striking how much can be gleaned from material evidence. Any material evidence that survives failure will provide mute but revealing testimony to product history. Material evidence will often prove to be the key to unlocking a train of events to failure, particularly when linked to documentary or witness evidence. However, material evidence has to be viewed in context and *not* in isolation, as there are occasions when failure was as a result of the incident, rather than being the cause of it. Above all else it is the attitude to evidence, along with a common-sense approach, that is the clue to unravelling product failures and accidents.[62,63] In effect, the investigator will try to explain the failure by sifting evidence carefully so as to reveal key or critical parts that bear witness to how the incident occurred.[64] On many occasions, this may be limited to a bundle of documentary evidence, where the expert will be required to 'tease out' relevant information on the matter.[38] On the surface, this may appear to be a tedious and time-consuming exercise, however reviewing documentary evidence represents a large part of the normal and expected work pattern. At times it might be reviewing a fracture surface, which contains fatigue striations, in others, it might be the trace of rubber on a metal part of an aircraft. Every so often, the microstructure of a small particle will reveal its origin, and hence allow a sequence of events to be constructed. Commonly the sifting process will be long and detailed – particularly when a large and complex device, such as an aircraft, fails – but at other times, it may be the speed of reaction to unfolding events that solves the problem at hand. Thus a photograph taken 'just in time' may short-circuit an investigation, particularly when evidence is deliberately removed or tampered with at a later time.[38]

## 1.9.1 Reverse Engineering (The 'Weakest-Link' Principle)

As highlighted above, the forensic engineer will initially attempt to glean relevant information by meticulous investigation. Additionally, the practitioner may also take a reverse engineering approach if the necessity arises.[65] To recognise how a component or system failed, the engineer must first understand how it worked in the first place, and how it was initially manufactured. By reverse engineering, the forensic engineer will attempt to produce a complete 'top-level' description of the product that will allow him/her to work out how the product functions.[66] By stepping back through the transformation stages, the investigator will then be in a better position to determine the most probable or expected point of failure within a structure or assembly – 'the weakest-link' principle.[38] When considering loading patterns within a system, forces must be connected in a chain so as to form a path. When there are several or many different components in that path, the load takes different forms along the path. The idea of load path through more complex configurations of many different components will help to elucidate how specific parts came to break rather than others. As it is the weakest component that will govern the strength of the whole assembly, failure outside of the weakest link will require careful consideration and analysis. The outcome of each investigation will be a clear and concise report with sound findings and a conclusion that accurately describe what happened, why and what corrective measures could or should be implemented.

## 1.9.2 Associated Costs of Forensic Investigation

Although the cost of any failure investigation will increase as the complexity and depth of analysis become more sophisticated, the resultant benefit will be a more complete picture or understanding of the true origins of demise. However, the concept of 'cost' should not be limited to the 'monetary' value of the investigation. A wider view should be taken that will encompass any business aspects associated with the failure in question. Consumers expect higher-cost products to translate into higher quality goods that have a lower failure rate and longer lifespan than lower-cost equivalents. This will apply to components and materials alike. When components and/or materials fail to meet stated performance specifications or service requirements, the consequences can be loss of manufacturing time and customer business. In general, lost business revenue will far outweigh most analytical costing.[38,67] However, any failure occurring in-service may also suffer legal ramifications (possible litigation) in addition to business implications (loss of customer goodwill), along with potential loss of company reputation. Society in general is always looking for someone to 'blame' for any misadventure, and is far more eager to undertake litigation as a process to redress any perceived loss, damage or injury associated with failure of a product. Consequently, rapid and effective use of appropriate failure analysis at the outset of investigation can short-circuit potential market place pitfalls, thus facilitating positive customer retention along with continued company growth.[38]

## 1.10 Historic Failures

Why should the failure or forensic engineer spend time re-visiting historic failures/disasters? Simple perusal of specific accidents and failures from the recent past often reveals a degree of notoriety, simply as a result of the scale of the particular disaster.[68,69] The ensuing public outrage over such events is often exceptionally vocal, necessitating the establishment of a public inquiry to investigate the root cause. There are many examples of 'forensic' methods used to investigate early accidents and disasters, one being the fall of the Dee Bridge at Chester, England.[70] It was built using cast iron girders, each of which was made of three very large castings dovetailed together. Each girder was strengthened by wrought iron bars along the length. It was finished in September 1846 and opened for local traffic after approval by the first Railway Inspector, General Charles Pasley. However, on 24 May 1847, a local train to Ruabon fell through the bridge. The accident resulted in five deaths (three passengers, the train guard and the locomotive fireman) and nine serious injuries. The bridge had been designed by the eminent engineer Robert Stephenson, and he was accused of negligence by a local inquest.[70,71]

Although strong in compression, cast iron was known to be brittle in tension or bending, yet on the day of the accident the bridge deck was covered with track ballast to prevent the oak beams supporting the track from catching fire. Ironically, Stephenson took this precaution because of a recent fire on the Great Western Railway at Uxbridge, London, where Isambard Kingdom Brunel's bridge caught fire and collapsed. This act imposed a heavy extra load on the girders supporting the bridge, and probably exacerbated the accident.[70,71]

One of the first major inquiries conducted by the newly formed Railway Inspectorate was conducted by Captain Simmons of the Royal Engineers, and his report suggested that repeated flexing of the girder weakened it substantially. He examined the broken parts of the main girder, and confirmed that the girder had broken in two places, the first break occurring at the centre. He tested the remaining girders by driving a locomotive across them, and found that they deflected by several inches under the moving load. He concluded that the design was basically flawed, and that the wrought iron trusses fixed to the girders did not reinforce the girders at all, a conclusion also reached by the jury at the inquest. Stephenson's design had depended on the wrought iron trusses to strengthen the final structures, but they were anchored on the cast iron girders themselves, and so deformed with any load on the bridge.[68,69] Others (particularly Stephenson) argued that the train had derailed and hit the girder, the impact force causing it to fracture. However, eyewitnesses maintained that the girder broke first, and the fact that the locomotive remained on the track showed it could not have derailed at all.[72]

In addition to the Dee Bridge failure, other disasters on the railways come to mind, such as the Tay Bridge disaster of 1879.[73] Again, the failure was the result of utilising cast iron that was routinely subjected to fatigue loading (Figure 1.8). Marine disasters were also common in the Victorian era and are largely forgotten now, unless remembered through an imaginative journalist's pen, such as the mysterious abandonment of the *Mary Celeste*. But the tragedy of the sinking of the *Titanic* in 1912 was so enormous that it is remembered by every generation by a new film or book. Similarly, the destruction of the *Hindenburg* airship in 1937 seemed to presage or echo the widespread human misery suffered at the hands of the Nazis. Such disasters continue to fascinate the public (see Section 1.12.1, Popular Books), not least because of a degree of uncertainty about the precise causes.

Where the Dee and Tay Bridge disasters concerned failure of metals (the cast-iron girders and columns respectively), the *Hindenburg* (Figure 1.9A),

**Figure 1.8** The beautiful bridge on the silvery Tay, with the collapse of its high girder section clearly evident.

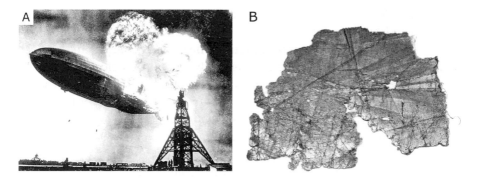

**Figure 1.9** (A) The *Hindenburg* disaster, 6 May 1937. The German passenger airship caught fire and was destroyed during its attempt to dock with its mooring mast at the Lakehurst Naval Air Station, New Jersey, USA. Of the 97 people on board, there were 36 fatalities, including one death among the ground crew. (B) Recent elemental X-ray analysis determined the fabric to be doped or reinforced with cellulose nitrate or acetate. (Public domain Wikimedia Commons.)

and the R101 airship of 1930, failed due to the low strength and flammability of their gas containers and outer envelopes.[74] They were composed of cotton fabrics reinforced with cellulose nitrate or acetate, and doped with rubber latex and other highly inflammable substances (Figure 1.9B).

Lightweight organic composites have been widely used in air- and spacecraft for a number of years, and have been involved in a number of disasters, the most notorious of which include loss of the Space Shuttles *Challenger*[75] and *Columbia*.[76] In the first disaster in January 1986, the O-rings on one of the *Challenger's* booster rockets failed, and the propellant exploded through the gap engulfing the entire structure shortly after take-off. The Viton rubber (a fluorinated copolymer) was very inflexible at the near-freezing temperatures at lift-off, and could not seal the booster sections correctly. The problem was well-known, but managers ignored the warnings. The second disaster, in October 2003, also involved damage during *Columbia's* take-off, and once again was recorded on cameras, as was the first disaster. The critical damage had been caused by a large piece of foam insulation falling away as the rocket rose, and striking carbon-fibre composite on the wing. The damage had fatally weakened the shuttle, but this was unfortunately missed at the time. When the craft attempted re-entry, the entire structure disintegrated. The aftermath of the accident exposed the testing regime at NASA as flawed; the problem had been appreciated, but tests were only conducted with very small pieces of foam, much smaller than that which actually flew off at launch.[77]

The final demise of Concorde, the supersonic airliner, was signalled by a horrific accident at Paris in July 2000.[78] It occurred when a tyre blew out during take-off, and a large fragment penetrated the fuel tank above. The jet fuel was ignited by the engine nearby and the Air France plane eventually crashed killing all on board. The root cause of failure was a length of titanium alloy that had fallen off a previous aircraft. The Concorde tyre was fragmented when it ran over the titanium, with the tyre fragments puncturing an unprotected fuel tank. The small fleet was grounded, but, after a short period, flights were resumed on the British versions. It transpired that tyre blow-outs were not uncommon on other Concordes, well before the Paris accident. So, after the fatal crash, the cross-ply tyres were replaced by a radial ply design, and the fuel tanks were reinforced by a rubber/aramid lining. However, the Concorde was finally retired in 2003, with a number of additional events coinciding with its final demise:

- In part, a change in the economic climate made the cost to fly transatlantic at supersonic speeds less viable.
- The airplane and its design were showing signs of age, approaching 30 years in commission.
- Due to the lack of competition, Concorde didn't benefit from many upgrades over the years so the technology ended up being very dated.

However, as dated as the engineering may have become over its life cycle, it remains a fact that the concept of a supersonic commercial airline, and the design that resulted from that concept, has not to date been surpassed. It could legitimately be said that technology and engineering receded with Concorde's demise as no viable replacement has been put in place.[79]

Material failure as a cause of major accidents and disasters has long been an unfortunate reflection of the lack of awareness of materials limitations. It is also unfortunate that the benefits of new materials are often overplayed and exaggerated when first introduced into the market. Engineers are then forced into the embarrassing position of explaining minor unanticipated failures, and being brought to account when lives are lost.

It can be seen that there is much to be gained from re-visiting historic failures/disasters and applying modern analytical techniques. The forensic or failure engineer will acknowledge that there are

- Engineering lessons to be learnt. For example,
    - New attitudes to, e.g., wind loading (Tacoma Narrows Bridge)
    - Demise of cast iron for bridge construction (Tay Bridge)
- Advances in understanding of the limits of materials, specifically how they were used and how they were applied 'in-service' (Tay Bridge, *Hindenburg* Airship)
- Opportunities to take a retrospective view of disasters that allows the use of modern analytical techniques that were not available at the time of initial investigation, such as the following:
    - Systematic gathering of evidence
    - Use of tools such as finite elements analysis and computational fluid dynamics
    - Use of modern analysis techniques for analysing fracture surfaces, along with detailed elemental chemical analysis

## 1.10.1 Computer-Aided Technology Limitations

At this point, it is prudent to raise a 'health' warning regarding current computer-aided technologies and design analysis. There is a popular misconception that, with current computing power, the modelling of component designs and their working environment will lead to the manufacture of products that will perform as specified throughout their working lives. This conception is totally misleading. Although there has always been potential for error in any design process or forensic/failure investigation, the ever-increasing use of computers raises a *new* failure potential. With engineering knowledge now effectively encoded by software programmes, the engineer or investigator can be led to a general (and totally false) assumption that the computer also holds subject-specific 'expertise'. The most basic of current computers can perform mathematic calculations and analysis at astounding speed, with excellent levels of accuracy. However, it is of vital importance that any judgemental decisions are made by the qualified engineer. There is, moreover, evidence that computing power and the vast range of available software is encouraging personnel with inadequate knowledge and training to engage in the process of design, failure and forensic analysis (the SCOSS report).[80] Moreover, the SCOSS report states:

> Those responsible in universities, the professional engineering institutions and government for the education of engineers and their continuing professional development should provide more guidance on understanding structural behaviour and its modelling for computer analysis, and on avoiding uncritical reliance on computer-generated results.

In layman's terms, there is an increased over-reliance being placed on computer generated output, with little or no appreciation of its actual meaning. It is of utmost importance to consider that the output from any computer is only ever as good as the data inputted in the first instance. A classic example that will illustrate this point is a mirror failure in Hubble Space Telescope:

> In 1990, the orbiting Hubble Telescope sent its first photographs back to Earth. However, its images were unexpectedly fuzzy and out of focus. NASA determined that the problem was the result of a human error made years before the launch: the telescope's mirror had been ground into the wrong shape. The mirror, alone, was tested prior to launch and found to be functioning properly. Nevertheless, the manufacturers did not actually test the mirror in conjunction with the other components. Rather, total reliance was placed on computer simulations to determine that the separate components would work together. Conversely, the simulation did not take into account the possibility of a misshapen mirror.
>
> As a result of the Hubble problems, NASA learned 'a great lesson' about 'the merits of actually testing a system rather than depending upon theory and simulation', explains Doran Baker, founder and vice-president of Utah State University's Space Dynamics Laboratory.[81] The problems with the Hubble Space Telescope taught engineers more than just to be wary of depending on theory and simulation. According to Doran Baker, those involved in the project also learned a great deal in their efforts to fix the telescope. After receiving those first blurred photos, space system engineers had to devise ways to perform the appropriate failure analysis on the telescope, even though the instrument was only accessible via radio signals from space. Once they analysed the problem, they had to plan and execute missions in which astronauts could remove and replace the faulty component. The result was that the engineers learned much more from correcting the telescope than they would have had the telescope not needed repairing.[81]

Engineers, working on the Hubble fix, faced problems as they arose, developing each solution as they went along. Reliance on the power of computers to both predict and counteract failure could be considered the weak link of this story. However, this is not a 'one-off' incident: another classic example is the demise of the Team Philips catamaran.[82] The loss of the craft (and no end of national pride and prestige) was directly attributable to a failure to correctly model loading conditions of rough seas, leading to structural failure and ultimate loss of the vessel.

Yet another example to emphasise the dangers of total reliance on computer-generated design is adequately illustrated by the loss of the Sleipner A oil platform.[83] The Sleipner A was a Condeep-type platform (abbreviated from concrete deep-water structure), built for Statoil in Norway by the company Norwegian Contractors (Figure 1.10). The Condeep-type platform contained two units, the hull and the deck. The gravity base hull included support pilings and concrete ballast chambers from which three or four shafts rise and upon which the deck would sit. Once fully ballasted, the hull would lie on the sea-bed. In the case of the Sleipner A, there were 24 chambers, of which four formed the 'legs' supporting the facility on top.

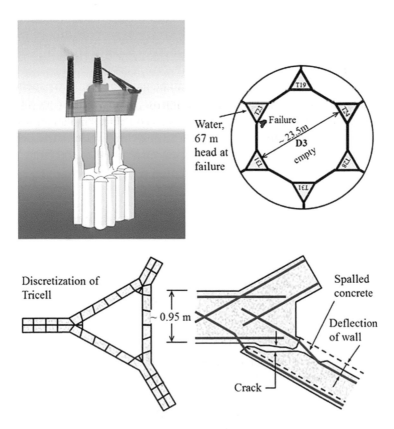

**Figure 1.10** Schematic explanation of failure of Sleipner A Oil Platform, Norway 1991. (Courtesy of Wikimedia Commons, licensed under Creative Commons.)

In August 1991, prior to the mating of the hull and the deck unit, the hull was towed into Gandsfjord. There it was to be lowered in the water – at a rate of 1 m per 20 minutes – in a controlled ballasting operation. As the hull was lowered to the 99-m mark, rumbling noises were heard; this was followed, immediately, by the sound of water pouring into the unit. A major crack had developed in the cell wall, with seawater gushing in at a rate that was too great for the de-ballasting pumps to handle. Within minutes, the hull began sinking at a rate of 1 m per minute.[84] As the structure sank deeper into the 220-meter deep fjord, the buoyancy chambers imploded, with the rubble striking the floor of the fjord and creating a Richter 3.0 magnitude shock wave that was recorded in a local seismograph station.

The post-accident investigation discovered that the root cause of the failure was the result of inaccurate finite element approximation during calculations in the design of the structure. Fundamentally, stresses on the ballast chambers had been underestimated by 47%. As such, the design of some concrete walls was far too thin. At a particular pressure, the walls failed, allowing seawater to enter the ballast chamber at an uncontrolled rate, eventually sinking the base unit. The base structure was subsequently redesigned, using hand calculations, and the Sleipner A Platform was successfully completed in June 1993.[84,85] Probably the greatest lesson to be learned from this case study is: *never treat computer analysis as a black-box process.*

It is clear that there are valid lessons to be gained from retrospective failure investigation (and that those lessons are *still* not being learned). Consider the Millennium Bridge. When the bridge opened in June 2000 (Figure 1.11A), it was expected to be a source of pride for the city of London. The design was a striking, slender, single arch construction, providing spectacular views of the city. However, the bridge had to be closed within days of opening due to

**Figure 1.11** (A) The Millennium Bridge, a pedestrian link across the river Thames, being the first new river crossing in London for over a century. (B) The torsional oscillation of the Tacoma Narrows bridge deck immediately prior to total collapse. (A&B Courtesy of Wikimedia Commons, public domain.)

excessive torsional swaying behaviour. Its design engineers curiously attributing this response to 'synchronised walking'. When comparing its slender design to that of the Tacoma Narrows Bridge collapse of 1940, an example of 'déjà-vu' would appear to exist. The torsional mode shape of the Tacoma Narrows design effectively divided the bridge into two halves that vibrated out of phase with one another (i.e. one half rotating clockwise, while the other half rotated anti-clockwise – Figure 1.11B). The bridge collapsed during this torsional excitation mode when a 600-ft length from the centre span broke loose from its suspension cables and fell 190 ft into the water below. As would appear to be the case, lessons from the past continue to be overlooked. However, on reflection, the Tacoma Narrows collapse most certainly provided the impetus for incorporating aerodynamics into bridge design.

As a point of interest, other suspension bridges of the period also suffered from excessive motion – for example, the Golden Gate Bridge (1937), New York's Bronx–Whitestone (1939) and Maine's Deer Isle Bridge (1939). These bridges also had a tendency to undulate in the wind, all of which required a retrofit with extra cables and/or stiffening devices.

## 1.11 Analytical Techniques

To deduce a train of events leading to the failure of a component or product, the forensic investigator will have a range of analytical techniques at hand. Each investigative method will be unique in its own right and also complementary to the others. The use of appropriate methods should reveal a complete picture of failure modes or mechanisms that had been at play at the point of failure. The experienced professional will already have a breadth of knowledge in techniques complementary to their core skills, while individuals new to the field of analysis must gain a good overview of the range and application of the many available techniques. An implicit understanding of available analytical 'tools' will allow more informed decisions regarding the most appropriate technique(s) to use in the course of investigation, thereby reducing both monetary and time cost.

A philosophy that will be advocated throughout this book is that failure analysis is primarily a problem-solving exercise, undertaken within a legal framework, being based on the knowledge and skill of the investigator. Therefore, the body of this book will expressly focus on the kinds of predicament associated with both metallic and non-metallic product failure. Chapter 4 will be devoted to a review of a 'typical' range of techniques and associated instruments that can be called upon by the investigator, and will include

- Techniques and instrumentation for materials characterisation
- Techniques that are best used in different applications

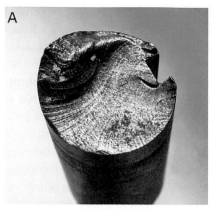 

**Figure 1.12** Two views of a drive shaft failure. As a fracture surface is a 'finger-print' of failure, with just the naked eye it is possible to state that the failure was mechanical in nature, with a torsional fatigue crack having initiated at a corner of the keyway, propagated some three-quarters of the way into the load-bearing section prior to final overload failure. However, further investigation is required to elucidate all factors that led to this failure in the first place.

- The basic fundamentals of each technique
- How to better interpret the data to get the most information from any analysis
- Strength, limitations and compromise associated with each technique

However, it is worth reiterating that the cause or causes of a large percentage of all failures can be detected with a careful physical examination using low-power magnification and some basic physical testing. Basic visual inspection of the failure will generally reveal the type of forces involved (tensile, bending, shear, etc.), whether the load was applied cyclically (fatigue) or was a single over-load (traumatic), the direction of the critical load and the influence of external forces such as residual stresses or strains (Figure 1.12). Then, accurately know-ing the roots of the failure, the investigator can pursue probable human errors, any latent cause or causes of failure and possible consequence of failure.[86]

## 1.12  Dissemination of Knowledge and Experience

At the outset, the meaning of the words 'dissemination', 'knowledge' and 'experience' should be defined as follows:

- In the context of forensic engineering teaching, 'dissemination' rep-resents more than just a knowledge transfer – it should literally rep-resent 'sowing seeds'.

- In a similar vein, knowledge is more than information; it is an attribute of the individual, resulting from processes of learning, understanding, reflecting, comparing, selecting the right information and critical and strategic thinking.
- Experience can be defined as knowledge and skill gathered through practice, with the individual ultimately evolving into an 'expert' in a particular sphere of activity.

It can therefore be said that a fundamental aim of forensic or failure investigation should be the ultimate dissemination and transfer of any such knowledge gained into routine engineering practice. Attention will now focus on a number of possible routes for the dissemination of such knowledge.

## 1.12.1 Popular Books

There is a seemingly endless public appetite for tales of disaster, with many authors attempting to fill the need. British journalist Nicholas Faith became one such author, having published five books, *Derail*, *Black Box*, *Blaze*, *Mayday* and *Crash*. Furthermore, *Crash* and *Black Box* were an integral part of and follow-on for two television documentaries. Starting with *Derail*, Nicholas Faith[87] presents an analysis of the most dramatic, bizarre and terrifying rail accidents over the past 150 years. He asks what lessons have been learned from past disasters and what steps remain to be taken. In *Black Box*, his focus is on air accidents,[88] written from the point of view of the crash investigation agencies. The book describes the investigations surrounding air accidents, with the conclusions reached and subsequent events. With *Blaze*, he continues by delving into the forensics of fire,[89] the process of investigating how a fire starts and grows. The narrative captures the often glossed-over details of the science of fighting fires. With *Mayday*, Nicholas Faith explores the world of 20th-century maritime safety and the causes behind tragedies.[90] He reports on incidents such as the sinking of the *Herald of Free Enterprise* in 1987, and of the *Estonia* eight years later. He queries why these disasters happen, why so many lives were lost and *what has been learned from so many mistakes*. The focus of *Crash* is on the automotive industry, and the book discusses how the technology and social acceptability of accidents have changed over time.[91] The book poses four vital questions:

1. Can we do more to reduce the number of lives ruined or ended by the all-conquering motor car, or
2. Is the quest for car safety doomed to failure?
3. In the future will technology save us by building safer, perhaps even un-crashable cars and safer roads, or
4. Will our fatal attraction to the car remain an obstacle to finally ending accidental deaths on the road?

Other popular books contain detailed investigation of disasters, with modern thinking and analytical techniques being retrospectively employed. In his book *The Beautiful Railway Bridge of the Silvery Tay*,[92] Peter Lewis concludes that fatigue was the most probable cause of failure, although this opinion differs from those of previous investigators. However, even today, fatigue still causes major railway accidents. In 1998, 101 people were killed when the wheel of a German high-speed train failed by fatigue. In the year 2000, the Hatfield derailment killed four people through a fatigue failure of the rails. This failure, along with a lack of understanding regarding fatigue of railway lines, caused major disruption to the whole rail network in the UK. Furthermore, repercussions eventually led to the demise of Railtrack. It is clear that we still have a lot to learn, with analysis of past disasters providing clues to the prevention of future catastrophes.

*Wheels to Disaster* re-evaluates the Shipton, Oxfordshire railway accident of 1874 and provides another example of retrospective failure analysis.[93] The outcome of the investigation was that the accident had been caused by a fatigue crack growing from a rivet hole in a wheel tyre. The origins of the accident at Eschede, Germany lay in the way the tyre was separated from the wheel centre by rubber supporting pads. A few days after the Eschede, an East Coast Mail Line train derailed in the UK. Failure had initiated at a point where a hole had been drilled through the wheel web. Not surprisingly, a fatigue crack grew, causing the wheel to collapse – a case of '*déjà vu*' when considering the Shipton accident?

As a final point here, it is clear that retrospective forensic engineering is a useful tool that will allow the current crop of budding engineers to re-learn the lessons of the past.

## 1.12.2 Event Reporting

Forensic engineering can therefore be considered as an *essential* design and management tool that will assist the prevention of disasters, accidents and failures. Furthermore, as a tool it will also *inform* the processes of design, manufacture, operation and maintenance. One of the best ways of preventing future accidents or disasters is to publish details of previous incidents or case-work. The collapse of a block of flats provides a classic example:

Ronan Point, May 1968[94]
For a while, high rise blocks of flats seemed like the perfect answer to the post-war housing shortage in the UK, providing cheap, affordable prefabricated housing. However, tenants were not always happy with the high-rise solution, as they felt isolated, and missed the sense of community that they had been

used to in terraced streets. The Ronan Point tower block was built using a technique known as Large Panel System building (LPS). LPS construction involved casting large concrete prefabricated sections off-site, then bolting them together to form the building. However, in May 1968, a gas explosion caused the 'blow-out' collapse of one complete corner of the building that failed like a stack of playing cards (Figure 1.13A). The structural failure was responsible for the deaths of three people inside the building. Subsequent investigation determined just how fragile the LPS manner of construction was.[95] The relatively low overpressure from the gas explosion should have led to localised damage at most, not a partial progressive collapse and the loss of four lives in total. The evaluation also found that the building was unusually vulnerable to ordinary wind and fire loading.[96] The partial collapse of Ronan Point led to major changes in the building regulations.[97] However, lessons are hard to learn, as similar failures have continued to occur, as illustrated by the progressive collapse of an apartment building on 27th February 2012, in Astrakhan, Russia (Figure 1.13B).

There is a long tradition of publicising the cause or causes of failure, but usually only when that failure resulted in a severe in loss of life or property, leading to it being classed as a disaster or catastrophe. Lesser failures have either become such a common occurrence, attracting little or no attention (car accidents), or receive no publicity whatsoever by being deliberately suppressed.

Over the last 20 years, there has been a rapid growth in both failure analysis and forensic science literature, with more volumes of case studies, journal papers, dedicated conferences and, increasingly, information posted on the Internet. Being the first international conference series of its type, the International Conference on Engineering Failure Analysis (ICEFA) has proved to be an ideal platform to raise the profile of forensic engineering failure research to an international audience. In addition, there is a growing body of published literature documenting investigation into a wide range

**Figure 1.13** (A) Ronan Point high rise progressive collapse, May 1968. (Courtesy of Wikimedia Commons, public domain.) (B) Apartment building failure on 27th February 2012 in Astrakhan (Russia) due to progressive collapse.

of product failures. The academic journal *Engineering Failure Analysis*[98] is devoted to publishing case studies of failed products, but it remains relatively isolated when compared to the abundance of strictly academic journals. Several volumes of papers taken from the journal are available,[99] with the majority reporting metal product failures. There is a long line of major disasters, within living memory, that inflicted enormous damage to the workforce at the affected sites, in addition to damaging surrounding public buildings and homes. Flixborough (1974), Bhopal (1984) and Buncefield (2005) are a few of the disasters which will be remembered by the wider public.[100] This and other accidents in chemical plants are discussed by Kletz[101] and in the British Petroleum training booklet.[102] Furthermore, *Loss Prevention Bulletin*[103] specialises in failures occurring in chemical plants, focusing on a wide variety of failure modes and their effects on the companies concerned.

Civil engineers have a long and distinguished record of publicising failures, reflecting the safety-critical nature of large structures such as bridges, dams, buildings and tunnels for example. When any failure occurs, it is likely to be both dramatic *and* life-threatening.[104–109]

The aircraft industry provides an outstanding example of an industry that takes reporting of any failure or near miss as a legal responsibility. As an example, the Geneva-based Bureau of Aircraft Accidents Archives (B3A), is an on-going statistical source for aviation accident data for aircraft capable of carrying more than six passengers, not including helicopters, balloons or fighter airplanes. Although not making good reading, Table 1.2 shows a record of air accidents and fatalities between 2006 and 2017. Browsing the site case-load reveals an excellent example of the record-keeping to be found at B3A.

In the seven-month period between January and August 2018, for example, there are 64 reported incidents, with a death toll in excess of 745 persons. These statistics include five commercial accidents that claimed 555 lives of passengers and crew. Along with far more in-depth detail, a chronology of

**Table 1.2  The Bureau of Aircraft Accidents Archives (B3A) for Fatal Air Accidents**

| Year | Fatalities | No. of Accidents | Year | Fatalities | No of Accidents |
|------|-----------|------------------|------|-----------|-----------------|
| 2017 | 399 | 101 | 2011 | 828 | 154 |
| 2016 | 629 | 102 | 2010 | 1,130 | 162 |
| 2015 | 898 | 123 | 2009 | 1,108 | 163 |
| 2014 | 1,328 | 122 | 2008 | 952 | 189 |
| 2013 | 459 | 138 | 2007 | 981 | 169 |
| 2012 | 800 | 156 | 2006 | 1,298 | 192 |

*Note:* Statistics are for all aircraft types. however, there were no *passenger* jet crashes in 2017, it being the safest year in the history of commercial airlines.

these disasters can be found on the B3A site. Three such events for commercial aircraft in 2018 are summarised below for information.

29 October 2018: A Boeing 737 Max, operated by Lion Air, crashed into the Java Sea shortly after taking off from Jakarta, Indonesia. All 189 passengers and crew were killed. A volunteer diver also died in the subsequent recovery operation. At the time of writing, investigators suspect that the plane may have suffered technical issues with its Manoeuvring Characteristics Augmentation System (MCAS). The entire fleet of Boeing 737 Max aircraft have been grounded world-wide.

18 May 2018: A Boeing 737 passenger plane crashed shortly after take-off from Jose Marti International Airport in Havana, killing 112 people. In August 2018, the Commission of Inquiry into the causes of the crash reported that a process of this magnitude (the crash investigation) requires an analysis of multiple factors and has not yet been completed.

11 April 2018: An Ilyushin Il-76 military transport aircraft of the Algerian Air Force crashed shortly after take-off from Boufarik Airport, near the Algerian capital Algiers, killing at least 257 people on board, including ten crew members. Most of the dead were military personnel and their relations. The Chief of Staff of the Algerian Army ordered an investigation into circumstances surrounding the accident[110] in an attempt to determine the cause of the incident. Russian authorities stated their intent to assist in the investigation.

Legislation is in place to force reporting of all incidents to inspection authorities, with appropriate remedial measures immediately implemented. Furthermore, these actions are undertaken under direct public scrutiny. Near misses of flying aircraft, for example, are widely publicised in the press.

The railways too are obliged to report SPADS (signal passed at danger or red), incidents which do not result in any accident but are an indicator of a potential problem. Just such a problem became reality in the Ladbrook Grove crash of October 1999, when a local train passed a red signal and collided with a fast express, with high loss of life.

### 1.12.3 Textbooks

There are many standard texts which are useful background for analysing structural, product and systems failure. When considering failure analysis *per se*, written text predominantly covers the field of metallic materials.[25,30–33,38,43,54,111–114] There is a shortage of case study compilations on non-metallic materials, but some substantial works have been published in the last decade or so. Among them are the books by Ezrin,[115] Wright[116] and Hays.[117] A recent work discusses numerous case studies of both metal and polymer failures from a forensic viewpoint, including intellectual property.[118] Several reviews exist in the literature concerning design in polymers, product failure and high-performance polymer fibres.[119] There are numerous texts dealing with mechanics and materials generally, including the latest update to Peterson's classic work on stress concentrations[120] and Roark's Formulae.[121]

### 1.12.4 Forensic Engineering Teaching

When considering forensic analysis, there is a growing need for both undergraduate and post-graduate courses in the area, so that the new generation of engineers is made aware of the potential downside to new product development and the serious consequences of product failure in both the marketplace and the courts of law. More importantly, they should be made aware of the range of both existing and new analytical tools available for studying failed products, and the methods available to tackle design deficiencies prior to market launch. While many such methods have been developed for safety-critical products in aerospace, their use is fast spreading to other safety-critical products such as medical devices and electronic products.

However, although there is little or no disagreement that forensic engineering courses would be beneficial to include in an engineering curriculum, on a worldwide scale there are few universities offering classes in forensic engineering.[122] Of the universities that do offer such classes, it is found that they are presented as a master's degree syllabus.[123] A recurring reason given for the neglect of engineering programmes to offer failure analysis modules is inadequate or inappropriate teaching materials.[124] Furthermore, it was found that many lecturers have difficulty adequately planning real-world assignments.[125]

One possible route to circumvent limitations of inadequate teaching materials and assignment planning is to raise the profile of failure case studies.[126] Such cases are real-life engineering situations, which reflect a wide range of engineering activity (both failures and successes), old and new techniques and theoretical and empirical results. The integration of case studies into the engineering curriculum will present a medium through which learning (e.g. analysis, knowledge, reasoning, assumptions and drawing conclusions) takes place. A good engineering case study is taken from real life and will include sufficient data for the student or reader to independently analyse the problems and issues. Content should allow understanding of possible conflicts, postulation of a sequence or train of events and determination of the most probable failure scenario. Furthermore, the forensic aspect should make it possible to elucidate any potential legal ramifications contained within the study.

## 1.13 Case Study Themes

Themes can be pursued, for example, marine disasters, power station disasters, colliery disasters, bridge failures, etc. However, it is important that the selection is representative, and to include the widest range of products, materials and failure modes. Examples of case study themes are presented in Chapter 8. In addition, specific groups of failures are presented as case studies within Chapters 9–13 including manufacturing faults, road traffic accidents, criminal activity, insurance fraud and agricultural accidents.

### 1.13.1 The Sayano-Shushenskaya Power Station Accident of 2009

An accident at Russia's Sayano-Shushenskaya Hydro Power Plant in August 2009 claimed the lives of 75 workers, when Turbine 2 catastrophically broke apart (Figure 1.14A). In addition to the loss of life, the turbine hall and engine room were flooded, the ceiling of the turbine hall collapsed and nine out of ten turbines were damaged or destroyed (Figure 1.14B).

Immediately after the accident, Russia's Federal Service for Ecological, Technological, and Nuclear Supervision (Rostekhnadzor) launched an investigation. The official report, released on October 3, 2009, blamed poor management and technical flaws for the accident.

According to the report, repairs on Turbine 2 were conducted from January to March 2009, at which time a new automatic control system (meant to slow down or speed up the turbine to match output to fluctuations in power demand) was also installed. The repaired turbine resumed operation on 16 March 2009. However, further operational issues emerged, with vibration amplitude increasing to an unsafe level between April and July. The unit was again taken offline until 16 August when circumstances at play forced managers to prematurely return the turbine back into operation.[127]

However, once back in service, Turbine 2 vibration was recorded at 'four times the maximum design limit. As the control system decreased the turbine's output on the morning of 17 August, the vibrations actually *increased*'[128], in a manner similar to a car being shifted down in gears whilst on a hill. The accident was therefore primarily caused by turbine vibration, which led to fatigue damage of the 'mountings of turbine 2, including the cover of the turbine. It was also found that, at the moment of accident, at least six nuts were missing from the bolts securing the turbine cover. After the accident 49 recovered bolts were investigated, of which 41 had fatigue cracks.' On eight bolts, fatigue damaged exceeded 90% of the total cross-sectional area.

The manufacturer had designed the turbines with a 30-year service life. At the time of failure, Turbine 2 had completed 29 years, ten months service

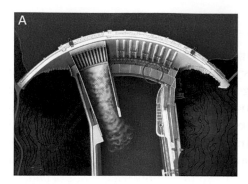 

**Figure 1.14** (A) The Sayano-Shushenskaya dam accident, with the destroyed turbine hall circled in red. (Courtesy of Wikimedia Commons, no license.) (B) Devastation of the inside of the turbine generator hall. (Courtesy of AP Photo/ Rossiiskaya Gazeta.)

lifetime. Investigators determined that total power failure after the initial explosion had knocked out the safety system that should have shut down the plant. Therefore, a malfunction turned into a catastrophe.[127,128]

On 11 November 2014, full renovation and repairs were finally completed, five years after the accident

## 1.14 Research-Enriched Teaching

As well as being a vital database for everyone involved in the safe design, construction, operation and maintenance of structures and components of all kinds, engineering failure case studies can be used to great effect in the teaching of engineering (Chapter 8). It can be safely said that the integration of case study analysis in the engineering classroom is a classic example of 'research-enriched' teaching. In this context, the failure or forensic practitioner will have compiled a bank of forensic failure data that would be a fertile source for current (and more importantly, relevant) case study input to almost all engineering courses. From personal experience, these data provide feed-stock for continuous assessment, examination papers and course content for post-graduate engineering courses, in addition to providing a classic example of research-enriched teaching. Furthermore, journal promulgation and attendance at national and international conferences (the International Conference of Engineering Failure Analysis series, for example) will allow the practicing professional to open his/her case-book to a wide audience. The publication and dissemination of engineering failure case study material will not only raise the general profile of engineering and provide feedback into the design process, but will also hold the potential for a significant contribution to the training of future engineering professionals.

## 1.15  Concluding Remarks

Forensic investigation casts the engineer in the role of the detective, using many analytical methods for investigation. He will apply those techniques in ways that avoid flawed reasoning while delivering convincing reconstruction scenarios. The forensic detective can therefore pinpoint where things went wrong, and thus provide valuable information that may prevent similar failures in the future. As everything physical is made from a material of some form or other, the professional working in the forensic failure analysis arena can find themselves involved with an incredibly wide range of applications. Their brief can range from investigation of the Space Shuttle Columbia accident, through the Potters Bar railway disaster, to a collapsed step ladder. Furthermore, forensic or failure analysts are not simply confined to the laboratory or the field. Unwanted failure of a structure or component can lead to injury, loss of life, loss of money, loss of company reputation or an environmental disaster. As such, the forensic engineer will often find him/herself involved within the legal process, acting as an expert witness in both civil and criminal cases. It is this particular aspect that differentiates a forensic analysis from that of failure analysis. Failure investigation tends to be restricted in that the centre of focus is on material selection, design, product use/abuse, manufacture and failure mechanics within the part itself. Subsequent reporting therefore presents a somewhat narrow outlook, not fully exploring causes of failure or identifying potential ramifications that may ensue. Forensic engineering is therefore the application of the art and science of engineering, undertaken within the science or philosophy of law. Requiring the services of legally qualified professional engineers, forensic engineering usually includes investigation of physical causes of accidents and other sources of claims or 'litigation, preparation of engineering reports, testimony at hearings and trials in administrative or judicial proceedings, and the rendition of advisory opinions to assist the resolution of disputes affecting life or property'[129,130].

The work of the forensic engineer demands a wide range of expertise and experience (both practical and academic), along with a constant exposure to people from many other walks of life. Additional pre-requisites demanded of the practitioner will include firm professional ethics, an open mind and a commitment to continued professional development. Forensic engineering (and failure) analysis can therefore be considered an essential design and management tool that will help prevent disasters, accidents and failures. As a tool, forensic analysis will provide an iterative feedback loop that will further inform the processes of design, manufacture, operation and maintenance. There is, however, immense scope for improvement in the arena, particularly when considering the increased transparency of publicising poor product performance. Probably one of the best ways of preventing future accidents or disasters is to broadcast details of previous incidents or case-work. In the last 20 years, there has been a rapid growth in both failure analysis and forensic science books[131-139] and literature, with more volumes of case studies, journal papers, dedicated conferences and, increasingly, information posted on the internet. There is no doubt that journals such as *Engineering Failure Analysis* have made great strides in publishing revealing and incisive case studies from investigators. Its expansion in recent years is a reflection of the growing desire to publish detailed case studies so as to limit further re-occurrence of identical failures, if not eliminate them altogether.

When correctly and accurately documented, forensic (or failure analysis) case studies can be considered 'stories' with an educational message.[140,141] It is recognised that introducing case study analysis to the classroom will provide a wealth of practical, real-life examples that can be used to contextualise theoretical concepts, in addition to enhancing the overall student learning experience. Case studies have an important role, in both the skills and knowledge base development of students, by exposing them to real-world issues with which they may be faced. Moreover, it is now documented that students can learn more effectively when actively involved in the learning process.[142-144] It logically follows that there is a continued and growing need for both undergraduate and post-graduate courses in this arena. A forensic curriculum will allow a new generation of engineers to be made aware of the downside to new product development, and the serious consequences of product failure in the marketplace. Perhaps more importantly, there is a need to raise the profile of a growing archive of failure case studies, as discussed above. As well as being a vital tool for anyone involved in the safe design, construction, operation and maintenance of structures and components of all kinds, engineering failure case studies are increasingly being used to great effect in the teaching of materials science and, in particular, engineering.[145]

Direct practical experience of failed products shows that there is still great room for progress in preventative engineering, and raising awareness among product designers and manufacturing personnel of the need to examine and analyse failures utilising current tools which have been developed to a sophisticated level over the past decade.

# References

1. Petroski, H., ed., *Design Paradigms: Case Histories of Error and Judgment in Engineering.* 1994, Cambridge: Cambridge University Press.
2. Leonards, G., Investigation of structural failures. *Journal of the Geotechnical Engineering Division, ASCE*, 1980. 50(2): pp 187–246.
3. Gdoutos, E.E., K. Pilakoutas and C.A. Rodopoulos, eds., *Failure Analysis of Industrial Composite Materials*. 1st ed. 2000, McGraw-Hill.

4. Petroski, H., *To Engineer Is Human: The Role of Failure in Successful Design.* 1992, Vintage.
5. Schlager, N.Y., ed., *When Technology Fails: Significant Technological Disasters, Accidents, and Failures of the Twentieth Century.* 1994, New York: Gale Group.
6. Schlager, N., *Breakdown: Deadly Technological Disasters.* 1995, Detroit: Visible Ink Press.
7. *Code of Hammurabi.* 2008, http://au.encarta.msn.com.
8. Hale, M., *The History of the Common Law of England,* Gray, C.M., ed. 1973, Chicago: University of Chicago Press.
9. Arnold, E.A., ed., *A Documentary Survey of Napoleonic France.* 1993, Lanham, MD: University Press of America.
10. Reichhardt, T., NASA braced for culture shock as Columbia inquiry reaches verdict. *Nature,* August 2003. 424(6951): pp. 863–864.
11. Whittingham, R.B., *The Blame Machine: Why Human Error Causes Accidents.* 2003, Butterworth-Heinemann.
12. Oliver, A., *Engineers Fall Victim to Blame Culture.* 17 July 2003, New Civil Engineer (NCE).
13. Chiles, J.R., *Inviting Disaster: Lessons from the Edge of Technology.* 2002, New York: Harper Business.
14. Bond, J., The blame culture—an obstacle to improving safety. *Journal of Chemical Health and Safety,* March–April 2008. 15(2): pp. 6–9.
15. *Working Paper DGCA/06-WP/11,* in *International Civil Aviation Organisation, Proposals for Further Improvement of Aviation Safety Worldwide.* 22 March 2006: Presented by Austria on behalf of the European Community, Montreal.
16. Kádárová, J., K. Teplicka and K. Teplicka, Product life cycle costing. *Applied Mechanics and Materials,* 2015. 816: pp. 547–554.
17. Gullo, L.J. and D.G. Raheja, *Design for Reliability.* 2012, John Wiley & Sons, ISBN: 9781118310038.
18. Wallace, M.M. and J. Stephenson, Design for reliability, in *Design for X: Concurrent Engineering Imperatives,* G.Q. Huang, ed. 1996, London: Chapman and Hall. pp. 245–267.
19. Cherkaev, E. and A. Cherkaev, Optimal design for the worst case scenario, in *III European Conference on Computational Mechanics, Solids, Structures and Coupled Problems in Engineering.* 5–8 June 2006, Lisbon, Portugal.
20. Cherkaev, E. and A. Cherkaev, Principal compliance and robust optimal design. *Journal of Elasticity,* 2003. 72(1–3): pp. 71–98.
21. Dolan, T.J., Risk and failure analysis for improved performance and reliability, in *Proceedings of the Twenty-fourth Sagamore Army Materials Research Conference.* 1977, Raquette Lake: Plenum Press.
22. Jorma Salonena, P.A., Olli Lehtinenb and Mikko Pihkakoskic, Experience on in-service damage in power plant components. *Engineering Failure Analysis,* 2007. 14(6): pp. 970–977.
23. Williams, S., Forensic engineering. *Proceedings of the Institution of Civil Engineers,* 2018. 171(1): pp. 12–26, https://doi.org/10.1680/jfoen.17.00015.
24. Reboh, Y., S. Griza, A. Regulya and T.R. Strohaeckera, Failure analysis of fifth wheel coupling system. *Engineering Failure Analysis,* 2008. 15(4): pp. 332–338.
25. Petroski, H., *To Engineer Is Human: The Role of Failure in Successful Design.* March 1992, Vintage.

26. Darlastona, J. and J. Wintle, Safety factors in the design and use of pressure equipment. *Engineering Failure Analysis*, 2007. 14(3): pp. 471–480.

27. Le May, I. and E. Deckker, Reducing the risk of failure by better training and education, in *Third International Conference on Engineering Failure Analysis.* 13–16 July 2008, Sitges, Spain.

28. Cooke, N.J., *Stories of Modern Technology Failures and Cognitive Engineering Successes.* 2007, CRC Press.

29. https://en.wikipedia.org/wiki/San_Bruno_pipeline_explosion.

30. Esaklul, K.A., *Handbook of Case Histories in Failure Analysis.* 1993, ASM International.

31. Tawancy, H.M., A. Ul-Hamid and N.M. Abbas, *Practical Engineering Failure Analysis.* 2004, Marcel Dekker.

32. Shipley, R.J. and W.T. Becker, *Failure Analysis & Prevention.* ASM Handbook Vol. 11. 2002, ASM International.

33. Esaklul, K.A., *Handbook of Case Histories in Failure Analysis.* 1993, ASM International.

34. Hull, D., *Fractography: Observing, Measuring and Interpreting Fracture Surface Topography.* 1999, Cambridge University Press.

35. Gagg, C. and P. Lewis, Environmentally assisted product failure – Synopsis and case study compendium. *Engineering Falure Analysis*, 2008. 15(5): pp. 505–520.

36. Gagg, C.R., P.R. Lewis and C. Tsang, Premature failure of a vacuum pump impeller rotor recovered from a pitch impregnation plant. *Engineering Failure Analysis*, 2008. 15(5): pp. 606–615.

37. Mahgereteh, H., An analysis of the gas pipeline explosion at Ghislenghien, in *Belgium 40th Loss Prevention Symposium (T5).* 2006.

38. Lewis, Peter and Colin Gagg, *Forensic Polymer Engineering: Why Polymer Products Fail in Service.* 2010, Cambridge: Woodhead Publishing Limited, ISBN-10: 1439831149, ISBN-13: 978-1439831144.

39. Evans, R.R. and W.M. Lindsay, *The Management and Control of Quality.* 2010, South-Western Cengage Learning, ISBN: 0538452609, 9780538452601.

40. Birsch, D., *The Ford Pinto Case: A Study in Applied Ethics, Business, and Technology*, J.H. Fielder, ed. 1994, State University of New York Press.

41. O'Hare, D., M. Wiggins, R. Batt and D. Morrison, Cognitive failure analysis for aircraft accident investigation. *Ergonomics*, 1994. 13(11): pp. 1855–1870.

42. Zotov, D.V., Reporting human factors accidents, in *27th International Seminar of the International Society of Air Safety Investigators.* 1996, Auckland, New Zealand.

43. Nishida, S., *Failure Analysis in Engineering Applications.* 1992, Butterworth-Heinemann.

44. Gano, D.L., *Apollo Root Cause Analysis-A New Way of Thinking.* 3rd ed. 2007, Apollonian Publications, LLC.

45. Dugan, J.B., K.J. Sullivan and D. Coppit, Developing a low-cost high-quality software tool for dynamic fault-tree analysis. *IEEE Transactions on Reliability*, March 2000. 49(1): pp. 49–59.

46. Grubišića, V., N. Vulic and S. Sönnichsen, Structural durability validation of bearing girders in marine Diesel engines. *Engineering Failure Analysis*, June 2008. 15(4): pp. 247–260.

47. Russel, B.M., J.P. Henriksen, S.B. Jørgensen and R. Gani, Integration of design and control through model analysis. *Computers and Chemical Engineering,* 2000. 24(2–7): pp. 967–973.

48. Plumbridge, W.J. and C.R. Gagg, The mechanical properties of lead-containing and lead-free solders—meeting the environmental challenge. *Proceedings of the Institution of Mechanical Engineers, Part L: Journal of Materials: Design and Applications,* 2000. 214(3): pp. 153–161.

49. Plumbridge, W.J. and C.R. Gagg, Effects of strain rate and temperature on the stress–strain response of solder alloys, *Materials Science: Materials in Electronics,* 1999. 10(5–6): pp. 461–468.

50. Ramachandran, V., A.C. Raghuram, R.V. Krishnan and S.K. Bhaumik, *Failure Analysis of Engineering Structures: Methodology and Case Histories.* 2005, Materials Park: ASM International.

51. Dennies, D.P., *How to Organize and Run a Failure Investigation.* 2005, Materials Park: ASM International.

52. Bhaumik, S.K., An aircraft accident investigation: Revisited. *Journal of Failure Analysis and Prevention,* 2008. 8(5): pp. 399–405.

53. Gagg, C., Failure of stainless steel water pump couplings, in *Failure Analysis Case Studies III,* D.R.H. Jones, ed. 2004, Oxford: Elsevier UK. pp. 225–236.

54. Brown, S., Forensic engineering: Reduction of risk and improving technology (for all things great and small). *Engineering Failure Analysis,* 2007. 14(6): pp. 1019–1037.

55. Lewis, Peter Rhys, K. Reynolds and Colin Gagg, *Forensic Materials Engineering: Case Studies.* 2004, Boca Raton: CRC Press.

56. Noon, R.K., *Scientific Method: Applications in Failure Investigation and Forensic Science.* 2009, Boca Raton: CRC Press.

57. Lewis, G.L., *Guidelines for Forensic Engineering Practice.* 2003, American Society of Civil Engineers (ASCE).

58. Neale, B.S., *Forensic Engineering.* Institution of Civil Engineers (Great Britain). 1999, Thomas Telford Ltd.

59. *Rule 35, Civil Procedure Rules (the Woolf Report).* 26 April 1999, http://www.open.gov.uk/lcd.

60. Houck, M.M., ed., *Forensic Engineering (Advanced Forensic Science Series).* 2017, Academic Press, ISBN-10: 0128027185, ISBN-13: 978-0128027189.

61. James, P.D., *Death of an Expert Witness.* 2002, Gardners Books, ISBN 0571204201.

62. John, G., T. Edwards, A. Wright, M. Broadhurst and C. Newton, Learning lessons from forensic investigations of corrosion failures. *Proceedings of the Institution of Civil Engineers - Civil Engineering,* 2009. 162(5): pp. 18–24.

63. Neale, B., *Forensic Engineering: A Professional Approach to Investigation.* 1999, Thomas Telford.

64. Sherman, R.M., *Point of Impact: Case Studies of Forensic Engineering in Personal Injury Lawsuits.* 2001, Tucson, AZ: Lawyers and Judges Publishing Company Inc.

65. Thilmany, J., Working backward: Reverse engineering gets a role in detecting and preserving, *The American Society of Mechanical Engineers (ASME),* 2005. 127(6): pp. 36–38.

66. Raja, V.F. and J. Kiran, *Reverse Engineering - An Industrial Perspective*. 2008, Springer.
67. Brealey, Richard A. and S.C.M., Franklin Allen, *Principles of Corporate Finance*. 9th ed. 2008, McGraw-Hill Higher Education.
68. Lawson, D., *Engineering Disasters- Lessons to Be Learned*. 2005, London: Professional Engineering Publishing.
69. Heller, S., *Design Disasters: Great Designers, Fabulous Failure, and Lessons Learned*. 2008, Allworth Press.
70. Lewis, P. and C. Gagg, Aesthetics –v-function: The fall of the Dee bridge 1847. *Interdisciplinary Science Reviews*, 2004. 29(2): pp. 171–191.
71. Lewis, P., *Disaster on the Dee: Robert Stephenson's Nemesis of 1847*. 2007, Tempus Publishing.
72. Petroski, H., Success syndrome: Collapse of the Dee Bridge. *iCivil Engineering ASCE*, April 1994. 64(4): pp. 52–55.
73. Lewis, P. and K. Reynolds, Forensic engineering: A reappraisal of the Tay bridge disaster. *Interdisciplinary Science Reviews*, 2002. 27(4): pp. 287–298.
74. Toland, J., *The Great Dirigibles: Their Triumphs and Disasters*. 1972, New York: Dover Publications.
75. Vaughan, D., Autonomy, interdependence, and social control: NASA and the space shuttle challenger. *Administrative Science Quarterly*, 1990. 35(2): pp. 225–227.
76. *Report of Columbia Accident Investigation Board*. 2003, http://www.nasa.gov/columbia/caib/html/start.html. Columbia Accident Investigation Board.
77. Mahler, J.G. and M.H. Casamayou, *Organizational Learning at NASA: The Columbia and Challenger Accidents (Public Management and Change)*. 15 April 2009, Georgetown University Press.
78. Orlebar, C., *The Concorde Story*. 2004, Oxford, UK: Osprey publishing.
79. Trubshaw, B., *Concorde: The Complete Inside Story*. 3rd Rev Edtion ed. September 2004, The History Press Ltd.
80. Structural safety, 1997–1999: Review and recommendations, *The Standing Committee on Structural Safety*. 1999, The twelfth report of SCOSS.
81. Hendley, V., *The Importance of Failure*. http://www.prismmagazine.org/october/html/the_importance_of_failure.htm.
82. Marsh, G., Re-launch test for team Philips. *Reinforced Plastics*, December 2000. 44(12): pp. 32–33+35.
83. Jakobsen, B. and F. Rosendahl, The Sleipner platform accident. *Structural Engineering International: Journal of the International Association for Bridge and Structural Engineering (IABSE)*, 1994. 4(3): pp. 190–193.
84. Sleipner, A. http://members.home.nl/the_sims/rig/sleipnera.htm, n.d.
85. Selby, R., F.J. Vecchio and M.P. Collins, The failure of an offshore platform. *Concrete International*, 1997. 19(8): pp. 28–35.
86. Lewis, P.R., K. Reynolds and C. Gagg, *Forensic Materials Engineering Case Studies*. 1st ed. 2003, CRC Press, SBN-10: 0849311829, ISBN-13: 978-0849311826.
87. Faith, N., *Derail, Why Trains Crash*. 2000, London: Channel 4 London.
88. Faith, N., *Black Box: Why Air Safety Is No Accident*. 1997, Zenith Press.
89. Faith, N., *Blaze*. 2000, St. Martins Press.
90. Faith, N., *Mayday*. 2001, London: Channel 4 Books.
91. Faith, N., *Crash: The Limits of Car Safety*. 1997, Boxtree, Macmillan Publishers Ltd.

92. Lewis, P., *The Beautiful Railway Bridge of the Silvery Tay*. 2004, Tempus Publishing ltd.
93. Lewis, P. and A. Nisbet, *Wheels to Disaster: The Oxford Train Wreck of Christmas Eve 1874*. 2008, Tempus Publishing.
94. Pearson, C. and N. Delatte, Ronan point, apartment tower collapse and its effect on building codes. *Journal of Performance of Constructed Facilities*, May 2005. 19(2): pp. 172–177.
95. Griffiths, H., A.G. Pugsley and O. Saunders, *Report of the Inquiry into the Collapse of Flats at Ronan Point, Canning Town*. 1968, London: Her Majesty's Stationery Office.
96. *The Ronan Point Apartment Tower Case*. https://eng-resources.uncc.edu/failurecasestudies/building-failure-cases/. Summarized from Rouse and Delatte, Lessons from the Progressive Collapse of the Ronan Point Apartment Tower, Proceedings of the 3rd ASCE Forensics Congress, October 19–21, 2003, San Diego, CA.
97. Harding, G., Revised part A of the building regulations: Technical update. *The Structural Engineer*, 18 January 2005. 83(2): pp. 18–19.
98. Jones, D., ed., *Engineering Failure Analysis*. 1995, Elsevier.
99. Jones, D., ed., *Failure Analysis Case Studies*. Vols. 1, 2 and 3. 2004, Elsevier, ISBN-10: 9780080444475.
100. Evans, D.J. and J.M. West, An appraisal of underground gas storage technologies and incidents, for the development of risk assessment methodology. *British Geological Survey for the Health and Safety Executive*. 2008, Norwich, UK: HSE Books.
101. Lindberg, A.K., S.O. Hansson and C. Rollenhagenab, Learning from accidents – What more do we need to know? *Safety Science*, 2010. 48(6): pp. 714–721.
102. Gil, F., *Integrity Management: Learning From Past Major Industrial Incidents*. Vol. Book 14. 2004, Sundbury on thames UK: British Petroleum.
103. Donaldson, T, ed., *Loss Prevention Bulletin*. Rugby, England: IChemE., https://www.icheme.org/lpb
104. Thornton, C.H. and R.P. DeScenza, Construction collapse of precast concrete framing at Pittsburgh's midfield terminal, in *Forensic Engineering: Proceedings of the First Congress*. 1997, ASCE.
105. Feld, J. and K. Carper, *Construction Failure*. 2nd ed. 1997, New York: John Wiley & Sons.
106. Kaminetzky, D., *Design and Construction Failures: Lessons from Forensic Investigations*. 1991, New York: McGraw-Hill.
107. Mindeguia, J.-C., P. Pimienta, H. Carré and C. La Borderie, Experimental analysis of concrete spalling due to fire exposure. *European Journal of Environmental and Civil Engineering*, 2013. 17(6): pp. 453–466.
108. Lew, H.S., S.G. Fattal, J.R. Shaver, T.A. Reinhold and B.J. Hunt, *Investigation of Construction Failure of Reinforced Concrete Cooling Tower at Willow Island*, W.V, U.S.D.o. Labor, ed. 1979.
109. Kodur, V.K.R., Spalling in high strength concrete exposed to fire — concerns, causes, critical parameters and cures. *Proceedings, ASCE Structures Congress*: Advanced Technology in Structural Engineering, pp. 1–9, May 2000
110. https://www.indiatvnews.com/business/news-a-chronology-of-major-air-disasters-in-last-20-years-477034.

111. Shipley, R.J. and W.T. Becker, *Failure Analysis & Prevention*. ASM Handbook. Vol. 11. 2002, ASM International, ISBN: 978-0-87170-704-8.

112. Tawancy, H.M., A. Ul-Hamid and N.M. Abbas, *Practical Engineering Failure Analysis*. 2004, Marcel Dekker.

113. Lewis, P.R., K. Reynolds and C. Gagg, *Forensic Materials Engineering: Case Studies*. 2004, Boca Raton: CRC Press.

114. Thilmany, J., *Working Backwards. Reverse Engineering as Applied to Failure or Forensic Investigation*. June 2005, American Society for Mechanical Engineers (ASME).

115. Ezrin, M., *Plastics Failure Guide: Cause and Prevention*. 2013, Hanser, ISBN 978-1-56990-449-7.

116. Wright, D., *Failure of Plastics and Rubber Products: Causes Effects and Case Studies Involving Degradation*. 2001, RAPRA.

117. Hays, M., D. Edwards and A. Shah, *Fractography in Failure Analysis of Polymers*. 2015, William Andrew, ISBN-10: 0323242723, ISBN-13: 978-0323242721

118. Lewis, P., *Designing with plastics. RAPRA Reviews*. 1993, Shrewsbury, UK: Rapra Technology

119. Lewis, P., High performance polymer fibres. *RAPRA Reviews*, 1999. 9(11). ISBN: 9781859571590

120. Pilkey, W., *Peterson's Stress Concentration Factors*. 3rd ed. 2008, New York: Wiley.

121. Young, W., *Roark's Formulas for Stress and Strain*. 8th ed. 2011, McGraw-Hill, ISBN: 9780071742474.

122. Delatte, N., An approach to forensic engineering education in the USA. *Proceedings of the Institution of Civil Engineers - Forensic Engineering*, 2012. 165(3): pp. 123–129, ISSN 2043-9903.

123. Delatte, N.J., *Beyond Failure: Forensic Case Studies for Civil Engineers*. 2009, American Society Civil Engineers, ISBN-10: 0784409730, ISBN-13: 978-0784409732

124. Bosela, P.A., Failure of engineered facilities: Academia responds to the challenge. *Journal of Performance of Constructed Facilities*, 1993. 7(2): pp. 140–144.

125. Hanna, A.S. and K.T. Sullivan, Bridging the gap between academics and practice: A capstone design experience. *Journal of Professional Issues in Engineering Education and Practice*, 2005. 13(1): pp. 59–62.

126. Jones, D.R.H., *Engineering Failure Analysis*. Elsevier B.V.

127. The act of technical investigation into the causes of accident at the Sayano-Shushenskaya HYDROELECTRIC STATION, 17 August 2009; In the branch of the open joint-stock company rushydro, Sayano-Shushenskaya GES p. s. Neporožnego; Federal service for ecological, technical and Atomic supervision; 3 October 2009 year. Source: http://www.gosnadzor.ru/news/aktSSG____bak.doc.

128. Sayano-Shushenskaya power station station accident. 2009, https://en.wikipedia.org/wiki/2009_Sayano–Shushenskaya_power_station_accident.

129. Kardon, J.B., *Guidelines for Forensic Engineering Practice*. 2nd ed. 2012, Forensic Practices Committee of the Technical Council on Forensic Engineering, American Society of Civil Engineers, ISBN: 9780784412466.

130. https://www.nspe.org/resources/issues-and-advocacy/professional-policies-and-position-statements/nspe-nafe-joint-position.

131. Carpenter, John, ed., *Forensic Engineering: Informing the Future with Lessons from the Past.* 2013, ICE Publishing, ISBN-10: 0727758225, ISBN-13: 978-0727758224.

132. *Beyond Failure: Forensic Case Studies for Civil Engineers.* e-book, Civilax – 5 August 2014, http://k2s.cc/file/ca92de095be28.

133. Zufelt, Jon E., ed., *Construction and Forensic Engineering, Congress on Technical Advancement.* 2017, Reston, VA: American Society of Civil Engineers, ISBN: 9780784481035.

134. Houck, Max M., ed., *Forensic Engineering (Advanced Forensic Science).* 2017, Wesport, CT: Praeger, ISBN-13: 978-0128027189, ISBN-10: 0128027185.

135. Hoffman, Conrad, *Forensic Engineering Experience & Example. The Forensic Engineer and Premises Liability.* December 2017, Pegasuze, ISBN-10: 0578196085, ISBN-13: 978-0578196084.

136. Blokdyk, Gerardus, *Forensic Engineering.* 3rd ed. 2018, 5STARCooks, ISBN-10: 0655308032, ISBN-13: 978-0655308034.

137. McLay, Richard W. and Robert N. Anderson, *Engineering Standards for Forensic Application.* September 2018, Academic Press, ISBN-10: 012813240X, ISBN-13: 978-0128132401.

138. Fiorentini, Luca and Luca Marmo, *Principles of Forensic Engineering Applied to Industrial Accidents.* December 2018, Wiley-Blackwell, ISBN-10: 1118962818, ISBN-13: 978-1118962817.

139. Zerbst, Uwe, Christian Klinger, Dirk Bettge and Richard Clegg, ed., *The Seventh International Conference on Engineering Failure Analysis.* 2018, Elsevier Ltd, ISSN: 1350-6307.

140. Baillie, C., *Effective Learning and Teaching in Engineering.* 1st ed. 28 October 2004, Routledge Falmer.

141. Grant, R., A claim for the case method in the teaching of geography. *Journal of Geography in Higher Education,* 1997. 21(2): pp. 171–185.

142. Raju, P. and C. Sanker, Teaching real-world issues through case studies. *Journal of Engineering Education,* 1999. 88(4): pp. 501–508.

143. Smith, C., Student written engineering cases. *International Journal of Engineering Education,* 1992. 8(6): pp. 442–445.

144. Raju, P.K. and C.S. Sankar, *Introduction to Engineering through Case Studies.* 2005, Anderson, SC: Taveneer, ISBN-10: 1930208782, ISBN-13: 978-1930208780.

145. Lewis, P.R. and C.R. Gagg, Post-graduate forensic engineering at The Open University, in *Proceedings of the International Conference on Innovation, Good Practice and Research in Engineering Education.* 24–26 July 2006, University of Liverpool, UK.

# Initial Aspects of Forensic Failure Investigation

# 2

## 2.0 Introduction

How will the forensic engineer approach a failure case? When an accident, component or system malfunction occurs, an engineering specialist may well be instructed to commence his detective work – examining the scene of failure, observing, documenting and characterising the incident (Chapter 1). Evidence gathered will include prior history of failure, environmental influences, failure modes and witness statements. Failure debris will require scrutinising to develop the most cost-effective approach to determining the actual failure mechanism at play, along with a root cause of failure. Once identified and analysed, the findings should lead to a series of proposals for corrective measures calculated to prevent future repetition of such failures. The more background data derived at the outset of investigation, the more effective the analysis will be in offering sound recommendations for remedial action.

At the start of a failure enquiry, the investigator will adopt an implicit assumption: that there are rational causes for failure and reasonably objective ways of determining those causes. The basis of investigation will therefore start (and then proceed) by a logical analysis of the way failure has occurred, with a range of possible causes (failure scenarios) tackled systematically to arrive at the true failure mode. There are a variety of initial basic methods that can be employed to determine why something failed. These methods are typically used (and taken for granted) in everyday life. Standard fault-finding guides for domestic equipment, for example, are usually tabulated as a checklist and suggest several possible causes for one specific symptom. It is up to the user to investigate each possible fault mode systematically, and in turn, to find the faulty component. By eliminating parts that are functioning correctly, the faulty part should be located quickly and remedial action taken. It is therefore often possible to evaluate failure using common-sense ideas and a simple checklist. Advances on the simple checklist are discussed in Section 1.4, where a range of qualitative, semi-quantitative and quantitative causal techniques are considered. However, there are an additional number of basic investigative methods that are frequently forgotten by textbooks *and* in the classroom. Starting with 'assessment' of the situation, three understated skills will be reviewed. Being somewhat abstract assets, these skills are considered essential tools for undertaking any investigation. The remainder of this chapter will then explore a range of more tangible assets, the use of which should ease the deductive process.

## 2.1  Three Essential 'Abstract' Assets for the Investigator

### 2.1.1  Assessing the Situation

At the outset, it can be said that the art of a forensic practitioner necessitates both logical reasoning and deduction. Therefore, the basis of any forensic investigation is deductive reasoning. This is an intuitive process of elimination by meticulous collection of all facts surrounding a particular incident or event. However, 'intuition' cannot be taught in the classroom; it will only develop with experience over time. In a similar vein, the power of 'deduction' is an additional abstract asset that will also be honed by both time and experience.

### 2.1.2  Initial Visual Observation as a Fact-Finding Exercise

There is a further abstract asset or tool that is often taken for granted: there is a tendency to neglect the vital importance of our eyes – the simple power of visual observation, so that others may see what we have seen and the importance attached to it. As the great (fictional) detective Sherlock Holmes said to his medical doctor sidekick, 'Watson, you see but you do not observe (perceive). The distinction is clear' (quote from *A Scandal in Bohemia*). When the forensic engineer examines debris from an accident he or she must look at much more than its general features. The aim must be to seek evidence elucidating what might have caused the debris to reach its current state, as illustrated by the following example.

> A result of heating aluminium to a point approaching its melting point, followed by an impact, will result in a phenomenon termed the 'broom-straw' effect or fracture (Figure 2.1). Although not fully understood, the mechanism at play is considered to be one of delamination between oriented microstructural grain boundaries – orientation being imparted by primary materials processing (sheet rolling for example). The degree of broom-straw formation will be dependent on exposure to heat and the force of impact. The presence of broom-straw, on any part of a post accident airframe structure, is considered highly indicative of in-flight fire – the assumption being that heating occurred in flight and stress occurred at impact. Although not a 100% certainty, evidence of broom-straw on aluminium airframe structures is highly indicative of an in-flight fire.

### 2.1.3  The Failure Scenario

After an accident or incident, the investigator will attempt to determine both the cause and the circumstances at play. Acute observation will allow

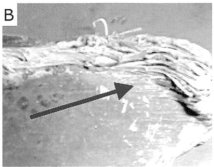

**Figure 2.1** (A) Broom-straw appearance of aluminium aircraft structure. (B) Closer view of broom-straw showing structural de-lamination. (Courtesy: Dudley Knox Library, Naval Postgraduate School, Monterey, CA.)

more than just scrutiny of the artefact/s under examination; it may also provide an insight into the situation that led to failure in the first instance. Close observations of this nature will allow assessment of the circumstances from which the physical evidence was gathered, thus triggering number of possible scenarios that may have led to the failure in question – the ultimate aim being an appropriate failure scenario that will accurately reconstruct the failure at hand. With this approach, the investigator will attempt to isolate events implied (or revealed) by available evidence. As the investigation evolves, further evidence may well be revealed, requiring adjustment to the initial failure scenario. As additional evidence is revealed, further re-iteration of the failure scenario will be required. The investigator will then test each iteration cycle against all evidence as it successively develops, to finally arrive at the most likely explanation of what occurred.

## 2.1.4 Summary of Understated Investigative Skills

As with deductive reasoning and observation, failure scenario development is not readily taught in the academic learning environment, practice (practical experience) being the paramount tutor. However, without doubt, descriptive case-study examples represent an important problem-solving tool, as reference material for both the student and practising professional.

The range of qualitative, semi-quantitative and quantitative causal techniques, considered in Section 1.4, provide a checklist (an 'aide-mémoire') pathway intended to assist investigative thinking. Prior to undertaking any laboratory analysis, there will be a raft of further actions to aid the investigator with his/her deductive process. Starting with visual observation, the balance of this chapter will review these more tangible assets.

## 2.2 Visual Observation

It should already be clear that the aim of all forensic investigation is to determine what actually happened during an incident or accident, based on *all* available material evidence. Section 2.1.2 above introduced the power of visual observation as a fact-finding exercise during the course of any forensic analysis. To take this point further, sharp observation will allow more than scrutiny of detritus or artefacts under examination: it may also provide an insight into the situation that led to failure in the first instance. Visual observation will allow assessment of the circumstances from which physical evidence was gathered, and provide insight as to a possible train of events (failure scenarios) leading to failure. Direct visual observation can be made by

- A visit to the scene of the accident or failure
- Selection of evidence for laboratory examination
- Inspection of 'scene of incident' photographs (and/or pertinent documentation) taken by bodies such as the police or factory inspectorate
- Consideration of witness statements from the incident or accident

Visual observation can be critical for crack detection. A human eye with 20/20 vision is able to resolve features as small as 75µm in size at a distance of 25 cm. It is possible under perfect conditions (on a mirror-polished surface) to detect a crack with a crack opening dimension (COD) as small as 10µm. However, the minimum detectable COD becomes much larger if the surface is rough or not perfectly clean. Surface features such as scratches and machining marks present visual 'noise' that will effectively mask any cracking. For this reason, other methods to aid crack detection will normally be required at a later date.

In many cases it is common practice to obtain an identical product or component (an exemplar unit) for direct comparison with its broken counterpart. Visual comparison can be undertaken as a simple method for checking if parts are identical and, if not, the reason for any divergence. Many manufactured articles are now identifiable from logos, date stamps and manufacturing codes either printed or embossed on the product, or in a concealed position within the product[1] (is it a counterfeit?). If the material of construction is unknown, it must be analysed for its constituent parts. However, although direct comparison with an unaffected product or component has a clear advantage, it is not always possible (air accidents for example). As a minimum, visual comparison should aim to identify the following product features:

- Overall dimensions
- Distortion in dimensions

- Fit with matching parts
- Surface quality
- Signs of environmental attack
- Traces of wear
- Identity marks
- Possible counterfeit product/component

## 2.2.1 Case Study: Visual Comparison

Visual observation can be used as a comparator, as illustrated by the following case study.[2] Here, simple comparative observations are used in a manner similar to fingerprint identification. A stolen Jaguar was recovered, and had sustained damage to the windows, roof and bodywork during the course of the theft. In addition, all four steel road wheels had their hubcaps missing. However, the owner stated that the steel road wheels had been substituted with his original alloy wheels, an expensive 'extra' for that make of vehicle. He submitted a substantial claim to his insurers for bodywork repairs, and for purchase of new alloy road wheels to replace those stolen from his vehicle. Similar to fingerprints, the observation of 'hub-prints' (the impressed markings from the wheel hub into the wheel rim, and vice versa) can identify mating surfaces that have been in intimate contact for extended periods of time. Examination was restricted to one side of the Jaguar. Road wheels on the opposite side were left untouched to facilitate a similar investigation by other interested parties at a later date.

Initial observation of the wheel rims revealed no finger markings in the wheel grime (Figure 2.2A). As the inner surface of the rim contains a deep well, it is natural to hold or grasp the wheel with your fingers in this well to take the weight of the wheel and provide a balanced lift. Lack of any markings in the road grime was indicative of wheels that had not been removed for some considerable period of time and travel. Arc lengths generated from bolting the wheel rim to the hub (Figure 2.2B) were measured and a match was observed

**Figure 2.2** (A) Undisturbed road dirt layer on wheel rim; (B) compare with impressed markings transferred from wheel hub to wheel.

between the flange and rim. In addition, machining marks from the flange had been impressed onto the wheel rim, and were clearly seen within the arc markings. The pitch of these machining marks was identical on both flange and wheel rim. Seven additional matching features were found on each wheel, verifying the initial observation that the steel road wheels had been on the vehicle and not removed for a substantial amount of time and mileage.[2] The claim was undoubtedly fraudulent. When advised of the criminal nature of fraud, the Jaguar owner rapidly withdrew his claim. Furthermore, the insurers advised him that they would not cover repair costs for damage sustained during the theft, nor would they insure him at any future time.

## 2.2.2 Case Study: Common (Engineering) Sense

The end of a shaft that connected to a tractor power take-off (PTO) coupling failed in service (Figure 2.3A). The shaft powered a cattle feed distributor, that is a hopper holding about 4 tons of feed, which is driven along a line of cows to deposit enough feed opposite each for a meal. The PTO output, from the tractor, rotated at 700 rpm, and was geared down by chains and sprockets to rotate the screw feeder at 20 rpm. The tractor would then drive slowly along the line of cows. Torsional force on the drive shaft would have been low, as a direct result of the speed reduction gearing. The feed distributor itself was about 14 years old.

The farmer stated that he was driving along the line of cattle, when a bucket fell off the top of the feeder bin and got caught in the screw mechanism. The feeder jammed and this shaft broke. However, the investigating engineer did not understand how sufficient torsional force could have been built up to break this shaft. The question that required an answer was: 'how *did* the shaft fail?' If it had been as a result of traumatic overstress then it would be classed as 'accident damage' and an insurance claim accepted. Conversely, if it had been the result of 'wear-and-tear', a claim would not be accepted. It was clearly

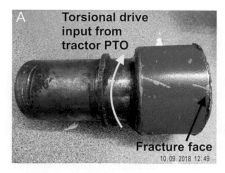

**Figure 2.3** (A) Cattle feed distributor shaft, broken at its tractor Power Take Off (PTO) connection. (B) Heavy post-failure fracture surface damage, with two undamaged low-lying sites circled in black.

apparent that both fracture faces had sustained extensive post-failure damage (Figure 2.3B). A heated debate ensued, as to whether fatigue striations were visible at two low-lying areas in the surface damage (circled on Figure 2.3B). However, appreciation of simple engineering principles silenced all raised voices. The fact that failure was sited across a diameter that was some 25% larger in cross-sectional area than that at the PTO connection totally ruled out torsional overload as a failure mechanism. Any torsional overload would result in failure of the *smaller* shaft diameter, and *not* the larger. This failure was therefore a direct result of fatigue (ie wear and tear) over a long period of time. As such, any insurance claim would be repudiated on the grounds of mechanical failure.

In summation, initial evidence at the scene of failure must be evaluated by careful visual observation, based on a common-sense approach, along with a refusal to jump to conclusions. This may entail a long and often tedious examination of the evidence left by an accident. Other debris may hide critical trace evidence, for example. However, information gleaned by simple observation may turn out to be highly significant and pre-empt the need for further or detailed and costly laboratory examination. An example of the strength of visual observation is given in Chapter 4 of a forensic engineering textbook,[2] where simply looking and listening revealed how the failure of a crane carrier bar occurred. Laboratory examination of the fractured carrier bar would *not* have identified the cause of failure, and would have been quite irrelevant, as the solution to the problem lay in circumstances at play *prior* to failure.

## 2.3 Forensic Photography

Gathering of any and all visual evidence must be recorded by camera or video – at the scene of an incident, or of a failed product, or at the dismantling of a failed device. 'A simple picture is worth more than a thousand words' is literally true, as well as being more convincing at a later date or in a courtroom. Forensic photography demands more than a 'point-and-shoot' approach to recording information. The general area in or around the vicinity of the incident is always important, but it is of little use photographing the surrounding area when the clue to failure lies in one small area or a surface feature that would pass unnoticed by a non-specialist onlooker. Careful examination and photography of the fractured artefact should follow, concentrating on any relevant details on external surfaces that are visible to the eye without any magnification. The photographic record should provide information on the size and condition of all pieces, and should show the relationship of any fracture to its component parts. As many pictures should be taken as is thought necessary to define and isolate key features on samples. At a later date such pictures, as an aide-mémoire, may well become invaluable.[2]

Lighting conditions are important in bringing hidden detail into picture. Shadows can often obscure such detail, and shadow-free photographs can be achieved only with very even illumination. The illumination source can be shielded either by an opaque screen (such as grease-proof paper) or a ring-illumination source used directly with the camera. The positive use of shadows is also very important, however, in highlighting relief on surfaces and is normally achieved with oblique or slanting illumination from a point source, such as the sun or a fibre-optic light.

When glass is broken, features are frozen into its fracture surface. As with metals, these fracture surface features provide a fingerprint of information regarding fracture initiation (origin), loading spectrum, crack propagation (crack growth and driving stress) and therefore the mechanism of failure. As an example of both lighting conditions and the importance of fracture surface features, Figure 2.4 shows macro-photographs of the fracture surface of a broken glass lens. Wallner lines are the rib-shaped marks seen in Figure 2.4, formed by propagating elastic waves reflected off free surfaces – thus creating a Wallner (lambda-shaped) line pattern seen on the fracture surface. Although Wallner lines do not show the *exact* crack front profile, they are curved in the same direction and can help in deducing which way a crack was running. For the lens in question, the crack propagation path ran from left to right in Figure 2.4, with a primary crack arrest mark arrowed in red on Figure 2.4A. This feature provides an indicator of where crack growth stopped (or was momentarily delayed), prior to resumption of crack propagation. The line curvature would suggest crack propagation accompanied by bending stress. Had the Wallner lines formed in a more perpendicular orientation, it could be said that thermal stress had contributed to crack propagation. The point being made here is that detailed observation of this glass fracture required good lighting (fibre-optic) and viewing conditions.

**Figure 2.4** (A) Photo-macrographs showing elastic waves reflecting off the free surfaces of a broken glass lens, creating Wallner (lambda-shaped) lines on the fracture surface as arrowed in blue. (B) A closer view of the lines. Note: crack running from left to right; primary crack arrest mark shown by red arrow; Wallner line curvature would suggest crack propagation accompanied by bending stress.

Of these conditions, an *acute* viewing angle was paramount, rather than one normal to the fracture surface, along with a dark, heavily shaded area beyond the glass component.

Rapid advances in digital imaging technologies have greatly improved many aspects of forensic photography. Digital imaging makes it possible to capture, edit and output images faster than processing conventional film. It is also possible to import individual frames of video for enhancement. Techniques that used to be applied in the darkroom through trial and error can now be used on a computer, and the results are immediately visible on screen. There are now several techniques available using digital imaging that were previously not possible using traditional photographic means (such as the ability to correct the perspective of an image). As long as it contains a scale of reference it is possible to take an image that was shot at an incorrect angle and correct it so that the scale is the same across the plane of focus. This has significance in cases where measurements must be taken of the evidence in an image. Further detailed information on a range of digital imaging techniques can be found in appropriate textbooks.[3] However, even now conventional film preserves more data than many high-resolution digital images, so it is still useful for record purposes. Furthermore, the analysis of old photographs in cold cases is an important area of research. Conventional film can of course be scanned to produce digital images, and subsequently enlarged to show details of interest.[4,5]

Considering both visual observation and forensic photography together, it can be said that they provide a useful combination for initial analysis and recording of failures. Such activity is essential before *any* laboratory analysis. Photography of the failed product, along with the actual context of failure, will provide a permanent record – frequently useful at a later stage of an investigation or in a court of Law.

## 2.4 Record Keeping

Observations made at the scene must be chronicled, usually with the aid of sketches, note taking and photography of the debris. The more complex the debris field, the greater will be the need for multiple images of the remains, thus ensuring that the final resting position of key components is accurately recorded. When compiling a case file, it is vital to ensure that record keeping is accomplished in an organised manner *and* initiated at the outset of investigation or instruction. The suggestion that this record might become part of a legal proceeding should be kept in the back of the investigator's mind at all times – simply as such a record will become the 'paper trail' for authenticating and tracing his/her actions, assumptions and opinion. A simple notebook or database with all relevant documentation, samples referenced and pertinent information should be kept. Information noted should include (but not be limited to)

- Relevant documentation bundles and witness statements
- Raw materials and any alloying or compounding
- Processing and any relevant standards
- Photographic record, to include date and time
- Signs of misuse or abuse and service environment
- Location from which sample was recovered
- Date and time of collection, along with number of pieces in sample if broken
- Any manufacturing or tooling marks on sample or design stress raisers

Furthermore, any evidence recovered from an accident or failure investigation site may eventually be presented and questioned in the courtroom. For evidence to be presented during the course of a trial, it will have made a convoluted journey from the initial scene of recovery, passing through the hands of a number of experts during the course of their examination, to finally reach the courtroom. This journey must be made in a validated and secure manner so that all involved can be assured that it (the evidence) has not been contaminated and that the evidence *was* that relevant to the investigation. In order to guarantee legitimacy of the evidence, the forensic engineer is required to follow a routine, termed the 'chain of custody', when collecting and/or handling items of evidence. A chain of custody of evidence will therefore represent a complete trail of where, when and by whom evidence was obtained. In addition, it will also clarify who had access to the evidence, and where it was stored prior to receipt in the courtroom. More detail regarding the chain of evidence can be found in Section 2.9.1 below.

## 2.5 Witness Evidence

If product failure has resulted in death, injury or damage to property, statements from those in the vicinity will often be available. The earlier statements have been gathered, the better. Memory fades, and the later a witness is asked to recall events, the greater the chances of error, particularly if the litigation process had been instigated. Bias creeps into statements, and there is usually a lack of technical detail, as the interrogator would generally be a lawyer having little or no technical expertise. If there are one or more victims, their memory may be affected by the accident.[1] For example, you can take two eyewitness statements of a particular accident/incident and, on reading them at a later date, they will appear to be recounting two totally different events. It is not that the individuals are lying, or trying to 'cover up' the truth; it is more to do with the brain's ability to recall events that occurred in a moment of shock, danger or extreme stress. As accidents generally occur very rapidly, the witnesses and/or victim often have great difficulty recalling the actual sequence of events. Therefore, material remains may often become a silent witness to an incident. Naturally, material evidence must be checked against any available witness statements.

## 2.6 Documentary Evidence

Documentary evidence is a vital source of facts surrounding a case, particularly if equipment is monitored regularly and automatically. In industrial cases, documentation is often copious as a result of health and safety legislation, and includes

- Maintenance sheets
- Design data
- Manufacturing records
- Quality control information
- Applicable standards

With such a wealth of information available, the problem is one of sifting records for the key fact or facts that will reveal how and why the incident occurred. For example, industrial processes are now usually automatically monitored by various sensors in the equipment. Variables such as time, temperature, pressure and volume of contents (die casting or injection moulding, for example) are measured and recorded remotely in computer databases. However, the data are usually indigestible until analysed, and visualised in the form of graphs or diagrams. Trends can then be seen to help interpretation of the facts of the matter at hand. A great advantage of such data compilations is their objectivity, with any bias being easily detected. Thus a faulty detector, such as a thermocouple, will show up when compared with other thermocouples in the system. A cross-check will be available from calibration records, etc. Systematic analysis will also expose whether or not the records are a fair representation of events. Sensors may be missing from critical parts of the system, although any absence can often be inferred by scrutiny of the data supplied. Computer records are not always infallible: computers crash, and data can be lost or transposed, as any PC user knows to their cost, but they are generally an invaluable source of unbiased measurements.

### 2.6.1 Case Study: Fatal Aircraft Crash

A case study that exemplifies the importance of record keeping, witness evidence, documentary evidence and relevant standards specification – along with meticulous analysis – can be illustrated by air crash investigation. On 4 October 1992 an EL AL Boeing 747-200 freighter crashed in Amsterdam, killing all four people on board and 51 people on the ground (Figure 2.5A). A schematic of the aircraft attitude prior to impact is shown in Figure 2.5B. The cause of the crash was determined to be the loss of the number 3 and 4 engines, which were stripped off the right wing, thus causing the Boeing freighter to crash as it manoeuvred towards the airport.

Subsequent investigation by the Nederlands Aviation Safety Board[6] found that the number 3 engine separation was as a result of breakage of the fuse pin, probably initiated by fatigue (Figure 2.5D). The pin was designed to break when an engine seizes in flight. As a result, the number 3 pylon and engine separated from the wing in such a way that the number 4 pylon and engine were torn off, part of the leading edge of the wing was damaged and the use of several systems was lost or limited. With both of the engines stripped off the right wing, the Boeing 747-200 freighter crashed as the pilots lost control. The final report also found the design and certification of the Boeing 747 pylon to be inadequate to provide the required level of safety. Furthermore the structural integrity inspection method that had been in place, clearly failed.[6]

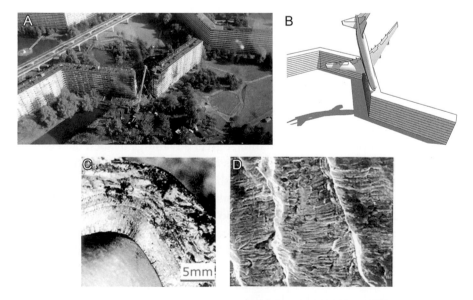

**Figure 2.5** (A) Aftermath of the 4 October 1992 EL AL Boeing 747-200 freighter air accident. (B) Schematic of aircraft attitude immediately prior to impact. (C) Optical micrograph of fuse pin fracture. (D) Scanning electron micrograph of fuse pin fracture surface. (Courtesy of Wikimedia Commons, licensed under Creative Commons.)

## 2.7 Product and Material Standards

Compliance with product standards and regulations is now a major issue for many manufacturers. Unless products comply with specific standards, manufacturers' ability to sell their products on the open market can be severely compromised. The American Society for Testing and Materials (ASTM) and British Standards Institution (BSI) have produced standards covering the safety, performance and reliability of most products, including the influence of mechanical and environmental factors. In addition, each standard is reviewed and updated periodically, thus ensuring continued relevance.

When considering any failure, an appropriate standard written to encompass the component or its application (e.g. wire ropes, chains, seating, etc.) may be applicable. Standards are an invaluable comparator against which failed components can be matched. Furthermore, equivalent standards exist in Europe (ISO), Germany (DIN) and Japan (JIS) – all are excellent sources of reference. As the majority of accidents occur within the home, domestic appliances are particularly well covered by appropriate standards. In the UK, statistics from the Royal Society for the Prevention of Accidents (RoSPA) and the Health & Safety Executive (HSE) show falls from ladders account for almost a third of all injuries involving falls from heights and cost the UK economy £60 million every year, as well as being the number one cause of workplace deaths.[7] A person falls, suffers injury and damage is found on the stiles or rung hooks, feet, etc. The user seeks to sue the supplier, alleging a manufacturing defect or flaw, poor construction, insufficient strength of alloy sections, etc. Such allegations are usually unfounded as there are stringent standards (BS EN131, 2018) covering the design, materials, construction and performance for industrial, trade and domestic ladders and steps. The principal reason for a ladder accident is that of ladder slipping, largely due to the user overreaching or because the ladder had not been properly secured. However, it is human nature to blame someone else for one's misfortune, so after suffering serious injuries while using lightweight step-stools or ladders, the user will frequently seek to establish that a manufacturing fault was responsible – as exemplified by the following case study of a step ladder failure.

### 2.7.1 Case Study: Step Ladder Failure

A stepladder was alleged to have failed while being used to paint the gutters of a single-storey house (Figure 2.6A). The user was a tall, heavy man but within the weight specified on the labelling of the steps for 'light domestic use'. The foot of the left stile had sustained a slight inward bend, whereas the right-hand stile had been bent inward to an approximate 45 degree angle. The stile is a rectangular box section aluminium alloy that has buckled inward at a

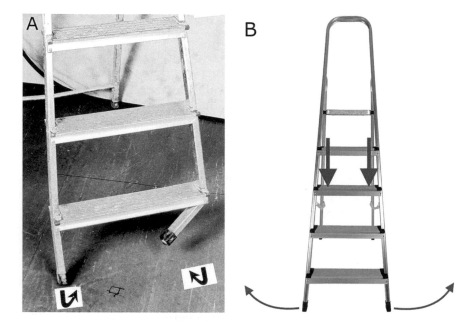

**Figure 2.6** (A) A stepladder as received for examination. Note the inward orientation of bending damage sustained by each stile. (B) A correctly used stepladder will generate compressive forces in the stiles that react against the floor, generating force components that act to spread the stiles outward.

point directly beneath the bottom tread, as shown in Figure 2.6A. The claimant alleged that the steps were of defective construction and/or that the box section was not strong enough or that there was a fault in the metal.

As in all designs of step stool ladders the front and back legs splay outward so that, whether viewed from the front, back or the sides, when they are opened out and placed on a firm, level floor the four feet are farther apart than any of the steps or side braces. Thus any load acting downward from someone mounting or standing on the upper treads generates compressive forces in the stiles that are reacted to by the floor and hence set up force components acting to spread the feet outward (Figure 2.6B). How then could the bottom of both stiles have been bent inward? Deformation imparted to both stiles has been caused by an inward acting horizontal force, with the lowest tread serving as a fulcrum. No sign of any similar mechanical distress could be found on any other part of the ladder structure. There was no evidence of faulty construction or material weakness and, even if there were, the major force that caused this failure was directed from the outside of the right stile foot.

The most likely way that a force system of sufficient magnitude could have been set up is if the stepladder had been lying almost flat on its left side, while a downward force was directed against the bottom of the right hand stile, as depicted in Figure 2.7. This would bend the left stile inward to some

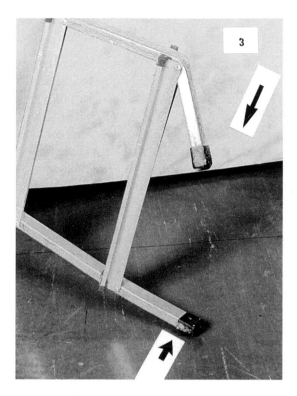

**Figure 2.7** Position of the steps when the stiles (legs) were bent, with the upper arrow representing an externally applied loading (the weight of the user's body hitting the stile) and the lower arrow showing a reaction force from the ground.

degree, whilst simultaneously imparting extensive inward bending of the right hand stile. From considering simple engineering mechanics, both stiles would rotate about the fulcrum of the bottom tread. How could such a situation arise? Simply by the user overreaching sideways while near the top of the steps, causing them to topple sideways onto their left side. On the way down some part of their anatomy struck the projecting foot of the right stile as the left stile lay nearly flat on the ground. The full weight of their falling body was applied to the foot of the right stile, producing exactly the form of damage shown in Figure 2.7.[2,8]

As a final word regarding product standards, much of the required background design and analysis will be included and will be of particular value in product liability disputes. All that may be required of the forensic investigator, or Single Joint Expert (SJE), is a comparative analysis between the relevant standard, and the product/component in dispute. It should quickly become apparent if a product did not conform to the required standard.

## 2.8 Patents

Patents are a rich and fertile source of information when undertaking a failure investigation. The inventor will claim a new product which professes to solve a particular problem and, as such, product failure will represent an important element of the problem. Furthermore, the failure of new products can be studied simply by referring to the patent that establishes that product. Registered design databases, such as that run by the UK Patent Office, are a valuable source of pictures of a design at the date of registration. The UK trademark database is useful for determining the ownership and identity of many commercial products. All such databases are readily accessible from the World Wide Web.

Among the foremost worldwide databases are those from Espace[9] and the US Patent Office.[10] They itemise patents from the principle patenting countries, primarily the USA, Europe and Japan, and complete patents can be downloaded for free. There are specialized databases for specific areas of failure, which are of great use in monitoring specific designs. The U.S. Food and Drug Administration (FDA) has a very large database of failures of medical devices at MEDWATCH, allowing an investigator to follow the failure history of specific hip joints, heart valves, stents and similar implants.[11] Since there is usually a plethora of different designs, identification by tradename or trademark on the compilation gives the required information.[9,10] Such databases tend to be more useful than statistical summaries such as that maintained by the UK Home Office for accidents in the home.

## 2.9  Surviving Remains (Detritus)

Material evidence that survives any failure may well provide wordless, but revealing, evidence of product history, often becoming a key to unlocking the way it failed. Such remains are normally preserved as the material evidence for further investigation, thus providing the proof positive of the cause of an accident and/or the justification of a case for compensation. However, there are occasions where a product or component can become broken or shaped as a *result* of the incident, rather than being its *cause*. So the material evidence has to be seen in its context, and not in isolation. For example, a broken plastic ladder tip started an investigation to determine the cause of a ladder slip,[2,8] where the user was severely injured by falling to the ground. Subsequent inspection of the accident scene showed that the tip had broken *after* the ladder had started slipping down the wall, and could not therefore be taken as the *cause* of the accident.[2,8] Trace evidence of its journey down the wall (scrape marks) provided visible evidence of how the tip was broken.[1,2]

### 2.9.1 Gathering of Evidence and the 'Chain of Evidence'

If the failure, accident or incident could subsequently end up as a criminal investigation, evidence gathered should be sealed in a tamper-evident bag or any other bag designed for that purpose (Figure 2.8) A range of differing size bags will allow the safe containment of recovered detritus prior to any further laboratory examination. Each bag should be clearly marked with information such as

- The incident
- Location from which sample was recovered
- Date and time of collection, along with number of pieces in sample if broken
- Name and organisation of the field investigator responsible for recovery
- Site of storage
- Adequate space (or integral card) to record its 'Chain of Evidence'
- Bar code (ideally)

Any material evidence recovered from an accident or failure investigation site may eventually be presented and questioned in the courtroom. For the evidence to be of use in a trial, it must make the journey from the recovery scene, pass through the hands of a number of experts for examination and may be subjected to a range of analysis (with agreement from all interested parties) to finally arrive in court. The course of this journey must be made in

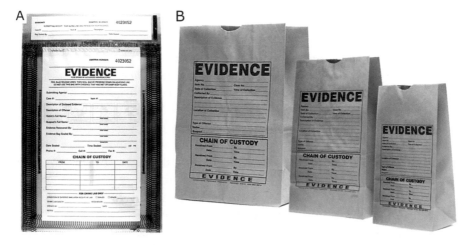

**Figure 2.8** (A) Heavy-duty re-sealable clear polyethylene bags providing protection for contents from air, moisture, dust and dirt. (B) Printed paper evidence bags allowing for evidence to dry. (Courtesy of Arrowhead Forensics, Lenexa, KS.)

a validated and secure manner so that all involved can be assured that it (the evidence) has not been contaminated and that the evidence was that relevant to the investigation. In order to ensure validity, the forensic engineer should follow a routine, commonly known as the 'chain of custody', when it comes to collecting and handling evidence. A chain of custody of evidence will therefore represent a complete trail of where, when and by whom evidence was obtained. In addition, it will also clarify who had access to the evidence, any testing undertaken, those in attendance at any testing and where it had been stored prior to receipt in the courtroom.

## 2.10 Product Liability

As discussed in Section 1.3, a product defect is a feature of a product or component that inhibits or prevents its correct operation. Features that may constitute a defect in a component can be latent or patent in nature, and can include

- Sharp corners in stressed areas
- Internal voids or inclusions that intensify stresses in a loaded product
- Deviations from proscribed dimensions, possibly caused by the sinking of surfaces of castings or mouldings
- Poor quality material, possibly caused by contamination
- Inherent cracks within components

Any such features can occur in combination with one another, and may also interact: a void near a sharp corner would be a serious defect in a product where the stress is at the corner. Not all such features may be defects by themselves. Whether a feature can be regarded as a defect depends on the product specification.[1,2] However, when any product or engineered system fails as a result of an inherent defect, the law of product liability may be enacted – particularly if failure results in injury.

Products liability refers to the accountability of any or all parties (along the chain of manufacture) for any perceived injury or damage caused by that product. Parties involved comprise: the product designer, the manufacturer (and his subcontractor/s), the product supplier and the retailer. Any marketed product that may contain an inherent defect capable of inflicting harm to the user, or to someone to whom the product had been loaned or given, will be open to product liability claims. While products are generally thought of as tangible personal property, the arena of product liability has expanded to include intangibles, pets, housing charts and scripts.

There are three types of product defects that incur liability in manufacturers and suppliers: design defects, manufacturing defects and defects in advertising and marketing. Design defects are inherent in that they exist before the product is manufactured. While the item might serve its purpose well, it could also be unreasonably dangerous in use by virtue of an inherent design flaw. Then again, any manufacturing defect is 'built-in' on the production line. It may be the case that only a few out of many products of the same type contain the flaw. However, with batch production methods, the probability exists that a considerable number of products may well be compromised. Defects in marketing deal with incorrect, scant or non-existent operating instructions and/or failure to warn customers of any latent danger contained within the product.

Under tort law, for any artefact that is found to pose a danger to the user or consumer as a direct result of a design or manufacturing defect, the manufacturer would be regarded as responsible, and thus liable.

### 2.10.1 Case Study: Premature Failure of a Presentation Cake Knife

A porcelain-handled cake knife failed as it was being used to cut a wedding cake. The failure caused a razor-type injury to two fingers of the bride's right hand. The knife in question was a presentation item from the product range of a well-known and reputable porcelain manufacturer. It comprised a steel blade set in a porcelain handle and was approximately three years old when the handle snapped. The knife had undergone minimal wear and tear within that period, usage being limited to approximately four occasions.

Observation of the broken parts showed there was no visible evidence of pre-existing cracks that might indicate a high degree of wear and tear.

Figure 2.4 shows the blade with part of the ceramic handle still attached to the tang, a porcelain sliver and the solid porcelain end of the handle. The knife tang and porcelain handle had been joined with an epoxy-based adhesive. Staining on the porcelain sliver (arrowed in Figure 2.9A) appears to be blood staining from the finger injuries mentioned above. Figure 2.9B shows the end of the blade and attached porcelain fragment, with the porcelain sliver temporarily put back in place.

There was no evidence of pre-existing cracking inasmuch as the entire fracture had formed at virtually the same instant and there was no sign of intrusions or staining such as would have occurred during use or when washing up. The fracture surface on the rear section of handle exhibits step and lip characteristics typical of combined bending and torsional overload failure, showing that the handle broke under heavy pressure consistent with cutting and twisting the blade, i.e. the failure of the porcelain handle was due to a combined bending and torsional overload mechanism that exceeded the strength of the porcelain at the tang end. It shows the tang should have been extended in length for the full length of the handle. The two design features raised by

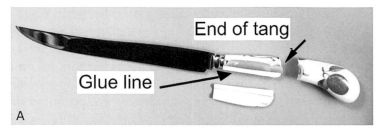

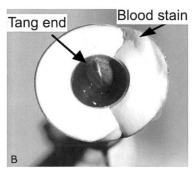

**Figure 2.9** (A) Parts of a broken cake knife. (B) End view of tang with porcelain sliver in place.

this failure should be readily foreseeable, from knowledge of basic engineering mechanics and materials principles. The unfortunate consequences of this breakage could have been avoided if the metal tang had extended almost the full length of the handle and if the end of the tang had not been directly in contact with the side of the hole in the porcelain. Perhaps more importantly, on no circumstances should a brittle material (including brittle polymeric materials) be used for fabricating a knife handle.

## 2.11 The 'Corporate' Environment

Associated costs, as a limiting factor in any forensic investigation, have been reviewed in Section 1.9.2. It was mooted that costs will rise in direct proportion to the complexity, depth and sophistication of analysis. However, incorrect or improper testing, analysis and reporting can also be directly attributed to the prevalent corporate environment. There exists a cost-driven work philosophy that dictates work to be performed by the lowest-cost personnel capable of performing the task – the so-called 'hire-and-fire' business environment. It is often the case that investigative personnel and/or test house staff are given 'just enough' training or worse still, too little training. Unfortunately, this business attitude can foster a situation in which personnel may prejudice or 'skew' a failure or forensic investigation by:

- Selecting the wrong sample
- Undertaking the wrong test
- Applying incorrect test parameters
- Performing the test incorrectly
- Recording results incorrectly
- Reporting findings incorrectly

All of which may be directly traced back to corporate attitudes to staff training (and salary) costs. A classic example can be recounted regarding bird strike testing on jet engines:

## 2.11.1 Case Study: Bird-Strike Testing of Aircraft Engines

Bird strikes have been a problem during aircraft take-offs and landings for many years, where a bird may actually be ingested into an engine. The engine must be able to handle a bird strike without uncontained failure (Figure 2.10A), fire or engine mount failure, and be able to shut down in a controlled fashion to pass this FAA requirement. The jet engine industry has a test to determine the effect of a bird strike on an engine – a chicken gun (Figure 2.10B). First used in the 1950s at de Havilland Aircraft in England, the chicken gun is a large-diameter, compressed-air cannon used to fire dead chickens at aircraft components in order to simulate high-speed bird strikes. On one particular day in the UK, a new technician running the test was apparently not told to thaw the frozen chicken before using it in the test. Needless to say, the frozen chicken hit the engine with the destructive impact of a bowling ball – the engine failed the test.

There remains confusion over the validity of this tale, relegating it to the realms of urban legend. However, one of the main users of the chicken gun is

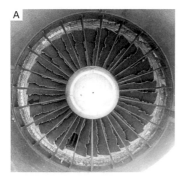

**Figure 2.10** (A) Bird strike damage to fan blades, contained within the engine nacelle. (Courtesy of Wikimedia Commons, licensed under Creative Commons.) (B) Engine stress testing to include bird strike simulation (Courtesy of Pratt & Witney).

Pratt & Whitney, the jet engine manufacturer. The chicken ingestion test is one of a series of stress tests required by the Federal Aviation Administration before a new engine design can be certified. In general, frozen chickens are thawed out and shot from a cannon into a running jet engine (on a test stand) to determine if the engine will stay within its nacelle.

At first reading, this example may be considered somewhat frivolous, or an urban legend. However, if this legend was found to be factual, it would recount a serious story of poor training attributable to cost saving as a direct result of the prevailing corporate environment. Therefore, the forensic or failure investigator must be aware of *all* compromise and limitations surrounding a failure event, investigation methods adopted and the prevailing environment (both corporate and system) during the course of his enquiries.

## 2.12 Abuse or Misuse

From a statistical point of view, it is not unusual to uncover failures that are a result of manufacturing defects. Added to this, operator errors and service conditions that exceed the design specifications will increase the likelihood of engineering component failure, compromising the manufacturer's reputation for reliability. In general, it is not possible to design a component or engineered system to withstand any foreseeable load. As a straightforward example, the bicycle wheel is perfectly capable of carrying the load of two or maybe three (drunk?) riders, provided that the loading direction is approximately in the plane of the wheel itself, where it is strongest. Nevertheless, a moderate sideways impact will readily cause buckling of the wheel. The direction and order of magnitude of a damaging force are sometimes all that is necessary to establish the cause of an accident. To expand on the step ladder case study reviewed in Section 2.7.1 above, steps and ladders are required to carry labels stating their maximum working load and pictorial 'Dos and Don'ts' for safe use. In their statements of circumstances of such a ladder accident, the claimants usually assert they had erected the steps properly on a firm, level floor. Conversely, if the ladder had been used correctly, and safety instructions were being properly observed, the very nature of damage sustained by the ladder may be such that it could *not* have been caused by normal forces generated by safe use of the ladder.

Occasionally the falling body may land on the step-ladder back frame and bend the leg, buckling the back-cross brace in a similar manner to that shown in the case study. In one particular accident, where the user's body landed directly above the cross brace, the brace was 'punched' through the back leg and penetrated the user's buttock – causing severe injury. He was a large man weighing some 130 kg (280 lb), and stated he had stood tiptoe on the very top of a set of steps to remove a suitcase from his attic and, as he pulled it out, the steps 'suddenly collapsed'.

Such damage as described following an accident is invariably the result of some kind of instability while the person injured was climbing or working from the steps near the top. Sometimes the instability is caused by overreaching, possibly to lift something heavy off a shelf, and other times by the steps not being opened fully and locked in position to prevent them folding forward, or by the feet not being set on a firm, level surface. With tall ladders, it is essential to have a second person standing on the bottom rung ('footing' the ladder) so that no rocking movement is possible while the user is working near the top. In fact, the stiles of ladders are usually so strong that the maximum recommended static weight of a user, acting at any angle, will not bend a stile. To cause ladder stile damage, dynamic forces are required. Such forces are generated only when the user's body is falling and these forces must react against something external to the ladder structure – generally the ground.

In spite of any built-in safety factor warning labels, and/or user's instructions, failure arising from inappropriate use, or service conditions, is common, and not just in the domestic environment. Equipment abuse within an industrial environment can be demonstrated by the following case, where a workshop technician sustained a severe eye injury.

## 2.12.1 Case Study: Abuse

A workman in a machine shop was removing a large drill, with a Morse taper shank, from a radial arm drilling machine. He said that he had used a soft (steel) drift to remove the drill from the machine, striking the drift with an ordinary engineer's hammer. This method was a normal everyday (and correct) procedure for drill removal. However, while undertaking the routine, the workman was hit in the eye by a flying metal splinter. The eye splinter was submitted to ascertain which of the tools being used was the likely source. The components and their respective compositions are shown in Figure 2.11. Because all four components in this case were made from different materials, it was decided to examine their respective microstructures and compare them with those of the splinter, in order to establish the source of the splinter. The microstructures are shown in Figure 2.12.

Direct microstructural comparison clearly reveals the eye splinter is a small piece of HSS, broken from one of the outer flutes of the drill, not the hardened steel hammer head, the medium carbon steel drift or the soft iron Morse taper shank. It was therefore a straightforward task to arrive at a sequence of events that caused the metal splinter to pierce the workman's eye. The simple answer was that the workman must have struck the drill flutes with the hammer when extracting the drill from the machine, and *not* used the drift as claimed. A trained workshop engineer would know full well that drill flutes are exceptionally hard, with little or no temper (therefore prone to splinter or shatter under impact loading). A hardened steel face of a hammer is perhaps the worst

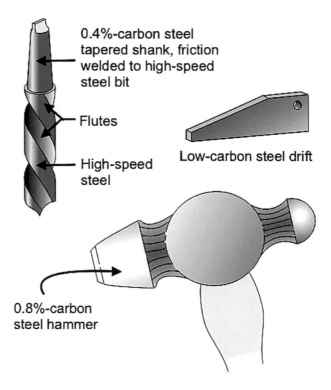

0.4%-carbon steel
tapered shank, friction
welded to high-speed
steel bit

Flutes

Low-carbon steel drift

High-speed
steel

0.8%-carbon
steel hammer

**Figure 2.11** Sketches of the hammer, drift and drill bit components, along with their respective compositions.

tool he could have use to knock the drill bit free. The resultant eye injury was therefore a direct consequence of the workman's irresponsible mistake and/or poor working practice, as well as a lack of any eye protection.[2]

## 2.12.2 Case Study: Misuse

On 28 December 2004 in St. Paul, Minnesota, MN, a natural gas explosion killed three people.[12] An MDPE gas pipe had worked its way out of its steel T-connector (Figure 2.13), which distributed gas from a main line through two MDPE pipes to neighbouring office buildings. Natural gas escaped from the failed coupling, seeping into the ground (which filtered out the foul odor), before the now-invisible gas leaked into the nearest office building some 40 feet away. At the instant of the explosion, an unidentified spark ignited it. Observation of the plastic pipe revealed that it had pulled out of the coupling over several seasonal freeze–thaw cycles. However the crux of this accident centred on the coupling type, where it was discovered that it had been designed for use with steel pipework and *not* designed to handle plastic pipe. The Director of Government and Public Relations for the area initiated an immediate investigation in an attempt to determine exactly how many incorrect couplings had entered service. This is a classic case of misuse (or inappropriate use) leading directly to catastrophic failure.

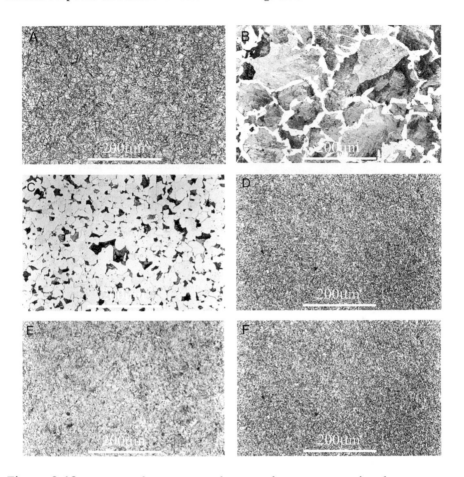

**Figure 2.12** Micrographic montage showing the structures of each component and the eye splinter. (A) martensitic structure of the hammer head. (B) Annealed 0.4% carbon steel drift. (C) 0.2% carbon steel tapered shank of the drill. (D) Splinter fragment removed from eye. (E) HSS steel drill tip fragment transverse section (longitudinal flutes). (F) HSS steel drill tip fragment longitudinal section (transverse flutes). (Courtesy of University of Cambridge DoITPoMS Micrograph Library, licensed under Creative Commons.)

Despite built-in safety factors, in-depth warning labels and user instructions, structural and equipment failure arising from service conditions is commonplace.[13,14] In general, there are five categories of unintentional service conditions that re-offend time after time. The five categories are as follows:

1. Reasonable misuse
2. Use of product beyond its intended lifetime
3. Failure of product due to unstable service conditions
4. Failure due to service condition beyond reasonable misuse
5. Simultaneous application of two stresses operating synergistically

**Figure 2.13** An MDPE gas pipe that had worked its way out of its steel T-connector (circled). The coupling type had been designed for use with steel pipework, not plastic pipe. An ensuing explosion killed three. (From Minnesota Pipeline Safety Report, Case 005217, public domain.)

Today, society is always looking for someone to 'blame' for any misadventure and, from the large corporation to the individual, are far more eager to undertake litigation as a process to redress any perceived loss, damage or injury associated with failure.[2] It is entirely normal or natural for anyone to make mistakes and the forensic investigator must consider if and how human factors had contributed to failure. Evaluation of potential human error is therefore an essential assessment during the early stages of enquiry.

Much the same approach that goes into designing a device or structure can be applied to gain an understanding as to why it failed. Therefore, to be in a position to identify the potential failure mode at play, the failure detective must first gain clear understanding of how that device or component was fabricated and, more importantly, the design specifications for service use. In order to find out why a part failed, the following questions must be answered:

- What manufacturing processes were used to produce the component or system in question?
- How does it work?
- What materials are in play and what are their properties?
- How was the part loaded?
- Where did the part fail?
- Did it fail at a recognised 'weak link'?
- Does it show evidence of anomalies in construction, conditions or use?

# References

1. Lewis, P.R. and C.R. Gagg, *Forensic Polymer Engineering - Why Polymer Products Fail in Service*. 2010, Woodhead, ISBN: 978-1-84569-185-1.
2. Lewis, P., K. Reynolds and C. Gagg, *Forensic Materials Engineering: Case Studies*. 2003, CRC Press.
3. Russ, John C., *Forensic Uses of Digital Imaging*. 2nd ed. 2016, CRC Press.
4. Lewis, P., *Beautiful Railway Bridge of the Silvery Tay: Reinvestigating the Tay Bridge Disaster of 1879*. November 2004, The History Press. p. 192.
5. Lewis, P. and C. Gagg, Aesthetics *versus* function: The fall of the Dee bridge 1847. *Interdisciplinary Science Reviews*, pp 177-191, 2004. 29(2): pp. 177–191.
6. Luchtvaart, Raad Voor De, Nederlands Aviation Safety Board, Aircraft Accident Report 92-1 1. EL AL Flight 1862, Boeing 747-258F 4X-AXG Bijlmermeer, Amsterdam, 4 October 1992, https://lessonslearned.faa.gov/ChinaAirlines1 862/ElAl_Accident_report.pdf.
7. Ladders account for around 40% of falls from height accidents in the UK, https ://www.laddergrips.com/ladder-accidents-in-the-uk, 2019.
8. Lewis, P.R., K. Reynolds and C.R. Gagg, *T839 Forensic Engineering, Open University Post-Graduate Course T839*. 2000, Department of Materials Engineering, The Open University.
9. http://www.patent.gov.uk.
10. http://www.uspto.gov.
11. http://www.fda.gov/medwatch.
12. Fatal gas explosion blamed on pipe fitting. *Minnesota Public Radio*, 2005, http://news.minnesota.publicradio.org/features/2005/04/27_ap_gas/.
13. Qua, Huck-Chye, Ching-Seong Tan, Kok-Cheong Wong, Jee-Hou Ho, Xin Wang, Eng-Hwa Yap, Jong-Boon Ooi and Yee-Shiuan Wong, *Applied Engineering Failure Analysis: Theory and Practice*. 2017, CRC Press, ISBN 9781138747869.
14. Otegui, Jose Luis, *Failure Analysis: Fundamentals and Applications in Mechanical Components*. 2016, Springer, ISBN-10: 331934773X, ISBN-13: 978-3319347738.

# A Framework or Methodology for Forensic Investigation

# 3

## 3.0 Introduction

The work of a forensic engineer is often compared with that of a detective in a criminal case, or the coroner undertaking an autopsy as a result of unnatural death. Based on both fiction and the cinema it is common knowledge how the investigator usually proceeds to solve those cases. In addition to the somewhat abstract assets discussed in Chapter 2, a substantial amount of *systematic* investigation of evidence, and an analysis of clues, are necessary. Principal steps, taken during the course of criminal (or medical) investigation, are closely shadowed by the failure or forensic engineer. In all cases, investigation normally entails considerable hard (and often monotonous) work by the individual, a team of specialists or by both. In general, the investigator will approach and undertake a failure investigation using both his intuition (developed primarily by experience) and a systematic path of analysis. However, there are no 'hard and fast' rules as to which investigative pathway forward is the more effective, or which (if any) should take precedence. The priority of any investigative method is simply an overriding requirement for immense care and effort, to ensure that the scientific and engineering integrity of the investigation is not compromised. Therefore, this chapter will present an overview of a somewhat specific path of investigation adopted by the writer.

## 3.1 Background to Failure Analysis Methodology

Publicly available literature, with a focus on failure analysis, usually deliberates on the technical aspects of failure, with the deep underlying reasons for failure often neglected, as demonstrated by the Challenger space shuttle accident of 1986.[1] It has been suggested that there should be a systematic way of performing failure analysis, with a methodology – similar to that of medical autopsies, or the criminal detective – being mooted.[2] A coroner will undertake an autopsy where an unnatural death has occurred, whereas the forensic engineer will examine parts, assemblies or structures that have been overtaken by premature failure. A similar analogy can be applied to the work of the police detective. In addition to intuition, a vast amount of substantial systematic investigation of evidence and an analysis of clues are necessary. The work of the failure or forensic engineer closely shadows the principal steps taken during the course of a criminal investigation. In both cases, investigation typically demands a substantial amount of background effort – that will often include the digestion of reams of paperwork – by all operatives engaged on the case.

## 3.2 An Investigative Path Followed by the Writer

A fundamental argument of any failure analysis will focus on an implicit assumption that there are rational causes for failure and reasonably objective ways of determining those causes. The generic framework for investigation described above may not sit well with all investigators, or with all situations. Individual techniques may have been honed by years of experience or by necessity within the individual working environment, i.e. the scope of an investigation may well be dictated by parameters such as cost/time constraints imposed by the specific consulting practice, the court of law, etc. Therefore, the balance of this chapter will focus on an investigative framework developed over time and experience. It is NOT to be taken as a blueprint for an unassailable investigation process; it is simply an account of what 'works best' for the writer. Indeed, there may well be readers that vehemently disagree with parts of the following methodology.

### 3.2.1 Intuition (Reasoning) and a 'Structured' Investigation Framework

It was mooted in Chapter 2 that the most comprehensive tools available to the investigator are his/her

- *Eyes*: the power of visual observation. As the great detective Sherlock Holmes said to his medical doctor sidekick, 'Watson, you see but you do not perceive'.
- *Ears*: almost as good as eyes – listen to what is being said.
- *Reasoning*: normally deductive – conclusions made on the basis of a single piece of data. Occasionally inductive – conclusions made on the basis of large amounts of data.

Unless the list of variables can be reduced or eliminated, the way in which many different factors can affect the strength of a product will make the task of failure investigation a difficult undertaking. However, reduction of variables is the exact procedure undertaken during the process of an initial examination of failure. It is an intuitive process of elimination by careful collection of all the facts surrounding a particular incident.[3-6] In this instance, it is suggested in Section 2.1 that 'intuition', or sound reasoning, will only develop with experience over time. The investigator will use a range of faculties to determine the most probable train of events that led to the root cause of failure in question. In addition, there exists a range of conceptual techniques (both creative and statistical) that will assist the investigator by providing a 'structured' pathway or guide to the root cause of failure.[7,8] Creative analysis techniques allow data to be assembled pictorially for ease of understanding. Types of creative or conceptual representation are presented in Section 1.4 (causal analysis), with computer-aided methods discussed in Section 1.4.1. However, of the qualitative and semi-quantitative methods reviewed, both fault tree analysis (FTA) and the cause-and-effect Diagram (CED) will provide a straightforward method or framework for the examination of a failure event. FTA and CED can help simplify very complex failures as both methods show the inter-relationship of many different possible causes producing a single symptom – thus providing an ideal framework for undertaking a failure investigation. Early Romans occupied the known world by using military manoeuvres to divide and conquer their antagonists. FTA and CED operate on the same concept. The failure is divided into individual root causes, and each one of those is divided further. It is easier to investigate and prove or disprove each root cause than the resultant failure. Although it may not be implicitly recognised by the individual, the route or structure most investigators follow is intuitive, and will closely align to an FTA or CED analysis. The CED engineering method in action can be demonstrated by the following simple example of poor-quality photocopies, as shown in Figure 3.1. An example of a computer-generated FTA solution is shown in Figure 3.2 (along with the usual caveat that the solution is only as good as the inputted data).

### 3.2.2 Individual Stages of Investigation

Having established an engineering/statistical method as the bones of an investigation, it can be 'fleshed-out' by the following sequence of stages or steps taken during the course of investigation or analysis of failure:

1. Accept an instruction or brief.
2. Most importantly: does the instruction and/or incident lie within your field of expertise?
3. Consideration of instruction/brief and known data/facts in play at that point in time.
4. Site attendance if required (evidence may be sent directly to investigator).
5. Initiate a photographic record of event.
6. Stop, look, talk, listen. Allow the scene/site/detritus to be assimilated.

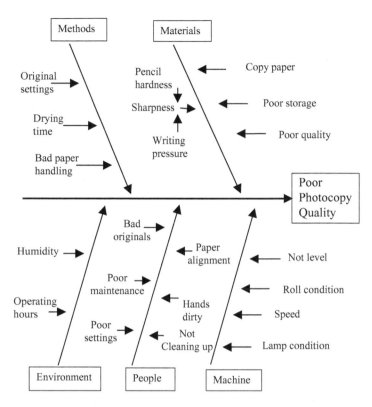

**Figure 3.1** A simple illustration of a cause-and-effect diagram (CED) for poor quality of photocopy.

7. Gather statements and any background data.
8. Allow the subconscious mind to begin the development of a train of events that would accurately reproduce the failure or accident presented before you.
9. Consider the weakest link of the failed system/component.
10. Selection, identification and preservation of appropriate components or samples for further laboratory examination. Evidence MUST be protected from further (secondary) damage e.g. plasticine or Blu Tack are an ideal medium for fracture surface protection; fracture surfaces must be protected from corrosion, including that generated by contact with finger grease.
11. For comparative purposes, select any identical parts that have not failed.
12. Organise appropriate packaging and transportation to laboratory.
13. Preliminary examination of the failed part(s), looking specifically at fracture surfaces, secondary cracks, other surface phenomena, signs of abuse/misuse, etc.
14. Laboratory visual examination and record keeping.

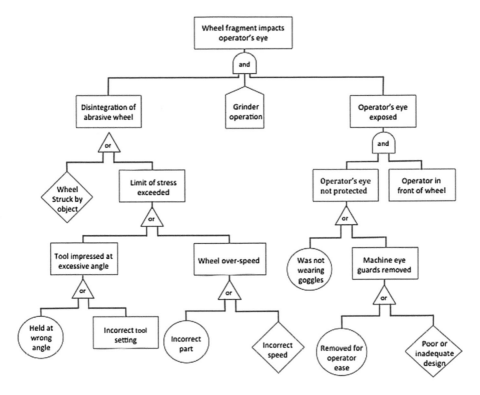

**Figure 3.2** A simple illustration of an engineering fault tree analysis (FTA) solution, developed by a commercial software company.

15. Macroscopic examination and photographic documentation.
16. Plan any required analysis, using a range of available tools, some of which are reviewed in Chapter 4.
17. Undertake any non-destructive testing.
18. Plan and undertake any microscopic examination and analysis. This may demand *appropriate* selection and preparation of metallographic sections, followed by appropriate metallographic examination and analysis.
19. Bulk chemical analysis for material composition.
20. Chemical analysis of surface corrosion products or deposits.
21. Undertake any mechanical testing. It must be recognised that no destructive testing should be attempted without full consent of *all* parties involved. With any product litigation case, the instructing solicitor or barrister should be consulted. It may be requested that all interested parties are in attendance, and observe all facets of any destructive testing
22. Appropriate mathematic calculations or mathematical modelling.
23. Determination of failure mechanism.

24. If required, initiate additional testing or reconstruction under simulated service conditions.
25. Re-evaluation of *all* evidence.
26. Formulation of failure mode and train of events in play at the instant of failure.
27. Consider potential legal ramifications, such as fitness for purpose (Consumer Rights Act 2015, UK), tort (an act that causes injury or harm to a third person), contract, criminal law, etc.
28. Draw final conclusions.
29. Compile the expert report.

Although presented in an ordered list, the sequence in which these (inter-related) stages are considered is relatively unimportant, as any one area may provide the key to failure in a particular situation. As a reminder, some of the earlier (and more abstract) points listed above are considered in more detail within Chapter 2. However, there are particular stages that demand expansion, as they contain intuitive technical understanding gained from experience rather than from a text-book. Therefore, the following paragraphs and sub-paragraphs will be dedicated to expanding those specific stages. The narrative will start with a basic equipment requirement for field investigation (Section 3.2.3), and initial approaches to investigation (Section 3.2.4). This will be followed by short reviews of background data collection (Section 3.2.5), sifting the evidence (Section 3.2.6), records (Section 3.2.7) and single items of evidence (Section 3.2.8). The final paragraph in this chapter will describe an understanding of the failure detective (Section 3.3), transformation stresses (Section 3.3.1), establishing a load transfer path to determine the weakest link (Section 3.3.2), computer-aided technologies (Section 3.3.4), the forensic engineering report (3.3.5), concluding remarks (Section 3.3.6) and references (Section 3.3.7).

## 3.2.3 The Field Investigation Kit

If at all possible, it is always advisable to undertake first-hand observations (a pre-requisite for expert evidence in any following litigation action). However, the failure scene *per se* is of utmost importance, requiring a philosophy of 'document and protect' as the guiding directive. If first-hand observation can be arranged, the investigator will require a basic field investigation kit. Although being relatively subject-specific (domestic product failure as opposed to large concrete structure failure, for example), field investigation of metallic component demise will require a field kit that will typically contain items as shown in Table 3.1. However, if a visit to the failure scene is not feasible, photographs and/or drawings of the scene both before and after the failure will undoubtedly be of some value in any further analysis.

**Table 3.1  Typical Content of a Field Investigation Kit**

| Item | Reason(s) |
| --- | --- |
| 1. The first and most important item is an unbiased, open and questioning mind. | Always be prepared, have questions to hand and always be prepared for the unexpected. |
| 2. Personal attitude. | Promote cooperation of, and assistance from, as many individuals as necessary. |
| 3. Professional manner. | Competent and organised manner will promote confidence, participation and value in your work. Work to your brief, and stay within your field of expertise. |
| 4. Camera – digital or conventional film. | Take as many pictures as you can of the item, the area and any supporting equipment. Be aware of photographing individual people. Be aware of colour temperature. |
| 5. Known colour chart or white card/paper. | With previous point in mind, use colour chart in the photo as a standard for colour temperature. |
| 6. Measuring tape and ruler – both steel and plastic. | Use tape/ruler in photos for scale. Being grey, steel rulers are ideal for light backgrounds. White plastic rulers perform well on dark backgrounds. Steel rulers will indicate if an item is magnetic (make certain the ruler itself is not magnetic), plastic rulers if the subject is magnetic. |
| 7. Magnets – both flat bar type and an extending 'wand' type. | Identify magnetic materials from nonmagnetic materials. The 'wand' can be useful for retrieval of items. |
| 8. Magnifying glass. | Sample and fracture surface observation. Magnification: 10–25×. Lower-power magnifier may also be helpful. |
| 9. Indelible ink marker (fine). | Identification/marking of sample items, bags, bottles, etc. |
| 10. Sample/specimen containers. | Generic evidence bags (paper and/or plastic). Plastic zip bags, pouches, etc. Plastic screw bottles, and containers. |
| 11. Swabs, cotton buds. | Sample collection. If used, always keep one for control sample. |
| 12. Illumination: flash light, fibre optic flexible probe inspection camera. | Flashlight for general illumination. Helpful to have a 90° bend attachment for observation in holes, crevices, etc. Flexible probe inspection camera for areas that are difficult to see with the naked eye, i.e. inside of cylinder chambers, corrosion checks, etc. |
| 13. Multimeter. | Check conductivity, resistance or continuity of surfaces, wires and components. |
| 14. Mirror, ball mounted. | Check around corners and under objects. |

*(Continued)*

**Table 3.1 (Continued)    Typical Content of a Field Investigation Kit**

| Item | Reason(s) |
| --- | --- |
| 15. Tool set: surface finish profilometer, scissors, screwdrivers, pliers, mallet, light hammer, punch, tweezers, blades, plastic toothpicks and pocket knife. Field investigation and recovery may well demand a full set of spanners, hexagonal keys, etc. | Various uses, particularly sample recovery. |
| 16. Pen, pencil, paper and lab book. | Writing utensils and writing medium. White paper may also be used for colour temperature (item 5). |
| 17. Technical information, hardness conversion charts, materials composition tables, etc. | Specification lists, design criteria, drawings, notes, etc. This is where incident-specific criteria will apply. |
| 18. Relevant addresses and phone numbers. | Assistance, information, etc. |
| 19. Mobile (cellular) phone. | Immediate 'outside' contact. Immediate information access. |
| 20. Eyewitness report forms. | To be filled out as soon after the event as possible. Therefore it is preferable to e-mailed or sent ahead. |
| 21. Laptop or palm pilot. | Computer provides access to spreadsheets, word documents, etc., in addition to the uploading of digital images from USB microscope. |
| 22. USB microscope (wireless) 10–200x. | Sample and fracture surface observation. Images directly downloadable to laptop PC. |
| 23. Fresh body (boiler) suit, rubber boots, wet wipes, surgical gloves, waterproof mat and paper towelling. | Personal protection and cleaning, in addition to site (or scene) contamination protection. |
| 24. Blu Tack or plasticine. | Fracture surface protection. |
| 25. Additional item(s). | Any additional investigation-specific items that an individual area of expertise may demand. |

## 3.2.4 Initial Approach to Failure Investigation

Armed with intuitive understanding and experience – along with a reference framework for undertaking an investigation – the forensic engineer will be in a better position to commence his/her enquiries. Just as for the criminal detective (Section 3.1), when initially approaching a failure the first course of action for the investigator is simple – do *nothing*. Visually study the evidence and surrounding scene, and let the subconscious mind assimilate the failure.

Stand back, use your experience and common sense and think about the failure. Key evidence from an accident scene has to be determined using simple observation, based on common sense, and a refusal to jump to conclusions. The power of visual observation is discussed in Chapter 2, where it is considered the single most important asset for the investigator.

As a matter of practice, the investigator will be looking for any indication of abnormal conditions or abuse in service, he will be asking questions (and listening to the answers), and will start documenting his observations, recording and sketching any and all important features, including dimensions. In addition, he/she will have at hand a range of additional 'tools' – both abstract and analytical – required for undertaking a thorough investigation. The somewhat 'abstract' tools are reviewed in detail throughout Chapter 2, and include

   (i)  Forensic photography
  (ii)  Record keeping
 (iii)  Witness evidence
 (iv)  Documentary evidence
  (v)  Product and material standards
 (vi)  Patents
(vii)  Surviving detritus
(viii)  Product liability
 (ix)  Abuse or misuse

Chapter 4 presents a review of a range of common analytical tools available to the investigator, along with the limitations and compromise associated with results gathered from each tool.

Any accident or failure can leave critical trace evidence hidden by other debris, so patience is a vital asset when first observing an incident or failure. Site observations must be documented, usually by recording sketches of the scene together with notes of the position of debris, and photographs taken from many different angles. The more complex the debris field, the greater the need for multiple images of the detritus so that the exact final resting positions of key components are recorded. Such activity necessarily must precede any laboratory investigation, where the physical evidence retrieved can then be subjected to analysis.

## 3.2.5 Background Data Collection

In addition to material evidence, a successful forensic failure investigation may often be dependent on gathering *relevant* background information. This can only be achieved as a result of thorough background research and painstaking fieldwork. Therefore, the first logical step to any failure analysis is to gather as much relevant background information as possible. Detail such as part information and identification, including pertinent drawings and specifications, should be collected and kept readily available for future reference. Any known service history of the component can be of importance, and should include manufacturing detail and service stress conditions that the component was subjected to during its lifetime. If investigating machinery failure, maintenance schedules and repair history records should be available for inspection. This form of documentation may well contain a record of any anomalous service events. Historic information relating to similar failures of the equipment will also be of importance, as it may provide evidence as to whether the failure was as a result of design or production shortcomings, or a result of abnormal service conditions. It is worth reiterating that a large percentage of any investigator's time is often spent on the more 'mundane' task of sifting through reams of documentation simply to tease out relevant and often vital information. This point has particular significance if and when the forensic engineer has been tasked with determining failure based entirely on evidence contained in a document bundle. This is often the case when given 11th-hour instructions (from a barrister, for example) to review a particular case file, and offer expert opinion on the failure contained within it.

### 3.2.6 Sifting the Evidence

What is deemed as evidence? The answer will be dependent on several (often many) different facts associated with the incident or accident. For example, whether having to consider metallic or non-metallic products (or composite products constructed of different materials). The following can be regarded as a minimum list:

- The testimony of any individuals who saw what happened before, during and after the accident, keeping the degree of witness statement variability in mind (Section 2.5).
- The circumstances surrounding the incident, such as the time and date, the environment, weather and so on.
- Technical records are an invaluable source of information and are often (but not always) available routinely for equipment failure (Sections 2.6 and 2.7).
- Material evidence itself is probably the most important focus of enquiry, followed by the details of its history and provenance, as illustrated by the following case study.

#### 3.2.6.1 Case Study: Failure of a New Design of Horse Bit

A company with a long and successful history of making horse bits was asked to produce a lighter version of an existing design (Figure 3.3A). However, the

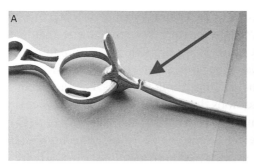

**Figure 3.3** (A) A new design of horse bit, showing fracture sited at a 'necked' section. (B) Fracture surface showing shrinkage cavities and hot tears.

new (lighter) bit failed during a race. As a result, the jockey was thrown to the ground and sustained injuries that prevented him from ever riding in a competitive race again. The new design had a slim section at each end of the bit. This narrowing resulted in the casting not being properly fed (i.e. cooling contraction could not be compensated for by the reservoir of molten metal), giving rise to shrinkage cavities and hot tears in the most highly stressed region of the bit (Figure 3.3B). The new design of bit failed simply because it was not strong enough to withstand the pull of the reins over the reduced cross-sectional area. The inherent casting weaknesses could easily be rectified by eliminating any taper at each end, and by changing the position of feed gates (the point of molten metal entry into the die cavity) to more appropriate positions. Legal action against the manufacturer resulted in substantial compensation being awarded to the jockey for his injuries.

## 3.2.7 Records

Quality standards such as ISO 9000 require systematic record-keeping of processes, materials and designs, and can give an insight into the past history of a particular product. Design-specific standards are a way of assessing compliance. However, as they are often drafted by a committee composed of manufacturers, they should be regarded with caution. Furthermore, they are often historic documents that may not have been modified to include the latest developments. Most contracts will specify compliance with one or more standards, so standards have an important status in the eyes of the courts. Section 2.7 contains a detailed overview of the relevance of product and materials standards in any forensic or failure investigation.

## 3.2.8 Single Items of Evidence

Single items of evidence always prompt the question: was this single product (or part) the only one to have failed? Occasionally, many similar broken products emerge, suggesting faulty design. Such a scenario would signify a more serious position for the product manufacturer, as rapid action should be taken to withdraw existing products in service, simply to prevent any further incidents. The result of any investigation may determine a product redesign requirement to withstand service conditions, or alternatives provided which are capable of resisting the working environment. Thus many fractured or leaking miners' lamps indicated one or more serious design flaws, and immediate action was required to provide alternative light sources to enable the colliery to keep working. Each broken product requires examination to provide a picture of the pattern of failure. If a common failure mode is found, details of each individual failure become unnecessary. Statistical analysis of many failures may provide further clues regarding any design flaw that may be causing failure, thereby helping the designer to improve the product. Design defects represent a serious challenge to the credibility of a manufacturer, and remedial costs can escalate rapidly. For these reasons alone, it is imperative to initiate an investigation as early as possible, simply to reduce any further failure escalation.[9,10]

Every product failure demands individual treatment, and usually starts with simple visual examination, careful measurement of its dimensions and determination of its condition compared with an equivalent intact component. Comparison is a simple way of checking if the parts really are identical, and if not, the reason for divergence. Many products are now identifiable from logos, date stamps and manufacturing codes either printed or embossed on the product in a concealed position of the product outer surface. If the material of manufacture is unknown, or degradation is suspected, it must be analysed for the constituent parts. The analysis should aim to be non-destructive, but if necessary, any sampling must be undertaken away from critical features such as fracture surfaces. Although direct comparison with an unaffected product or component is ideal (exemplar comparison), it is not always possible. Visual inspection aims to identify the following product features as a minimum:

- Overall dimensions
- Distortion in dimensions
- Fit with matching parts
- Surface quality
- Traces of wear
- Identity marks[10]

Only then can inspection of the damaged part or parts begin. Fracture surfaces require particular care during inspection, particularly in the way the product is lit. It may seem somewhat surprising that lighting should be important, but fracture surfaces can provide detailed information regarding the way the crack or cracks have started, grown and ended. Often key features of the fracture surface are only visible in oblique illumination, where tiny details are revealed by their shadows.

The search for key details does not stop at the fracture surface however. Cracks, which have not grown to completion, are one objective of the search. Such sub-critical cracks provide evidence of the way the component has been loaded in service, and might show why failure has occurred in the first place. Thus the discovery of sub-critical brittle cracks on a polypropylene storage tank showed the tank to be under-designed for its function. In the same vein, sub-critical cracks on an acetal plumbing fitting were an indicator that stress corrosion cracking (SCC) as a failure mode should be brought into the picture.[10] As so often found, such cracks frequently start at localised stress raisers, such as changes in design configuration, sharp section changes, etc, where the overall stress becomes magnified to levels in excess of the design brief.

## 3.3 The Failure Detective

As suggested in Section 3.1 above, the individual must train him/herself to become a 'failure detective'. The investigator must be aware of transformation stresses that can be introduced during the course of manufacture, and become familiar with both the manufacture and operation of the system in question. By undertaking a 'reverse engineering' exercise (Section 1.9.1), the forensic engineer will be in a better position to establish a load transfer path for the system, and become aware of the normal or expected location of a fracture or failure in any component part or the system as a whole (the weakest-link principle, Section 1.9.1). Any deviation from the normal or expected failure location must have been caused by additional (unknown) factors that have to be revealed.

### 3.3.1 Transformation Stresses

The art of manufacturing is simply a transformation process. However, any process that transforms the shape or properties of a material will introduce residual stresses, as seen in Figure 3.4. These stresses are universally present and are a complex function of the material or component processing and service history. As such, residual stresses cannot be determined by calculation or predicted from first principles with any great accuracy. However, for complete assessment of design integrity, knowledge of the total stress field is required. Nevertheless, in many circumstances, designers limit their assessments merely to consideration of stresses generated by externally applied loads that the component will be exposed to during its service lifetime. Although an essential requirement in the design process, the adoption of such a narrow focus holds the potential to miss a key integrity factor. Consideration of residual stresses will provide an indicator as to how close a material is to its ultimate limit. Such deliberation may also define performance under fatigue loading and can often influence the response to environmental conditions.

**Figure 3.4** Photoelastic stress analysis of injection moulded polycarbonate safety glasses showing residual moulding stresses. High fringe order zones indicate high levels of stress, whereas zones of low fringe order represent unstressed areas.

### 3.3.2  Establishing a Load Transfer Path to Determine the 'Weakest Link'

The forensic engineer may be presented with a broken component and the circumstances (documentation and/or photographs of the incident scene or wreckage) of an accident for which it may have been responsible. The first step in the investigation is to ascertain, from the state of the damaged parts, the direction and order of magnitude of the force(s) involved, so that these can be traced from the critical (broken) component through all the linkages to the source of the external loading. An external force applied to a system of linked components can only set up stresses in individual members if they remain linked so that each is able to react to the force acting on its neighbour. If one member is stressed to a level where it suddenly fractures (regardless of whether this is due to overstress or pre-existing weakness at that point), this interrupts the load transfer so that members beyond the break are no longer required to react to the external force. By estimating the stress levels in the various members of the linkages and comparing these with the fracture of the critical component it is then a straightforward matter to establish whether or not the critical component was causative of, or a consequence of, the accident.

The principle here is that if there is evidence of mechanical distress or damage on both sides of a fracture, then abnormally high forces had been transferred before the component broke, and it did so simply because it was the weakest part in that particular load transfer path. If, however, there is no sign of prior mechanical distress on either side of a break, then the fracture must have occurred before adjacent links in the system had reached stress levels capable of causing deformation, i.e. none was grossly overstressed.

Evidence should therefore be sought as to whether any inherent material or manufacturing fault had existed or some progressive deterioration could have been responsible for a spontaneous fracture under a foreseeable loading condition. If such a fault is identified in a linkage that played a vital role, for example, in the steering or braking system of a vehicle, then there is a high probability that it could have been the cause of loss of control and thus causative of an accident.

A simple example of load transfer path can be given by considering a roof truss (Figure 3.5A), which is designed to take the static weight of the roof (plus any snow or live loading), and transfer the forces vertically downward through the roof framing to the exterior walls, thus transmitting the load downward to the building foundation (and, ultimately, the ground). Tensile and compressive forces within the truss members are shown by respective arrows, which graphically illustrate load transfer paths within the structure.

The simple beam in bending gives a further example of how loading creates a combination of tension and compression through its thickness (Figure

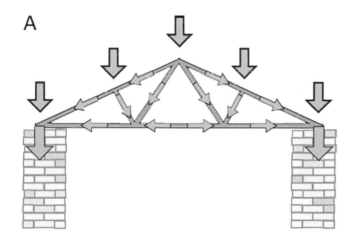

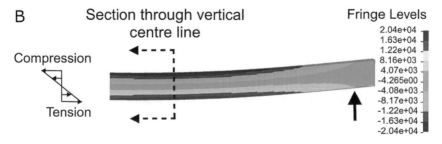

**Figure 3.5** (A) A simple illustration of load transfer paths within a roof truss. (Courtesy of LinkedIn Slideshare, licensed under Creative Commons.) (B) Stress distribution through a simply supported rectangular beam in bending.

3.5B). The idea of a load path through more complex configurations of many different components is helpful in elucidating how specific parts come to break rather than any other, as it is the weakest component which will govern the strength of the whole assembly (the 'weakest-link' principle).

### 3.3.3 Case Study: The Weakest-Link Principle

To demonstrate the 'weakest-link' principle, consider the failure of the simple shoelace (Figure 3.6). A cotton fibre shoelace will almost always fail at one of the two eyelets at the top of the shoe. The reasons for this can be elucidated by consideration of the sequences of events at play when tying a shoelace:

- Lace pulled tight by hand to provide enough lace to enable a knot to be tied, since when last undone, the lace will have been loosened to allow the user to remove the shoe from his or her foot.
- Fibre sides of laces rubs against all eyelets during tightening, but greatest wear occurs against the topmost eyelet, with wear concentrated in the softer fibres of the lace. Movement is greatest at the top eyelets, and the applied load is at a maximum here too.
- Knot tied in the free ends of the lace and pulled tight against the top eyelet. The lace is in tension, with bending loads both within the knot and at most of the eyelets.
- Use of shoes causes repeated oscillations of the lace, mainly concentrated at the top eyelet. Wear of the lace continues, causing loss of fibre fragments, and decrease of lace diameter at and near the top eyelet.
- When the lace has become thin enough at this position, it fails suddenly, often during tightening when putting on the shoe. This is when tension in the lace is greatest.

**Figure 3.6** (A) A simple illustration of the 'weakest-link' principle in action. Wear and failure of a shoe lace will be expected around top eyelets. (B) Wear is expected to be evenly distributed at both top eyelets.

In summary, the load in each lace is transmitted to the top eyelet, where the lace bends around, and is supported by, the eyelet. The load is only partly supported by the eyelet, as some of the load is transmitted to the adjacent leather. However, there is some friction between the lace and the brass (or plastic) eyelet, so that wear of the softer fibre of the lace occurs each time that the lace is pulled through. With repeated use, wear occurs and fibre is lost. This wear results in loss of shoelace thickness. The shoelace breaks when the section thickness (at the top eyelet) reaches the failure load of the lace. It is the weakest link in the load path between the user's hands and the shoe. So in this very simple example, the weakest link is the area of greatest wear, and this is where failure may be expected to occur. However, if the shoelace failed at another location, different conclusions could be inferred regarding the nature of the failure, the pattern of use or the quality of the shoelace. Any deviation from the expected failure could result from a number of reasons, relating either to the prior manufacture of the lace, to particular patterns of use or to environmental causes.

### 3.3.4  Case Study: Failure of a Backhoe Dipper Arm

A backhoe (Figure 3.7) was lowering large-diameter pipes into a trench, using its dipper arm. Without warning the arm dropped, with the sudden and unexpected jolt causing the load to slip and fall into the trench, seriously injuring one worker (who subsequently died from his injuries some weeks later), and narrowly missing a number of others. On investigation, the reason for the arm jerking was found to be a result of an unexpected loss of hydraulic pressure. The relevant hydraulic distribution block was quickly identified, recovered and taken for detailed laboratory inspection. As with all hydraulic systems, the backhoe had in-built pressure-limiting devices, installed as a safety feature designed to prevent failure (or explosion) by an over-pressure event. As the backhoe had two independent systems, installed at the hydraulic pump and within a line manifold, it was clear that the distribution block failure was not the result of excessive hydraulic pressurisation. Visual observation of the

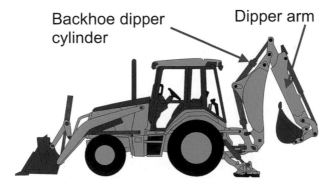

**Figure 3.7** Schematic diagram of a backhoe showing its dipper cylinder and arm.

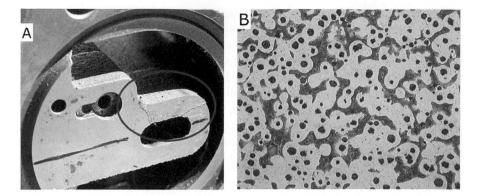

**Figure 3.8** (A) The hydraulic distribution block, with one of two identified sites of cracking circled in red. (B) Microstructure conforming to that of the specified nodular (spheroidal graphite) cast iron.

block uncovered extensive cracking present on two areas of the casting, one site being shown in Figure 3.8A.

Simple reference to operating manuals quickly established that there was no direct load train connecting the cracked zones, other than hydraulic pressure. To establish the material of manufacture, and for fracture surface observation, it was necessary to (metallurgically) cut a section from one of the cracked regions. This was only undertaken after gaining permission from all interested parties. The metallurgy confirmed three points:

(1) The material was an SG iron, conforming to the design specification (the graphite is in the form of spherical nodules. (Figure 3.8B).
(2) The section entered service in an unacceptably weakened condition by virtue of the inherent degree of gas porosity and shrinkage (Figure 3.9A).
(3) The presence of beach markings, often difficult to see in such a granular material (Figure 3.9B).

It could therefore be stated that the hydraulic distribution block failed as a result of latent defects introduced by the casting process. As well as reducing the load-bearing section, the porosity and shrinkage casting defects had also acted as stress raisers, sensitising the region to onset of fatigue cracking, with normal pressure pulsing of the hydraulic pump providing the cyclic loading input required for fatigue.

## 3.4 Computer-Aided Technologies

Over the past two decades, computer-aided technologies, such as computer-aided design (CAD), computer-aided engineering (CAE) and finite element

**Figure 3.9** (A) Unacceptable degree of gas porosity and shrinkage located at both sites of cracking. (B) Appearance of beach markings indicative of fatigue crack growth.

analysis (FEA), have become important (if not essential) tools in the product development process. These technologies embrace the use of computers in all activities from the design to the manufacture of a product. With the use of computer simulation, complex failure mode analysis becomes an easy undertaking before the manufacturer commits to costly prototyping. Computer modelling therefore allows for product designs that maximise the safe operating life of structural components. It is at the forefront of information technology, offering a greatly shortened design cycle along with lower product development costs.

In addition to enhancing the design process, computer-aided technology is also gaining ground in the forensic engineering arena. The use of this category of analysis will allow the comparison of simulated loading scenarios to actual failure detritus. If simulated cases deformations match those of the 'as found' damage, then causal factors which contributed to the failure or accident will become more obvious, as illustrated by the following case study of an open-ended spanner.

### 3.4.1 Case Study: Failure of an Open-Ended Spanner

An incident with failure of an open-ended spanner provides an ideal example to demonstrate the use of FEA, and also to demonstrate the 'weakest-link' principle in action: a heating and ventilation engineer was using a new chromium-plated 12-mm open-ended/ring combination spanner, when it suddenly failed across the open-ended jaw section. The unexpected (and sudden) failure caused the engineer to drive his right hand into an adjacent block wall, causing severe damage to both tendon and bone. The engineer testified that he was using the spanner in a normal manner and not subjecting the tool to any excessive force. Simple FEA analysis identified the sites of maximum stress (whilst being used to tighten a bolt) as being along the spanner shank (Figure 3.10A), and at one particular point of stress intensity on the jaw (seen as a fringe pattern concentration in Figure 3.10B). Under normal circumstances, failure would occur at one of these two points. However, the jaw had actually broken at a different point altogether, therefore not at the site of its 'weakest link' (Figure 3.10C). The jaw had broken with practically no prior deformation or loss of parallelism, indicating that the tool had failed in a brittle manner. However, it had only been subjected to light loading conditions over its short service lifetime. Under magnification, it was possible to see that the fractured jaw contained a myriad of fine cracks on its surface running parallel with the main fracture (Figure 3.10D). Along with a crack jump and brittle fracture features on the actual fracture surface, all three features were characteristic of hydrogen cracking, where hydrogen, evolved during the chrome-plating process, is absorbed by the metal. Clearly, the spanner had missed a final low-temperature post-plating baking process, designed to expel hydrogen from the electro-deposit. If this final bake process had been carried out after electroplating, it had not been effective in expelling all the hydrogen trapped within the chromium, thus directly leading to embrittlement of the

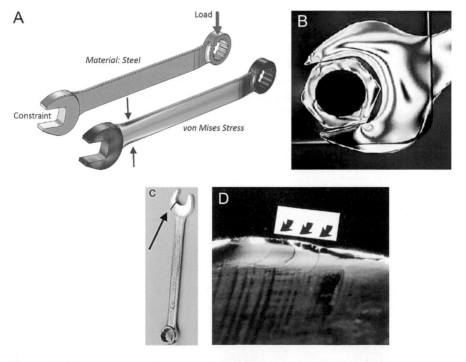

**Figure 3.10** (A) FEA analysis of a spanner showing site of maximum stress. (Image made using the COMSOL Multiphysics® software and is provided courtesy of COMSOL.) (B) Fringe pattern at spanner contact showing points of maximum stress within its jaw. (Courtesy of Prof. K. Ramesh.) (C) Actual site of jaw fracture. (D) Fine cracks on the jaw surface running parallel with the main fracture.

steel. Entering service in such an (hydrogen) embrittled condition, the spanner was completely unfit for its purpose – it would be expected to fail virtually the first time it was subjected to a heavy torque, either tightening or loosening, and failure would occur under quite modest service loading. The heating and ventilation engineer successfully sued the manufacturer, with damages being awarded for injury and lost income, assessed over the course of his recovery.

Generally, computer-aided simulations involved in such a comparison are highly non-linear, requiring advanced simulation techniques that include dynamic impact, contact, non-linear material properties and material fracture. The analysis will apply engineering calculations to determine stresses, strains and deformations of the component or structure in question.[11] Results of interest can then be plotted on a graph. In addition images and animations can also be further output to visualize the event, which can help with understanding how the failure or accident occurred. Throughout this book, and particularly within case study input, numerous examples of the use of computer-aided analysis, as a forensic tool, will be found.

To demonstrate the power of computer-aided analysis, the simple nut and bolt provides an ideal example. In general, it is almost always found that a bolt – having failed by fatigue – will have its crack initiation sited one thread root below its clamping nut. From observation of the FEA analysis shown in Figure 3.11, the site of intensified stress (arising from the clamping action of the nut and bolt) is sited at and around the first bolt thread below the nut face. Any site of localised stress intensification will sensitise the zone to the onset of fatigue crack initiation.[12]

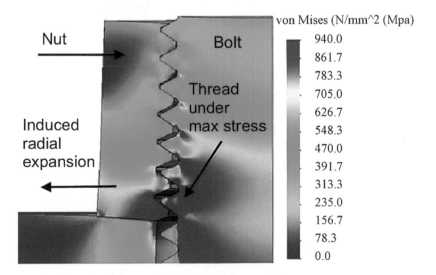

**Figure 3.11** A finite element analysis (FEA) of a simple nut and bolt, showing the point of maximum stress located one thread below the nut surface. Also, note the degree of induced radial expansion. (Courtesy of László Molnár, Károly Váradi and Balázs Liktor.)[12]

However, it should be emphasised yet again that *uncritical* reliance on computer-generated results should be avoided, as previously discussed in detail in Section 1.10.1.

## 3.5 The Forensic Engineering Report

The findings of any forensic engineering investigation will (generally) be promulgated as a report. However, two aspects need to be borne in mind when reporting any findings:

- First, the investigator *must* have examined any items relevant to providing answers to the questions within the brief. Any reporting must demonstrate what evidence and statements had been relied on, and how any deductions had been arrived at during the course of enquiry. The investigator must not step outside of the brief, and should on no account advance interpretations and explanations that are beyond their field of competence and experience – otherwise evidence presented is likely to be of no greater value than that of a lay witness.
- Second, the forensic report must be written so that a layperson will understand the basis for (and relevance of) the reported conclusions. In this respect judges, barristers, solicitors, insurance claims inspectors and so forth must be considered as laymen. However, they are experts in their jobs just as much as the forensic engineer will be an expert in his/her field. Ultimately, it is the expert who will have to convince the court. (It is often cynically remarked that it is not the side that is right, that will win a case in the civil court, but the one that tells the better story). Writing a forensic report is not like writing an in-house report on a service or process failure, where the readers will be technically competent, know the jargon, and do not need a great deal of basic background information regarding the product or the field of application. The report must, however, be accurate and written with an eye on the fact that it will eventually be disclosed to other parties and *their* experts, who may eventually seek to challenge both findings and explanations. In many cases, experts from all concerned parties may be asked by the court to prepare a joint (expert's) statement setting out:
  (a)  Those issues upon which the experts agree
  (b)  Those issues upon which the experts disagree
  (c)  A synopsis of the reason or reasons for disagreement on an issue
  (d)  Details of any issues requiring further exploration or laboratory-based investigation

In the UK, rules relating to experts and the expert report are set out in Part 35 of the Civil Procedure Rules (CPR), with Criminal Code rules being similar. All experts are duty-bound to be fully conversant with CPR 35, and to adhere to the practice and direction for form and content of an expert's report. Barristers often rely on expert responses, as a basis for cross-examination of other experts during a trial.

## 3.6 Concluding Remarks

All contributors to the literature agree that *robust* failure analysis will provide insight into failure mechanisms *provided* any analysis is thoroughly undertaken (being both vigorous and accurate), and all *necessary* tests are performed. If the analysis is incomplete, then inaccurate conclusions will be reached with possible serious future consequences and repercussions. It follows, therefore, that the forensic investigator is, in essence, a failure detective, and must

- Follow a totally independent pathway when undertaking an investigation and when offering or promulgating an opinion on the mode of demise.
- Resist any temptation to jump to conclusions. Every failure is unique, and should be treated as such. Any prior preconceptions may lead to neglect or overlooking of important details or facts.
- Recognise that conflicting evidence can often be elucidated by close scrutiny of the physical remains of failure. Physical remains will override any prior statements of condition or pre-conceived cause of failure.
- Not limit his/her observations to the broken artefacts alone. Study of a fractured surface in isolation will not allow a fully justifiable conclusion to be determined. For a more complete picture of failure, observation of the whole system *and* its surroundings is a necessity.

## References

1. Petroski, H., *Success through Failure: The Paradox of Design*. 2018, Princeton University Press, ASIN: B077JGFH99.
2. de Castro, P. and A.A. Fernandes, Methodologies for failure analysis: A critical survey. *Materials & Design*, 2004. 25(2): pp. 117–123.
3. Lampman, S.R., *Failure Analysis and Prevention*. ASM Handbook. Vol. 11. 2002, ASM International, ISBN: 978-0-87170-704-8.

4. Witherell, C., *Mechanical Failure Avoidance: Strategies and Techniques*. 1994, McGraw-Hill, Inc. pp. 31–65.
5. Carper, K., *Forensic Engineering*. 2nd ed. 2001, Boca Raton, FL: CRC Press. p. 401.
6. Komiya, S. and A. Hazeyama, Projecting risks in a software project through Kepner-Tregoe program and schedule re-planning for avoiding the risks. *IEICE TRANSACTIONS on Information and Systems*, April 2000. E83-D(4): pp. 627–639.
7. Wulpi, D. and B. Miller, *Understanding How Components Fail*. 3rd ed. 2013, American Society for Metals, ISBN-10: 0871706318, ISBN-13: 978-0871706317.
8. de Castro, P.M. and A. Fernandes, Methodologies for failure analysis: A critical survey. *Materials & Design*, April 2004. 25(2): pp. 117–123.
9. Lewis, P.R., K. Reynolds and Colin Gagg, *Forensic Materials Engineering: Case Studies*. 2003, Boca Raton, FL: CRC Press.
10. Lewis, P.R. and C.R. Gagg, *Forensic Polymer Engineering - Why Polymer Products Fail in Service*. 2010, Boca Raton, FL: CRC Press, ISBN: 978-1-84569-185-1.
11. Bremar Automotion. *Forensic Failure and Crash Analysis*. Prosolve Ltd. http://www.bremarauto.com/portfolio/forensic-crash-and-failure-analysis/, n.d.
12. Molnár, L., K. Váradi and B. Liktor, Stress Analysis of Bolted Joints Part I. Numerical Dimensioning Method. *Modern Mechanical Engineering*, 2014. 4(1): pp. 35–45.

# Analytical Methods 4

## 4.0 Introduction to a Typical Forensic Engineering 'Toolbox'

Various experimental methods are available to examine the physical remains of engineered products or systems that fail. Physical evidence of failure may include many pieces of detritus, or just a single sample. When observing a unique specimen, the approach taken is fundamentally constrained by the need to preserve and conserve it for possible scrutiny by others. Care must therefore be exercised in the choice of examination methods, particularly as items for study may be vital legal evidence that cannot be chopped up for examination. Forensic investigations are, therefore, fundamentally different from routine quality control, where thousands of components may be available for examination. The different approach required of forensics is biased to non-destructive methods of inspection, simply so that evidence is preserved. This is not to say that destructive or partly destructive methods cannot be used at all, but rather that they should only be used as the last resort. Indeed, all parties in a dispute are required to agree *any* destructive testing programme, normally with all interested parties in attendance.

After possible site inspection and gathering of relevant information (Chapters 2 and 3), there are a variety of techniques and methods available for in-depth analysis of any physical remains. The range of these tools will span from a simple eyeglass and macro-photography at one end of the scale (Chapter 2), to non-invasive or invasive analytical methods at the other. Each method will be unique in its own right and also complementary to the others. The use of appropriate methods should reveal a complete picture of failure modes or mechanisms that had been at play at the point of failure. The experienced professional will already have a breadth of knowledge regarding techniques complementary to their core skills, while the individual new to the field of analysis must gain a good overview of the range (and application) of the many available techniques. However, as a result of high capital equipment cost and complexity associated with the majority of analytical techniques, their use as an investigative tool will invariably be undertaken by a third party (accredited) test house. Nevertheless, the forensic engineer has a duty to be in attendance during the course of any analytical procedure, thus enabling him/her to appreciate, and be fully aware of, the exact sequence of analysis undertaken on the artefact in question. As it is the forensic expert who will be reporting all findings to the court, the forensic practitioner will then be fully conversant with both the procedure undertaken, and with any limitation or compromise associated with data gathered by that particular technique or tool.

Whilst acknowledging that potentially all engineering disciplines may be involved in forensic work, in a textbook such as this, it is impossible to include them all, or even the majority. So areas such as chemical, marine and electrical engineering, though vitally important, must be omitted. This book is primarily concerned with engineered products, namely, small or medium-sized manufactured artefacts used for machinery, automobiles, structures and domestic products. As such, the basic disciplines required for failure investigation are biased towards fractography, metallography and materials engineering, with substantial input from the disciplines of mechanical, production and structural engineering.

However, no matter what specific discipline of forensic engineering is being pursued, any analysis should always be approached using non-destructive (NDT) methods where ever possible, leaving any potentially destructive procedure as a last resort. Therefore, for clarity of presentation, available tools reviewed in this chapter will be assigned to three groups: non-destructive (NDT) methods (Section 4.1), microscopy and metallography (Section 4.2) and mechanical (usually destructive) testing methods (Section 4.3). Each group of tools will be considered in turn, thus forming the basis of a typical toolbox of analytical methodology available to the forensic (metallurgical) engineering investigator.

## 4.1 Non-Destructive Testing (NDT)

As should now be clear, the objective of any forensic engineering investigation is to determine the cause or causes of accidents or failures, whilst recognising and operating within a legal framework. However, forensic engineering is primarily a problem-solving exercise. As with any analytical assignment, there is a need to be aware of available tools, and the quality of information they provide. Although there is no pressing need to understand any deep underlying scientific theory associated with each tool, the investigator must appreciate any inherent limitations and 'trade-offs' that may be associated with a particular technique. For example, when driving a car, clearly, it is not necessary to know the way the car engine works. However, any car driver must be aware that excessively high revving of the engine, or driving like a 'boy (or girl) racer', will run the risk of excessive clutch wear, engine bearing damage and/or engine seizure, as well as probable loss of vehicle control.

Many failure problems will lead to litigation, possibly because there are several different possible interpretations of the damage. As an example:

### 4.1.1  Case Study: Failure of a Power Transmission Shaft

A marine engine power transmission shaft failed whilst in service (Figure 4.1A). It was argued that the shaft was under-designed for the power it was required to transmit; therefore the basic design of engine was at fault. However, on subsequent inspection, it was clearly evident that the shaft had succumbed to fatigue failure, as witnessed by the presence of fatigue striations (Figure 4.1B). Therefore, it could also be argued that the material of choice was not adequate in terms of its fatigue endurance limit. In reality, the shaft failed as a direct result of raised localised stress, sited at the root of a female thread. The thread had been formed with a root radius that was far too sharp, thus acting as a stress concentrator that had increased the normal working stresses by some 3×, hence the initiation of a fatigue crack. So, failure was the result of a latent manufacturing issue, rather than any design or materials shortcoming.

As there can often be cases where damage is open to different interpretations, initial examination must be non-invasive. For the forensic investigator, a first group of appropriate tools to be reviewed is that of NDT. NDT will embrace a wide group of analytical techniques used to evaluate the properties of a material, component or system without causing damage.[1] Almost certainly, the first universal method of NDT was that of visual inspection, with the power of observation being a basic asset for the forensic investigator, as discussed in Sections 2.1.2 and 2.2. At this point, it is worth reiterating that the cause of some failures can be determined by careful examination using nothing more than visual and/or low-power magnification. However, for the majority of situations that demand more in-depth analysis, the investigator will

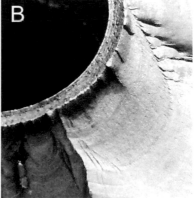

**Figure 4.1** (A) Marine engine splined power transmission shaft, clearly showing fatigue beach markings initiating from an internal screw thread. (B) Ratchet markings around the thread root periphery attest to the multi-start nature of fatigue, suggesting highly raised localised stress as a direct result of a particularly sharp thread root.

have at hand a variety of investigative procedures, unique to his/her special-ism. As such, the rationale behind this chapter is to outline a *typical* range of techniques and associated instruments available to the forensic engineer (the 'tools'), and highlight any limitation and compromise associated with each.

There is a range of NDT methods that can be used to diagnose and ver-ify the extent of any inherent defect, without being invasive, destructive or inducing mechanical damage. NDT integrates the use of current technology for *indirect* assessment of strength, integrity and fitness for purpose of mate-rials, components and structures. As non-destructive tools, they lie at the opposite extreme to destructive methods such as tensile and impact testing.[1,2] The types of defect that NDT may be employed to evaluate can be classified into three major groups:[2]

- *Inherent (liquid state processing) defects*: flaws that are established or introduced during the course of initial raw or feedstock material production.
- *Processing (solid-state) defects*: faults or imperfections that are in effect 'manufactured-in' at the time of further processing (e.g. heat treatment), or on the production line itself (e.g. machining of a com-ponent part).
- *Service (performance) defects*: introduced during the course of a nor-mally expected service lifetime for the material, part or device.

The range of defects or (micro-) structural differences that may be present within these three groups are cracks, both surface and subsurface; tears; porosity; casting, machining, rolling and plating defects; de-lamination; absence of sound interconnection; inclusions; segregation; poor weld pen-etration in both metallic as well as polymeric components; stress-raisers; fatigue failure/cracking; blow holes, shrinkage, limited adhesion of surface coatings (PVD), exceptionally high residual stresses, etc. For the investigator, each NDT method will have its own associated set of advantages and disad-vantages, with individual methods being better suited for a particular appli-cation than others. Therefore, a range of NDT methods commonly employed by the forensic metallurgical investigator will be reviewed, starting with what must be the most widely used NDT tool: that of visual crack detection.

## 4.2  Crack Detection and the Human Eye

As an initial approach to failure, the power of visual observation has been reviewed as something of an abstract or comparative 'tool', to be utilised at the outset of an investigation (Visual Observation, Section 2.2; Visual Comparison, Section 2.2.1). However, as a general non-destructive approach, visual observation can also be utilised in the context of crack detection. Cracks on the surface of a component can often be seen by eye, thus representing convincing evidence for structural failure. A crack is an opening of the material that must have been formed by plastic deformation (strain), no matter how small the crack. A human eye with 20/20 vision is able to resolve features as small as 75 μm in size at a distance of 25 cm.[3] It is possible under perfect conditions (on a mirror-polished surface) to detect a crack with a crack opening dimension (COD) as small as 10 μm. Conversely, the minimum detectable COD becomes much larger if the surface is rough or not perfectly clean. Surface features such as scratches and machining marks present visual 'noise' that will effectively mask any cracking.[3]

> *Advantages*: visual crack detection – no specialist equipment required. Instantly available, a cheap technology.
>
> *Limitations*: difficulty in observation of large areas or complex structures, requiring good access for the individual. Visual inspection is extremely slow, therefore visual inspection may rely on a 'representative' proportion of damage. The limitations of the survey method in performance terms are in speed and coverage. In addition, the reliability of visual inspection methods for detection of cracking or localised corrosion is questionable.

## 4.2.1 Surface Appearance of Common Cracks

The ability to visually detect a crack in a component will be dependent on both the type of crack, and its surface-breaking properties. Typical cracks found in service will include mechanical fatigue cracks, thermal fatigue cracks and stress corrosion cracks. Each category of crack will be formed differently, and exhibit different surface morphology.

- Mechanical fatigue cracks are formed when a material is subjected to repeated stress cycles. Cracks formed in this way are relatively straight and characteristically do not branch significantly, if at all. When the service stresses that formed the crack are removed, the COD shrinks considerably and may close almost entirely. There will be little or no corrosion within the crack, making observation somewhat difficult, even for the seasoned investigator.
- Thermal fatigue cracks can manifest in regions of components with cyclically applied temperature excursions. Thermal fatigue cracks will typically form in patterns. For example, under a tensile stress, multiple small thermal fatigue cracks often form perpendicular to the direction of applied stress. In a material without any predominant directional stresses, thermal fatigue cracks often form in a fabric-like

**Figure 4.2** Morphology of stress corrosion cracking.

or cobblestone pattern. There may or may not be oxidation inside the crack, and thermal fatigue cracks can have a very small COD.

- Stress corrosion cracking (SCC) is caused by a combination of stress, a sensitised material and a corrosive working environment.[4] Stress corrosion cracking will vary in appearance, ranging from a single crack to a series of cracks lying together in a slightly feathered pattern (Figure 4.2) but virtually always following the sensitised zone. Any internal residual stress or strain will provide the driving force needed to propagate SCC. Stress corrosion cracks typically have small opening dimensions.

Apart from visual observation, there are several NDT methods available for detecting hairline cracks. These techniques can be used to discover cracking in a worn component, such as a crankshaft, before regrinding or refurbishing. Such methods are routinely used for the inspection of safety-critical parts, such as bodies and wings of aircraft, pressure vessels and so on. Therefore, a range of other available crack detection techniques will be reviewed, being typical tools available to the forensic investigator.

### 4.2.2 Other Crack Detection Techniques

*Liquid penetrant testing (LPT)*: liquid penetrant (sometimes termed 'dye-penetrant') testing is one of the earliest forms of non-destructive testing. The technology can only inspect non-porous components and then only for surface breaking defects. However, it is one of the few inspection techniques that can be used to examine ceramics. The component is sprayed with a penetrating dye, which is allowed to soak into the cracks by capillary action. The surface is then cleaned off with a removing solvent. However, dye will remain in any crack-like defect. The component is then sprayed with a chalky powder that acts as a 'developer'. This will draw out the dye from the defect, thus highlighting its presence in the white developer layer, providing a visual indication of a surface crack defect.

A simple example of the method in action concerned a cycle crank arm that was returned to the supplier after a very short time in use (Figure 4.3). The owner had seen a crack emanating from the square taper axle attachment, and also suspected a smaller crack close to the pedal thread. Results from a dye penetrant inspection revealed a latent crack running from the square axle drive hole, indicating its precise location. The pedal thread feature showed no red line, indicating a surface scratch rather than an embryonic crack.

In July 2000 hairline cracks in parts of the wing structure of a British Airways Concorde were detected by routine inspection using dye-penetrant and ultrasonic methods. However, it has to be emphasised that this particular cracking issue was not relevant to the subsequent Paris disaster.

> *Advantages*: penetrant testing can be used on metal, glass, ceramics and many solid materials. It is relatively simple to use and does not require elaborate or expensive equipment.
> *Limitations*: the procedure will not detect subsurface discontinuities, and it may attack some plastic and composite materials.

*Magnetic particle inspection*: a technique only applicable to ferromagnetic materials. The component is painted with a white paint and then placed in a strong magnetic field. The attachment of strong horseshoe permanent magnets is often sufficient. Any cracks in the component produce local magnetic poles and when 'ink' consisting of iron, or iron oxide, particles in paraffin is

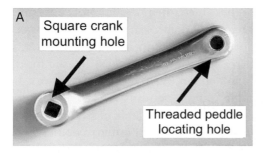

**Figure 4.3** (A) A bicycle crank arm with a suspected cracking defect. (B) After dye penetrant application, the presence of a latent crack is clearly evident.

sprayed on to the component the particles settle at these poles and highlight the crack. Once again it is only applicable to surface cracks in ferromagnetic materials, but is very quick and cost-effective. A similar technique is that of ultraviolet (UV) examination of fluorescent inks.

> *Advantages*: quick and relatively uncomplicated, providing immediate indication of defects. Will reveal both surface and near-surface defects, being the most serious stress concentrators. Can be adapted for both site and laboratory use, with no elaborate sample pre-cleaning requirement.
> *Disadvantages*: restricted to ferromagnetic materials – usually iron and steel. Cannot be used on austenitic stainless steel. Requires an electrical supply. Spurious, or non-relevant indications possible; thus interpretation is a skilled task.

*Ultrasonic testing*: a non-destructive method in which beams of high-frequency sound waves are introduced into a material being evaluated to detect hidden cracks, voids, porosity and other internal discontinuities. High-frequency sound waves reflect from flaws in predictable ways, producing distinctive echo patterns that can be displayed and recorded by portable instruments. Ultrasonic flaw detectors are suitable for test house, laboratory and/or field use. They generate and display an ultrasonic waveform that is interpreted by a trained operator, often with the aid of analytical software, to locate and categorise flaws in failure detritus.

When considering interference fit applications such as hubs or bearings on shafts, ultrasound is an ideal non-destructive route for determining contact pressures of any press-fit situation. During press-fit assembly, interference will induce a pressure at the interface. The magnitude of this contact pressure is of importance as it is a stress-raiser, and so can cause fatigue crack initiation and ultimate failure.

In the semiconductor industry, the use of ultrasonic microscopes for evaluating the integrity of the bond between adjacent surfaces has increased dramatically over the last few years. These instruments are rapidly becoming an acceptable diagnostic tool in the quality control laboratory. New specifications have been written incorporating them into military specification and standards (MIL) requirements. The technique is principally used for the non-destructive investigation of integrated circuits (ICs) and capacitors, for detecting de-laminations, cracks, voids and die-bonding faults.

There are two types of acoustic microscopes available to study and evaluate area array packages: the scanning laser acoustic microscope (SLAM) and the C-mode scanning acoustic microscope (C-SAM).[5] Both instruments employ high-frequency ultrasound signals to detect internal discontinuities within both materials and component packages. The SLAM

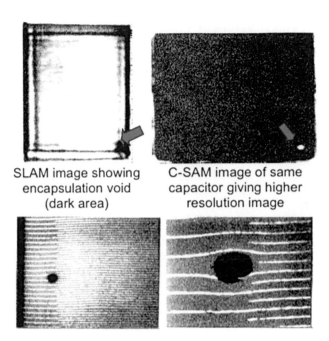

SLAM image showing
encapsulation void
(dark area)

C-SAM image of same
capacitor giving higher
resolution image

**Figure 4.4** Investigation of encapsulation void in a capacitor (MLCC). Acoustic microscopy requires a trained eye to interpret results. Investigation by scanning electron microscopy (SEM, bottom two images) gives a clearer picture but is a destructive technique.

operates in the through transmission mode, and the C-SAM is a reflection mode instrument.

> *Advantages*: small, portable, microprocessor-based instruments. Can be used on metal, ceramics, plastics or any other homogenous material that is compatible with sound transmission. This process can be used on the surface of parts or under water (immersion testing). Ultrasonic testing is completely non-destructive and safe in use.
>
> *Limitations*: include the moderately high initial cost of equipment and the need for highly trained personnel to interpret the results (Figure 4.4). A range of materials that are generally not suited to conventional ultrasonic testing include wood, paper, concrete and foam products.

## 4.3 Hardness Testing

A hardness test involves measuring the size or depth of an indentation created under pressure from an indenter, usually a diamond or a hardened steel ball. The size or depth of damage gives a measure of the tensile stress of the softer

Table 4.1  **Typical Values of Vickers Hardness Number, Hv, for a Range of Different Materials**

| Material | Hv | Material | Hv | Material | Hv |
|---|---|---|---|---|---|
| Tin | 5 | Limestone | 250 | Tungsten carbide | 2,500 |
| Aluminium | 25 | MgO | 500 | Polycarbonate | 14 |
| Gold | 35 | Window glass | 550 | PVC | 16 |
| Copper | 40 | Fused silica | 720 | Polyacetal | 18 |
| Iron | 80 | Granite | 850 | PMMA | 20 |
| Mild steel | 140 | Quartz | 1,200 | Polystyrene | 21 |
| Fully hardened steel | 900 | $Al_2O_3$ | 2,500 | Epoxy | 45 |

material. Hardness testing can also be used to estimate the resistance to wear (or abrasion) of a material, its tensile strength and the type of heat treatment that it has been subjected to. In addition, because it leaves only a tiny indentation on the surface of the specimen, hardness testing has the advantage of not markedly altering any evidence the surface may contain. Typical values of Vickers hardness for a range of different materials are shown in Table 4.1.

Although a large number of cases presented in this book utilise the Vickers hardness test (see Figure 4.5 for an example of such a test), it is important to remember that there are a *variety* of hardness test methods, as shown in Table 4.2. The Vickers test using a diamond pyramid can be used with very small loads (down to a few grams), and will give results comparable with standard tests that produce indentations measuring up to 1 mm across. In addition, there are also ultrasonic methods of hardness testing.

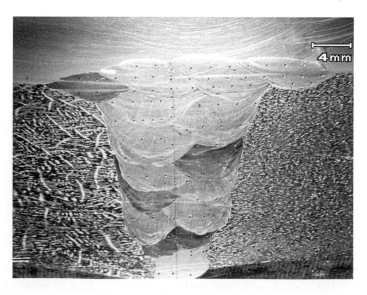

**Figure 4.5** An example of a Vickers hardness survey of a multi-pass weldment.

**Table 4.2    Flow Chart Showing the Range of Available Hardness Test Methods**

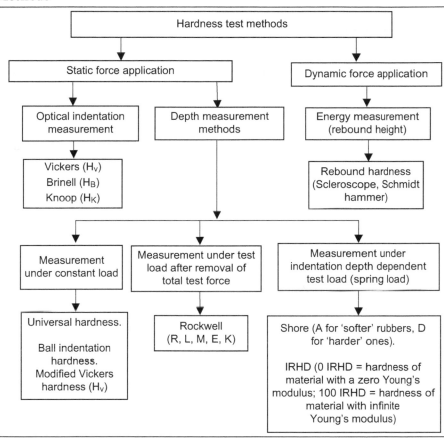

## 4.3.1  Case Study: Pin-Punch Splintering

A pin punch was being used by a DIY enthusiast. It was one of a set of punches that he had purchased from a local market. He was driving out corroded bolts on some equipment that he was attempting to repair. As he was hitting the striking end of the punch with a ball peen hammer, the end of the punch shattered causing splinters to embed in his cheek. Although not seriously hurt, he was shocked that the punch had failed in such a way when being (correctly) hit on its striking end. Longitudinal sectioning and a Vickers hardness survey of the pin punch showed an increasing hardness as the striking face was approached, having values of 390 Hv away from the striking face, increasing to 600 Hv as the striking face was approached. This result can be clearly seen in Figure 4.6, with hardness indents visibly getting smaller as they approach the striking face. As a result, the hardness survey was repeated using a Brinell ball indenter, producing identical results. These results were the reverse of those expected, showing that the striking face had mistakenly been hardened and tempered, rather than the punch end.[2]

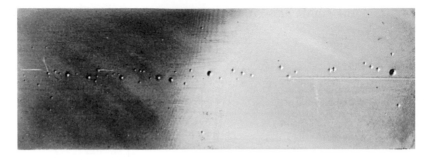

**Figure 4.6** Longitudinal polished section of a pin punch. Hardness indents (both Vickers and Brinell) clearly get smaller as they approach the striking face.

*Advantages*: Vickers hardness has one scale that covers the entire hardness range, with a wide range of test forces to suit every application. Rockwell hardness has a direct number readout and rapid testing time, usually less than ten seconds. With a direct readout, there are no questionable optical measurements required. When compared to 'other available hardness test methods, the Brinell ball will create the deepest and widest indentation. Therefore, the test will inherently average the hardness value over a wider contact area of material. As such, the test will more readily assimilate multiple grain structures along with any variability in consistency of the material. This method is ideal for obtaining the bulk, or macro-hardness, of a material – particularly for those materials with diverse (heterogeneous) micro-structures.'[2]

*Limitations*: samples must be clean and have a smooth test point to get good results. Possibility of error during the test sequence caused by permanent shifting of the specimen, and of other components that may be in the force flow during the test. Limited possibility of the test of edge zone hardened specimen, due to high test-load requirements. Sensitivity of any indenter to damage, thus raising a danger of erroneous measurements. When using a conical indenter, it may influence any hardness test result obtained.

### 4.3.2 Relationship between Hardness and Tensile Strength

It can be shown that the Vickers hardness test (Hv) bears a direct relationship to the tensile strength ($\sigma_{TS}$) of a material. However, the way in which a material responds will depend on the ratio of its yield stress to Young's modulus (E). For softer metals with a low $\sigma_{TS}/E$, it is found that

$$Hv = 3\sigma_{TS} \tag{4.1}$$

Where Hv is the Vickers hardness number. The relationship shows that yield strength in tension is about 1/3 of the hardness. To find the ballpark figure for the yield strength, convert the hardness number to MPa and divide by 3.

For example take the Vickers number, which has the dimension kg/mm$^2$, and multiply by 10 to convert it to /mm$^2$ (MPa), then divide by 3.

For materials with higher $\sigma_Y$ /E (including a wide range of glasses and plastics), the relationship between Hv and $\sigma_{TS}$ becomes more complex. However, there is a useful empirical relationship between tensile strength (in MN m$^{-2}$) and Hv (in Nmm$^{-2}$), namely

$$\sigma_{TS} = 3.2Hv \qquad\qquad (4.2)$$

When forensic specimens are small or when standard test bars for tensile or shear testing cannot be taken from material evidence, a hardness test may be the only means of obtaining a useful estimate of the tensile strength of the material.[6]

## 4.4 Indirect Stress/Strain Analysis

Hardness measurements are one way in which the material response to a given load can be measured directly in a standard way. The strain produced by application of the load changes the dimensions of a small part of the surface of the material. However, are there any other ways in which the strain to which a sample has been exposed can be determined? The question is important for forensic work as large deformations and strains are frequently found in failed specimens, such as car parts recovered from accidents. Strain monitoring techniques are well-known for the structural analysis of components at the design stage of product development. A standard load will be applied to such products, and the strain in different parts of the structure monitored by strain gauges.

There are a number of methods for indirect stress or strain analysis when investigating the possible or actual failure of a product or part. Failure can be from externally applied stress or from residual (manufactured-in) stresses.[3] Both external stress and residual strain (or a combination of both) can cause a part to fail prematurely. It is relatively straightforward to detect failure due to poor design, or excessive service loading forces.[3] However, residual stresses and strains are altogether different. Here, a simple manufacturing process can generate residual strain just about anywhere, anytime. As a tool, photoelastic inspection will allow the detection of frozen-in strains, allowing identification of failure, with the method revealing the actual levels of stress in the part. Two basic methods used to conduct an indirect stress/strain analysis are reviewed here:

### 4.4.1 Brittle Lacquer Technique

Brittle lacquers can be used to provide a simple and direct way of estimating the degree of strain in deformed bodies. The product is coated with a thin

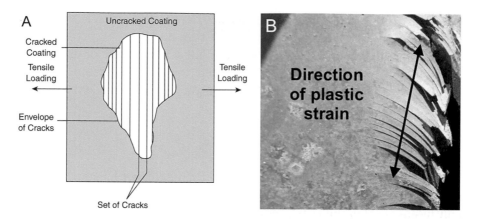

**Figure 4.7** (A) Schematic diagram of parallel cracks in a brittle lacquer under tensile loading. (B) Cracking of paint layer on a load frame fabrication. Note how the paint has acted as a 'brittle lacquer', clearly allowing the direction of plastic strain to be determined, as arrowed.

layer of brittle polymer, using a solution of the polymer in an organic solvent. After drying it forms a thin uniform film that is well-bonded to the underlying metal. When strained to a high degree, the metal deforms plastically, and the thin film also deforms. Because it is brittle, however, the film cracks at only very small strains, so that it is covered with a pattern of cracks that reflect the underlying plastic deformation of the metal product (Figure 4.7A). The cracks are oriented at right angles to the tensile strains in the metal, so the method allows easy identification of the plastic deformation. In other words, the crack pattern bears witness to the loading history of the product, as shown in Figure 4.7B, for example. Complicated shapes such as castings or large weld runs are not an issue for the technique, as the coating can be sprayed on. However, what is not generally recognised is that any paint coating on a structure can, and will, act as a brittle coating, thus providing historic evidence of loading stresses applied to a structure. Therefore, the basic idea can be used for identifying loading patterns in failed products, as described in the case where a cyclist was injured after riding into the back of a parked vehicle. This incident adequately describes the technique of brittle lacquer coating in action.

## 4.4.2 Case Study: Cycling Accident

A young man suffered injury when riding home on a bicycle, after working a night shift. It was raining and just starting to get light. He claimed a poor weld on the frame head tube broke, causing him to ride into the back of a van that was parked without lights. The bicycle frame had been sent to a laboratory,

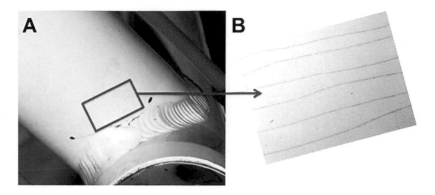

**Figure 4.8** (A) Head tube of a cycle frame (on the left) showing a network of cracks in its 'brittle lacquer' paint coating, introduced as a result of plastic strain prior to fracture. (B) Magnified view of cracking.

where the head tube was cut out for closer examination. Subsequently, the section was sent for independent opinion as to the cause of failure.

On receipt of the head tube section, the first observation was the clear presence of cracks in the white paint coating (Figure 4.8). On closer inspection, the paint coating adjacent to a weld on the head tube of the cycle frame revealed an interesting pattern of strain cracks (shown as an enlarged view of the red square zone). The paint had acted as a brittle lacquer, and bore witness to the fact that the steel tube had undergone extensive deformation under frontal impact before the weld failed.

The scenario can be reconstructed: the rider, head down, half asleep after a night shift, gritting his teeth as he rode into the rain, ran straight into the back of the unlit parked van; the weld failed under the impact. There was no evidence whatever of fatigue cracking in the fracture surface. If there had been a pre-existing crack, the paint film on either side of it would not have exhibited strain cracks.[6]

## 4.4.3 Photoelastic Stress Measurement

Some transparent plastics are birefringent and lend themselves to photoelastic stress analysis. The part is placed between two polarising mediums and viewed, in the crossed polar position, from the opposite side of the light source. Fringe patterns are observed without applying external stress, thus allowing observation of moulded-in or residual strains in the part.[3] Figure 4.9A shows the stress fields present in a seven-member model bridge truss, centrally loaded and simply supported. A high fringe order indicates areas of high stress level whereas low fringe order represents an unstressed area. Furthermore, close spacing of fringes will represent a high stress gradient, whereas uniform colour will be an indicator of uniform stress in the part. The plastic part may be subjected to stress by applying external force to simulate

**Figure 4.9** (A) Stress fields present in a centrally loaded seven-member simply supported model bridge truss. (B) The distribution of residual stress of transparent plastics, in the form of fringed pattern, can be seen on this plastic DVD jewel case. The DVD case was illuminated with a polarised light source and photographed through a polariser filter. (Courtesy of Wikimedia Commons, licensed under Creative Commons.)

'in-service' conditions. Both applied and residual stress fields can be exposed using models of structures in photosensitive material placed between polarising filters in the crossed polar position.[3] Figure 4.9B shows a polycarbonate injection moulded protractor containing residual moulding stresses that are clearly visible under the cross polarised photoelastic viewing method.

Photoelastic coating can also be used to analyse opaque plastic component parts. The part in question must be covered with a photoelastic coating, service loads are applied to the part, and coating illuminated by polarised light from a polariscope. Although any 'frozen-in' residual stress (introduced during initial moulding process) cannot be detected by this technique, it is possible to fabricate the component part out of a transparent plastic material.

> *Advantages*: areas of high stress concentration will easily be pinpointed by observing changes in fringe patterns brought about by external stress.
>
> *Limitations*: residual stresses cannot be readily observed with this technique.

## 4.5 Conventional (Contact) Radiography

Radiography is another NDT technique that could have been discussed in Section 4.1.2, other crack detection techniques. However, conventional radiography is not always successful in detecting cracks unless the line of the crack is parallel or nearly parallel with the X-ray beam, Figure 4.10. Radiographs will reveal whether defects or discontinuities are present in castings, forgings and weldments, in addition to polymeric and other non-metallic parts.

'A source of radiation is directed toward a sample, with a sheet of radiographic film having been previously placed behind the object (Figure 4.11A).

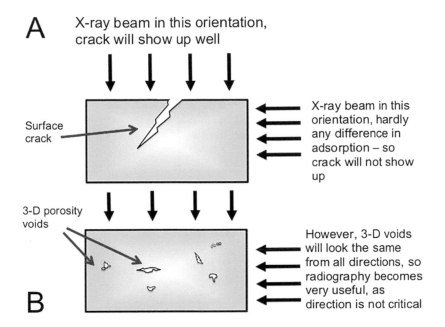

**A** X-ray beam in this orientation, crack will show up well

Surface crack

X-ray beam in this orientation, hardly any difference in adsorption – so crack will not show up

3-D porosity voids

However, 3-D voids will look the same from all directions, so radiography becomes very useful, as direction is not critical

**B**

**Figure 4.10** (A) X-ray orientation for crack detection. (B) Three-dimensional voids.

The density of the image formed on the film is a function of the quantity of radiation transmitted through the object, which in turn is inversely proportional to the atomic' weight, density and thickness of the object.[3,6] Therefore, radiographic images directly correspond to density, atomic number and thickness variations of static, solid objects as shown by the camera X-ray, Figure 4.11B.

Information that can be gathered from radiographs would include surface cracking, internal voids, reinforcing flow patterns, density of reinforcing and fibre orientation (particularly at critically stressed points). As an illustration, a contact radiograph of injection moulded vehicle panel containing 22% by volume of 13-mm diameter glass fibres is shown in Figure 4.11C, along with a photographically enlarged radiograph of the same panel at a different point of flow (Figure 4.11D).

A particular case concerned the failure of a semiconductor. Semiconductor devices are almost always part of a larger, more complex piece of electronic equipment. These devices operate in conjunction with other circuit elements and are subject to system, subsystem and environmental influences. Common problems which can arise include de-lamination of plastic packaged components due to moisture ingress and exposure to soldering heat, lead plating defects, die damage and contamination issues. However, if a single component fails in-service, catastrophic failure of the device or system may result.

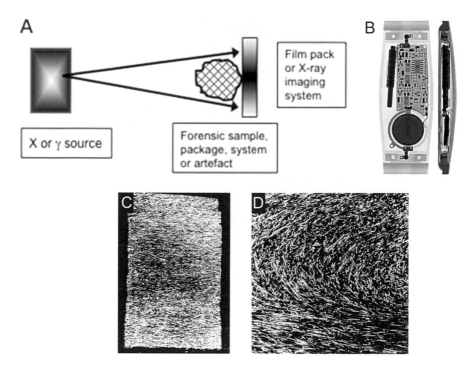

**Figure 4.11** (A) Schematic of conventional contact radiography. (B) Contact radiograph of a digital camera, clearly revealing internal mechanism and discrete components. (Courtesy of Nick Veasey, licensed under Creative Commons.) (C) Contact radiograph of injection moulded vehicle panel containing 22% by volume of 13-mm diameter glass fibres. Transverse section. (D) Longitudinal section.

### 4.5.1 Case Study: Failure of a Vehicle Motherboard

A motherboard from a vehicle management system failed, causing the vehicle to crash. Although there were no serious injuries to the driver or passenger, they were shaken enough to pursue the reason or reasons for failure. Simple probe inspection of components mounted on the motherboard isolated a glob-top encapsulated chip-on-board (COB) semiconductor as the mal-functioning device. Subsequent X-ray inspection revealed two broken lead wires, as seen in Figure 4.12. Further investigation of the materials used for fabricating the device revealed the cause of failure.

Under normal operating conditions, the dice, wire bonds and board laminates all expand and contract at different rates with changes in temperature. The coefficients of thermal expansion (CTEs) range from a low of 3 ppm/°C for the silicon wafer to a high of 17 ppm/°C for the copper board tracks. The best reliability is achieved by closely matching the CTE of the encapsulant to the CTEs of the device and the board. Although normal epoxies have CTEs of 50 to 70 ppm/°C, standard encapsulants achieve CTEs of less than 25 ppm/°C through high loading levels of silica filler.

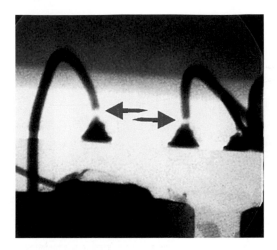

**Figure 4.12** X-ray micrograph of broken lead-wires within a chip-on-board (COB) device.

In the original design, the CTE of the epoxy-based encapsulant material had not been well-matched to the CTEs of the other materials in the assembly. During changes of temperature experienced during field operating conditions, the encapsulant expanded excessively with respect to the other materials and stressed the bond wires. This led to thinning, and ultimate fracture of the bond wires, as illustrated by arrows shown on the X-ray micrograph of Figure 4.12.

The technique has further advantages for failure investigation in that information obtained from a radiograph is often of value when choosing a location for sectioning. For example, radiographs can be utilised to determine a site that will provide optimum information for metallographic sectioning and examination. More to the point, it will allow the choice of a sectioning site that will not destroy useful information.

### 4.5.2 Case Study: Heavy Goods Vehicle Fire

A heavy goods tractor unit (Figure 4.13A) was overtaken by a fire in its engine compartment (Figure 4.13B). Subsequent inspection by a fire expert identified a back-to-back idler bearing unit as one of two possible sources of ignition (Figure 4.13C). A radiographic image of the recovered bearing unit was taken with an X-ray machine, as seen in Figure 4.13D. It became clear that the back-to-back bearings had been exposed to heat at a level where both the lubricant and plastic rolling element retaining spacers had simply carburised. The forward-facing bearing had a single rolling element that had moved out of registration (arrowed in red on Figure 4.13D), whereas, the rear bearing had managed to retain rolling element registration. This would suggest the forward-facing bearing had been exposed to higher levels of heating than that of the rear bearing i.e. the forward bearing was facing the heat source. Loss of registration

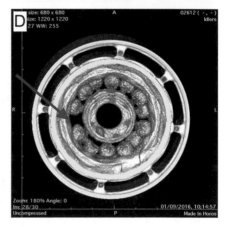

**Figure 4.13** (A) Heavy goods tractor unit, overtaken by a catastrophic engine fire. (B) The extent of fire damage, shown immediately prior to inspection. (C) A back-to-back idler pulley bearing unit (red arrow) identified as a potential source of ignition. (D) Comparative medical diagnostic radiogram of the idler pulley bearings, with loss of rolling element registration in the forward unit, clearly seen as a gap, arrowed in red.

in the forward-facing bearing would have occurred as a result of gravity acting over time on the rolling element, as the lubricant became progressively more fluid from the increasing temperature. Furthermore, as there was no evidence of rolling element or rolling track breakdown, it could be concluded that loss of registration of the bearing element occurred as a result of a fire in the subject vehicle, rather than being the cause or source of that fire. Conclusive evidence was later found, which confirmed a short circuit, in a high-tension cable, as the source of ignition.

This method is often of vital importance in personal injury cases, such as when a young man suffered injury when using a circular saw. A tooth came off an almost new blade and went right through his eye, lodging close to his brain. The final position was so delicate that surgeons could not risk removing it. However, X-rays taken at different angles allowed identification as a tungsten carbide tooth of the same dimensions as those on the failed blade, two of which were found to have broken off due to faulty brazing.

As a final point, digital radiography (DR) is a more specific portable X-ray method, using a source of X-rays (or gamma rays) and a digital detector. The result is a digital X-ray image, displayed on a computer screen. The use of radiographic testing includes pipe inspection (welds, wall loss, pipe under insulation, valves), aircraft component testing, testing of welds on ships and marine structures, defects in power lines, casting defects during the production process (e.g. automotive parts), etc.

> *Advantages*: non-destructive, as the object should not be damaged by either X-rays or gamma radiation. An ideal technology for detecting internal defects such as voids, cracks or inclusions in materials and assemblies. Easy to establish internal clearances between parts in an assembly (such as electronic chip packages) and to check for displacement of internal components (such as reinforcing in concrete).
> *Limitations*: laminations, tight cracks or cracks parallel to the film plane are not readily discernible. Use of equipment carries inherent radiation risk.

## 4.6 Summary of NDT Inspection

As a final word on NDT inspection, the techniques reviewed above do not represent the limit of the NDT principles available to the failure or forensic engineer. Electrical potential drop, infra-red, acoustic emission and spectrography have been used to derive information that the above techniques have been unable to yield. Furthermore, development across the board is a continuing process; for example resonant inspection (RI) is a novel NDT technique that was originated by scientists at Los Alamos National Laboratory in the USA.[7] It has recently been developed for industrial applications as a whole-body resonance inspection system. Being particularly suited to inspecting smaller mass-produced hard components, one test will inspect the complete component without radiation, the need for scanning, immersion in liquids, chemicals, abrasives or other consumables.

## 4.7 Forensic Optical Microscopy

The second group of tools to be reviewed is that of microscopy. Both visual inspection and microscopy are important parts of the examination of failed components. However, visual resolution is limited and subjective (Section 2.1.2, Section 2.2 and Section 4.1 above). After an initial inspection of the morphology of a specimen, and the recording of appropriate images, higher magnification may be required to establish causation.

## 4.7.1 Macroscopic Examination

Macroscopic examination involves observation of a failed component, fracture surface or section under low magnification (10 to 50×), preferably using a stereo-microscope.[6] For field observations, a simple eyeglass is all that will be required. However, today, just about everyone will be carrying a cell or smartphone in his or her pockets. Recognition of this fact has led to the development of low-cost attachments that turn a smartphone into a reasonable 'magnifying glass', with a 5–20× magnification. There are also a large number of third-party lens attachments for smartphones available, with a choice of fisheye, wide-angle, macro or telephoto lenses. If he were around today, Sherlock Holmes may well lament the demise of the hand-held magnifying glass – his most favoured tool. However, I do feel he would also agree that this simple development has proved to be particularly practical when undertaking an on-site investigation. One particular advantage this advance has, over that of the conventional magnifying glass, is the added ability to take, save and manipulate images of objects under scrutiny.

In addition, the amount of information that can be obtained from examination of a fracture surface under low magnification is surprisingly extensive, exposing information representative of the entire piece[6], such as size and distribution of inclusions, uniformity of structure and grain size, segregation and presence of fabricating defects, which will all be readily discernible. Macroscopic examination of fracture surfaces can reveal information relevant to the failure mode (fatigue, shear, brittle or ductile), indentations and surface damage suffered in service (some of which may have caused a fatigue failure) and environmentally enhanced failure (corrosion, stress corrosion cracking, hydrogen embrittlement, etc.).

Fracture characteristics such as direction of crack growth and origin of failure can be determined, as in the fatigue failure shown in Figure 4.14. Here, a ten-month-old combine harvester was overtaken by failure of a drive shaft. Observation of the mating fracture surfaces immediately revealed that failure was as a direct result of classic multi-start rotating bending fatigue – a totally mechanical failure mechanism. A number of fatigue crack initiation points could be determined, along with the area of final failure. A point of interest was that, contrary to expectation, the keyway was not a major influence in initiating this failure. When the product design (service) lifetime was considered, it can be said that the shaft failure was a premature mechanical event, covered by a manufacturer's and/or dealer's warranty. Therefore, advice to the combine owner was to pursue this avenue for recovery of his losses.

**Figure 4.14** Fracture surface of a combine harvester drive shaft, showing multi-start rotating bending fatigue failure. Points of secondary fatigue crack initiation are arrowed in red, with the prime point of initiation arrowed in blue. Note: shaft diameter: 50 mm (approx. 2 in.).

## 4.7.2 Examination under Magnification: Optical Microscopy

More advanced magnification tools include reflected light microscopy and scanning electron microscopy. For non-metallic materials, it is often useful to use either a low-vacuum scanning electron microscope or an environmental scanning electron microscope. The latter two differ slightly in their mode of operation and the information that is obtainable but offer the advantage that the specimens can be examined without the need for a conductive coating. A major advantage of scanning electron microscopy is that it can be used in conjunction with energy dispersive X-ray analysis to obtain information regarding the surface chemistry of a fracture surface, component, etc.[3,6] The advantages and limitations of different microscopy techniques are outlined below.

### 4.7.2.1 Universal Serial Bus (USB) Microscopes
In terms of magnification, a range of USB microscopes will eclipse the cell phone magnifying glass described earlier. Basically, a USB microscope is a digital microscope powered by a laptop computer, or similar device, by

means of a USB cable connected to one of the USB ports. Images can be viewed, recorded and stored similarly to a digital camera by using the software that comes with the microscope. The magnification is typically in the range 5–200× magnifications and the microscope comes with built-in LED lights for illumination of the subject.

### 4.7.2.2 WiFi-Enabled Microscopes

A major disadvantage with the USB microscope is that it has to remain attached to the computer or device by means of an umbilical (power/communication) cable. However, there are now a number of WiFi-enabled microscopes that are becoming popular for fieldwork. Instead of having an umbilical cable, USB or dongle type of connection, these microscopes have a WiFi access point built within its camera enclosure. Once turned on, it will create a new hotspot for the camera, so any tablet or smartphone can (wirelessly) connect to the camera and access the microscope's video output. Coupled with an appropriate image-capturing application, this type of instrument will provide all the necessary features to capture, store and manipulate microscope images captured in the field.

> *Advantages*: exceptionally low cost. Compact device. Non-destructive. Provides a portable solution that is ideally suited to field investigation. Ideal for field work, so of particular use in remote areas, where the location dictates that access to scientific instruments is limited or non-existent.
>
> *Limitations*: USB microscopes have a lower level of magnification compared to other forms of optical microscopy. Low device resolution may not be adequate for detailed imaging of the subject. Limited depth of field, as with all optical microscopes. Best results are obtained when working with flat objects.

### 4.7.3 Reflected Light Microscopy

'Reflected light microscopy is used to study the microstructure of opaque materials. Contrast in the viewed image, is the result of differences in reflectivity of the microstructure. The maximum magnification achievable is limited to around 1000×'[3,6]. The primary purpose of a microscopic examination is to reveal detail that is too small to be seen by the naked eye or by macroscopic examination. Magnifications would usually range from 20 to 1000× for a standard optical microscope. Microscopic examination of etched sections can be used to reveal grain structure such as grain size, inclusions, phase distribution, cracks, porosity, internal defects, surface coatings, etc., as well as thermal and mechanical history and external features such as corrosion. In short, optical microscopic examination can reveal a great deal about the past history of a specimen and how it will (or did) react in service. However, it should be remembered that the technique will reveal detail about a particular (and often small) portion of the specimen and therefore may or may not be representative of the entire article. Further details of the techniques involved can be found in standard textbooks on microscopy.[7]

As an example of optical microscopy, Figure 4.15 shows a polished and etched steel micrograph, showing a large quench crack. During the hardening and quenching process, there are two stresses involved: thermal stresses arising from rapid cooling, and transformation stresses due to an increase in volume as the structure changes from austenite to a martensitic structure. The severity of such transformation stresses can cause excessive distortion or even cracks within the section, as seen in Figure 4.15.

### 4.7.4 Stereo Microscopy

A stereo microscope exploits 'the brain's ability to superimpose two images from different angles, and perceive spatially accurate 3D objects. In the stereomicroscope this is achieved by transmitting two images from the sample inclined by a small angle (10–12°) to yield a stereoscopic image when the sample is viewed through the eyepieces.'[3] Although stereomicroscopes allow images to be obtained with excellent depth perception, they only have a lim ited resolution. Practically, the maximum working magnification tends to be of the order of 90–125×. The images can be subsequently recorded utilising a digital camera, then saved onto a personal computer (limitation: the image is taken through a single camera and thus the 3D effect is lost). However, stereomicroscopes are a valuable tool for detailed examination of fracture surfaces.[3]

**Figure 4.15** Polished and etched section showing a quench crack through the section. The different colour around the crack in the micrograph is an artefact of the etching process. (Courtesy of University of Cambridge DoITPoMS Micrograph Library, licensed under Creative Commons.)

*Advantages*: magnifies objects up to 1000×. Cheap to purchase and operate. Small and portable. Natural colour of the specimen can be observed and recorded.

*Limitations*: specimen preparation may introduce anomalies. The depth of the field is limited, particularly at higher magnifications, where the image can become increasingly distorted and blurry. Requires a light source, particularly at high magnifications. Specimen surface must be flat.

## 4.8 Metallography

When metallic composition or bulk structure needs to be checked or inspected, a polished section can be prepared by cutting a small sample from the object under observation. This form of observation is by necessity a destructive technique, so care is needed in the choice of sectioning site. Further, it has to be remembered that the specimen itself may become evidence, and so must be carefully conserved for possible viewing by others. The cut and polished samples may be lightly etched by chemical swabbing to outline the crystal structure. This is the technique termed metallography. Metallographic examination provides an indication of the class of material involved and whether it has the desired/required structure and how it had been manufactured and heat-treated.[8,9] It is particularly useful for seeking internal defects in welded or brazed assemblies.[10,11] In brief, the microstructure of a material can be equated to a fingerprint taken during the course of a criminal investigation. The case of the counterfeit coin of the realm, reviewed in Section 12.1, is an ideal example to illustrate the unique nature of material microstructures.

Metallographic examination can be used to reveal grain structure such as size, inclusions, phase distribution, cracks, porosity, internal defects, surface coatings, etc., as well as thermal and mechanical history and external features such as corrosion. In short, microscopic examination can reveal a great deal about the past history of a specimen and how it will (or did) react in service.[12] In other words, the microstructure of a material can be equated to a fingerprint taken during the course of a criminal investigation. However, it should be remembered that the technique will reveal detail about a particular (and often exceptionally small) portion of the specimen that *may* or *may not* be representative of the entire article.

As an analytical technique, metallography is relatively simple to undertake, consisting of the following four steps:

- Selection of section
- Surface preparation

- Etching
- Examination under magnification

However, as metallography is a necessarily destructive operation, care is required in the choice of section (the section itself will become evidence, and so must be conserved for possible viewing by others). Care is also required to ensure that the chosen section will reveal material that will be representative of the metal under examination. The actual position of sectioning should always be recorded.

The orientation of the section may also be of importance if the section is selected to show, or is suspected to possess, directionality. Figure 4.16 shows micrographs of a rivet that was sheared off in a collision. The two views shown have been taken at 90° to each other, and show the same fracture at the end of a rivet. Failure of the rivet occurred either prior to, or during, a collision between two heavy goods trucks. An examination was undertaken to determine the exact nature of failure. Unfortunately, the investigator only looked at section (A) and duly concluded that it had failed in tension by mechanical overload, prior to collision. But when the complementary section (B) was subsequently examined, failure had clearly been in shear, having been 'sliced' off by the collision impact.

The case demonstrates how easy it can be to misinterpret microscopic evidence by selecting an inappropriate section. Without knowing the direction of sectioning, Figure 4.16A gives no indication that the fracture was the result of transverse shear forces whereas Figure 4.16B clearly shows the direction of shear and the grain distortion. A further micrograph to illustrate microstructural flow as a result of mechanical influence is shown in Figure 4.16C. Here, an extruded 6061 aluminium bar had been sheared at one end, as evidenced by the clear microstructural flow.

Examination of metallographic sections can be invaluable to the forensic engineer. As a tool, this technique can reveal important information such as case depth of a case-hardened or induction-hardened product (Figure 4.17A), thickness and adhesion of plated coatings or crack initiation in heat-affected zones (HAZ) of welds (Figure 4.17B), all of which may have a direct bearing on the cause of failure. By observation under magnification, the metallographic section may also reveal information regarding the method of manufacture (rolling, machining, casting, hot forging, cold drawing, etc.) (Figure 4.18), and thermal history (either intentionally during manufacture or accidentally during service) of a component.

*Advantages*: can select a representative sample to be evaluated (a critical step for successful analysis). The relationships between material structure and mechanical properties thus allow

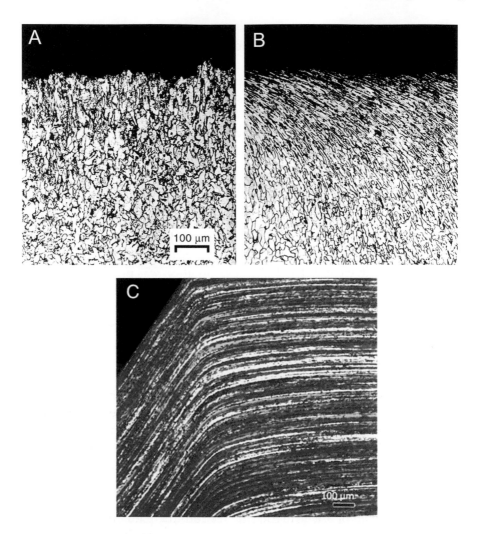

**Figure 4.16** Microstructure of a broken rivet. (A) A transverse section of fracture surface. (B) A section parallel to line of action of force showing clear deformation of the grain structure. (C) An unrelated example of microstructural flow in an extruded aluminium bar that had been sheared at its end. (Courtesy of George Vander Voort.)

metallographic characterisation to be used for materials specification, quality control, quality assurance, process control *and* failure analysis.

*Limitations*: destructive. No universal technique to embrace all demands of metallographic specimen preparation. Dangerous chemical etchants are often required. Specimen preparation may introduce unwanted artefacts or anomalies. Possibility of preferential etching of phases, in multiphase materials.

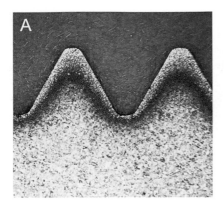 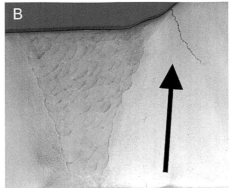

**Figure 4.17** (A) Transverse section through the tooth of a gear where question had been raised as to its depth of hardening. The case depth can be clearly seen and measured at approximately 200–300 µm (0.2–0.3 mm), being more than adequate for the intended use. (B) Crack initiation at the HAZ of a weld, with crack propagation taking a path into the parent material.

## 4.9 Scanning Electron Microscopy

Developments in electronics and instrumentation have brought micro-analytical techniques out of the realm of research and into the hands of everyday failure analysis. One of the best-known instruments is the scanning electron microscope (SEM), with fractography (observing the fracture surface of materials) and microstructural analysis (identifying the material microstructure) being two of the most popular uses of the SEM. To illustrate the potential of SEM for forensic examination, Figure 4.19 shows a montage of three common modes (types) of fracture: brittle, ductile and fatigue. The fracture surface of each mode is shown at two different magnifications, thus allowing direct comparison of each mode of failure. The ability to expose and compare surface features that are impossible to discern with the naked eye or the optical microscope has revolutionised forensic investigations.

Operation of a scanning microscope is more complex than that of optical instruments, due to the need to observe specimens in a vacuum. However, the benefits of SEM are numerous, and include

- Large depth of field, so that features of a rough surface are all in focus
- Exceptionally high degree of resolution (down to 0.5 × 10–6 m)
- Non-destructive for conducting materials
- Elemental analysis using energy dispersive X-ray analyser (EDAX)

The case study presented in Section 3.3.4 highlights a requirement to distinguish between different forms of cast iron that was dependent on observing

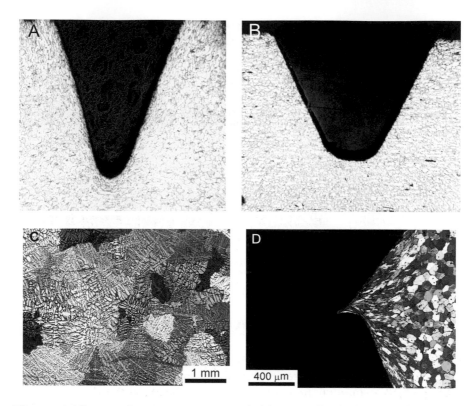

**Figure 4.18** Manufacturing route revealed by metallography: (A) rolled screw thread on alpha brass. (B) A machined brass screw thread on alpha + beta brass. (C) As-cast wrought (worked) aluminium alloy viewed in cross-polarised light. An etched oxide layer produces 'colours' dependent on grain orientation and oxide thickness. (D) The effect of machining deformation on a previously equiaxed structure. Shear strain causes the elongation of grains with subsequent etching (anodising) producing 'mottled' grain colours. (C, D courtesy of University of Cambridge DoITPoMS Micrograph Library, licensed under Creative Commons.)

the microstructure of a failed casting.[7] So, a brief review of types of cast iron microstructures will now be given, simply to further demonstrate the power of the SEM for microstructural characterisation.

Figure 4.20A shows a typical micro-section of a grey iron casting, and Figure 4.20B shows that of a ferritic SG iron, both polished and photographed at a linear magnification of 75×. By making a casting quickly solidify in the mould, the carbon does not appear as graphite but instead separates as iron carbide ($Fe_3C$), rendering the casting extremely hard and brittle. Figure 4.20C shows such a structure, the section polished and etched but again at a linear magnification of 75×. If a casting with such a microstructure is broken it exposes a silvery white, sparkling fracture, which is why these types of casting are called 'white' cast irons. For certain applications where high hardness and maximum abrasion resistance are required, for example, the nose of a

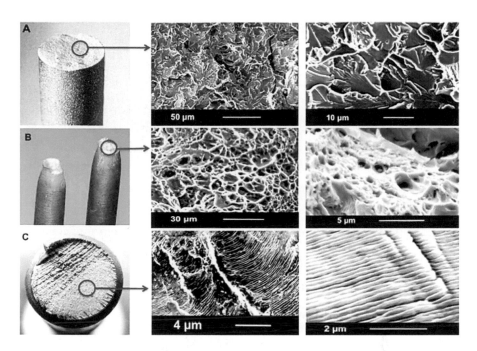

**Figure 4.19** Montage of fractographs to illustrate the depth of field and magnification capabilities of scanning electron microscopy (SEM). Three failure modes are shown at two different magnifications: (A) a ductile overload fracture. (B) A brittle fracture. (C) A fatigue fracture. For all three cases, the material of manufacture is steel. The power of the SEM to differentiate between different fracture mechanisms is clearly demonstrated.

plough-share or the surface of a roll for cold working metals, castings are often made with chills inserted in the moulds so that the metal that solidifies in contact with them develops the iron carbide microstructure. Finally, there is a fourth group, the malleable cast irons, which have to be cast as wholly white iron and subsequently heated for several hours at temperatures where the iron carbide breaks down to form clusters of graphite. Strength is determined by whether they are produced with a ferritic or a pearlitic matrix after the heat treatment. Figure 4.20D shows such a structure, where the graphite appears as the ragged black areas and the matrix is equivalent to a medium carbon steel. A casting with this type of microstructure will be strong, tough and reasonably ductile, hence the name 'malleable' cast irons. However, the application of such malleable castings has been largely ousted by the SG irons which are cast directly into the moulds and require virtually no subsequent treatment.

One of the drawbacks of scanning electron microscopy is that generally, the sample needs to be electrically conducting in order to prevent imaging artefacts from charge build-up on the specimen surface. Charging can be a particular problem with non-metallic materials that are generally insulating.

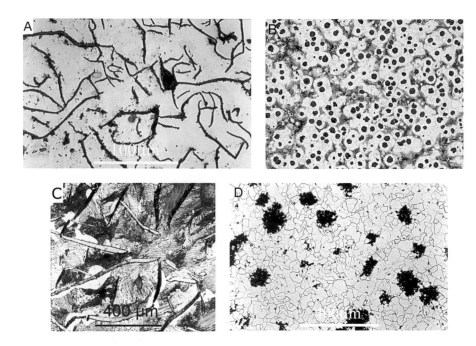

**Figure 4.20** Microstructures of four different types of cast iron. Linear magnification: 75×. (A) Flake graphite ('grey') cast iron. (B) Nodular (spheroidal graphite) cast iron. (Courtesy of University of Cambridge DoITPoMS Micrograph Library. Creative Commons Attribution Non-commercial Share Alike 2.0.) (C) Chill ('white') cast iron where carbon is present as $Fe_3C$. (Courtesy of George Vander Voort.) (D) Malleable (blackheart) iron.

The traditional method to avoid charging has been the use of a conductive coating of either carbon or gold. However, this can be a drawback for forensic investigation, where it is often necessary to preserve the sample in its original state. An alternative strategy is to use low-voltage imaging. At low voltages, there is a threshold typically between 1 and 5kV where the emission of secondary and back-scattered electron yield is high and gives a natural suppression of charging or 'zero-charge' imaging. However, with the advent of modern field emission electron microscopes, which are optimised for low-voltage performance, it is now feasible to image in this accelerating voltage range without the use of conductive coatings. An alternative approach is to use either low-vacuum or environmental scanning electron microscopes.

### 4.9.1 The Environmental Scanning Electron Microscope (ESEM)

A drawback with conventional SEM is the need to coat the sample surface of non-conductors with carbon or gold to bleed away 'incoming electrons. If this is not achieved, the electrons build-up on the sample surface, and so prevent or inhibit image formation. A very high vacuum is needed in conventional SEM, as air molecules scatter and absorb the electron beam.'[3,7] However, the environmental scanning electron microscope '(ESEM) allows a small bleed of gas to pass over the sample being examined without entering the main column of the instrument. As the primary electron beam hits the surface, any electrons that stay on the surface are neutralized rapidly by reaction with positive ions formed by interaction of the primary beam with the gas molecules bled into the microscope. This enables the electrons to be carried away from the surface of the sample, preventing the harmful build-up. The prime interest in these instruments is for observing living things, or materials that would deteriorate rapidly at low pressures.'[3,7] Thus water-absorbent fibres and wood lose water very rapidly at low vacuum, suffering severe damage. Care is still needed for all polymers, however, as the highly energetic electron beam (accelerated through 20,000 V, typically) can itself cause direct damage to samples by chemical reaction with the polymer chains.

Application of the ESEM during the course of an investigation into a hit-and-run road traffic accident is an ideal illustration of its use as an investigative tool.

### 4.9.2 Case Study: Hit-and-Run Accident

A car struck a wooden post after knocking a cyclist off of his bicycle. However, the driver of the vehicle did not stop he just drove on – a typical 'hit-and-run' accident. A witness saw the car drive off but was unable to get the vehicle registration number. Nevertheless, the witness was able to provide a detailed description of make, colour, etc. of the suspect vehicle. A vehicle matching the description was located sometime later, and a sliver of wood was found behind one of its headlights. The vehicle owner 'recalled' that he had run into his own wooden fence, which had damaged the front wing. His explanation was that the wooden sliver must have become trapped behind the headlight as a direct result of his 'bump'. Comparative observation on an environmental scanning electron microscope (ESEM) revealed a different story. A sample taken from the suspect's fence proved the wood to be birch (Figure 4.21A), whereas the sliver recovered from behind the headlight was pine (Figure 4.21B). Samples from the post that had been struck at the accident scene showed that it was also pine. Thus a simple comparison of respective microstructures revealed that the car in question was the vehicle involved in the hit-and-run incident.[12]

*Advantages*: a wide-array of applications. Excellent depth of field. Detailed three-dimensional and topographical sample imaging. Also provides morphological and compositional information. Versatile information gathered from a range of different detectors. As a direct result of advances in computer technology and related software, SEM operation is relatively user-friendly. Data output in digital form. Once prepared, SEM samples will require minimal upkeep thereafter.

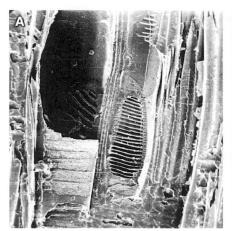 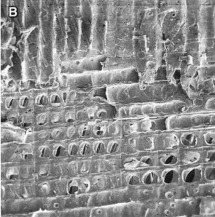

**Figure 4.21** Environmental scanning electron micrographs of recovered samples of (A) birch, and (B) pine.

> *Limitations*: high-priced, costly servicing contracts, large and requires housing in an area free of any electric, magnetic or vibration interference. Maintenance entails ensuring a steady voltage/current to electromagnetic coils, along with continuous cool water circulation. Specialist training is required to prepare SEM samples. Gathered images will be in in grey scale. Preparation of samples can result in artefacts. There is no absolute way to eliminate or identify all potential artefacts. Limited to solid, inorganic samples small enough to fit inside a vacuum chamber under moderate vacuum pressure.

## 4.10 Optical Versus Scanning Microscopy

Ease of use is the single most attractive feature of optical microscopy. Furthermore, samples can be analysed in air or water; the images are in natural colour with magnifications of up to 100 to 1,000 times. With the advent of modern electronics, charge-coupled devices (CCD) allow image processing. However, optical microscopy is not the dominant investigative technique, primarily because of its limited depth of focus and resolution capability. As a result of these limitations, the SEM outshines the optical microscope as the tool of choice. This is clearly demonstrated in Figure 4.22, where a transistor was being examined to determine a cause of failure of the device. A striking difference can be seen between the series of optical and SEM micrographs at the same magnification. When considering the optical micrographs, only a section of the transistor is in sharp focus. In the lower SEM images, the whole specimen is in focus.

Microscope Image

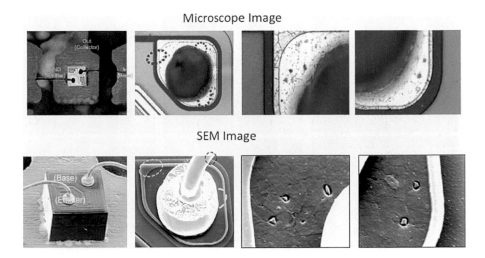

SEM Image

**Figure 4.22** Electrostatic discharge (EDS) damage to a transistor, with optical images along the top row and SEM images along the bottom. Note superior depth of field and focal length of SEM, along with its '3-D' appearance.

For the optical microscope, the depth of focus is the distance above and below the image plane over which the image appears in focus. As the magnification increases in the optical microscope the depth of focus decreases. The three-dimensional appearance of the specimen image is a direct result of the large depth of field of the SEM. It is this large depth of field in the SEM that is the most attractive feature of the technique.

### 4.10.1  Case Study: Printed Circuit Board (PCB) Failure

A newly configured printed circuit board (PCB) was found to be regularly failing its quality control. The consensus of opinion was that there was an electrostatic discharge issue. To determine the cause of failure, a specific integrated circuit was dis-encapsulated by the investigator, and observed under optical and SEM magnification. As in the previous case, optical observation did not give a clear image of the damage, as illustrated by the top row of images on Figure 4.23. However, SEM observation clearly revealed an over-voltage melting of the tracking. It later transpired that a test station operative was inadvertently applying 50 volts to the wrong pin during a hot connection test.

### 4.10.2  Case Study: Prototype PCB Failure

A further case concerned a printed circuit board (PCB) subcontractor who undertook the manufacture of a prototype board, utilising a production line that had been recently shut down because of a lack of firm orders. However, it very quickly became apparent that the prototype was prone to unacceptably high failure during its 'burn-in' stage. Electrical testing and observation under

**Figure 4.23** Integrated circuit (IC) track melting as a result of 50 volts being inadvertently applied to the wrong pin during hot connection at a PCB test station. Optical images along the top row, SEM images below.

low power optical magnification located the problem in a series of ceramic chip capacitors (Figure 4.24A). Subsequent examination of a metallographic section, taken through one of the problem components (Figure 4.24B), suggested that there was no obvious fault with the discrete component. However, closer examination of the interface between the joint solder and copper board tracking on an SEM revealed internal cracking just above the solder intermetallic line (Figure 4.24C and D). EDAX analysis of the solder showed that it was within specification (Figure 4.24E) and was not the direct culprit. Thermal profiling of the board showed a localised hot spot in the region of failure, generated by local power components. However, cracking of the type observed was a direct result of thermal fatigue. This failure mode had been induced by the cyclic nature of board burn-in combined with a mismatch of thermal expansion coefficients of the ceramic capacitor and FP4 resin circuit board. The fatigue mode had been exacerbated by having a localised hot spot.

## 4.11 Energy Dispersive X-Ray Analysis (EDAX)

Both the SEM and ESEM may be coupled to an energy dispersive X-ray analyser (EDAX) for quantitative, or semi-quantitative analysis. EDAX analysis can be performed in an SEM to gain an understanding of the chemistry of materials. When the electron beam in the SEM strikes the sample, it excites the atoms of an element, producing X-rays characteristic of the element. The scan is typically displayed in graphical format, with the X-axis showing photon energy and the Y-axis indicating the number of times energy of that photon keV has been generated (Figure 4.24E). Section 9.3 (Figure 9.5) shows an

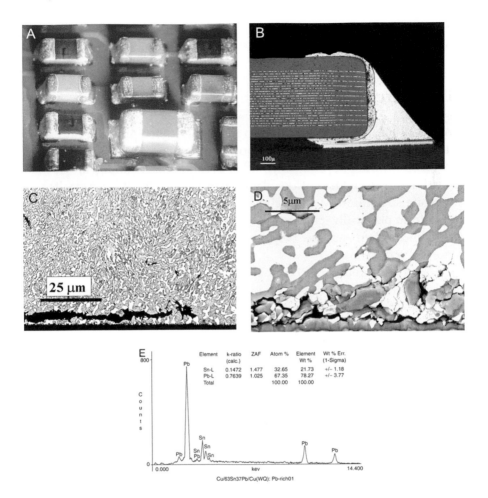

**Figure 4.24** (A) Ceramic capacitors on printed circuit board (PCB). (B) Section through a failed ceramic capacitor. (C&D) Cracking within solder joint. (E) EDAX analysis of solder.

elemental scan for an aluminium alloy rivet that had been recovered from a failed container. The EDAX scan showed that the rivet was of the correct composition. No sample was destroyed or altered microstructurally to make this analysis. It is not as precise as a chemical analysis but is more than adequate for checking material specifications.

The SEM will provide fractographic and microstructural information, whereas the EDAX will provide accurate quantitative information on local variations in material composition. In addition to resolving chemical composition, EDAX can be utilised to ascertain unusual micro-constituents or contaminants within the material, even down to a single inclusion particle. An example that readily demonstrates the use of the SEM, coupled with the power of EDAX, is presented in Section 12.4.1 and is concerned with the theft of a large quantity of copper alloy. A further example to highlight the use of SEM in the course of forensic examination can be found in Section 9.5.

## 4.12 Destructive Testing Methods

When considering a metal in relation to an engineered product, the emphasis tends to be on its mechanical response to service loading. Most materials have an elastic limit beyond which other events occur. A totally brittle material will fracture suddenly (like glass) or progressively, like composites and composite structures. Most engineering materials do something different; they deform plastically (change their shape in a permanent way). It is important to know when and how they do this, and to be able to correlate the data to normal service loads.

Component failures can occur by overloading (car accidents, hoist chains, cables and almost anything that is broken in collisions) or 'the result of poor design, incorrect material selection, manufacturing defects or environmental factors. Mechanical properties of a failed component are, therefore, of prime interest in any failure investigation, as they will provide an insight as to how the component would' be expected to perform under 'service' loading conditions.[3,7] As discussed in the introduction above, standard mechanical testing methods for measuring different material properties will be limited, simply as most forensic investigations are fundamentally constrained by the need to preserve and conserve any and all evidence for possible examination by others. Therefore, conventional tensile, torsion and fatigue testing will not be reviewed in any academic depth. If required, there are a number of excellent books available that describe these procedures in detail, Professor Kyriakos Komvopoulos's *Mechanical Testing of Engineering Materials* (January 2017, Cognella Academic Publishing) being an ideal example.

### 4.12.1 Tensile Testing

Component failures can occur by traumatic overloading events, as a result of poor design, incorrect material selection, manufacturing defects or environmental factors. Mechanical properties of a failed component are, therefore, of prime interest in any failure investigation, as they will provide an insight as to how the component would perform under 'service' loading conditions. Some of the most widely quoted mechanical properties are those determined by a tensile test. Although generating a wide range of mechanical data (tensile strength, modulus of elasticity, elastic limit, yield point, ductility and breaking strength), the tensile test does not tell the whole story, and further mechanical information may be required to permit a satisfactory conclusion to any failure inquiry. The tensile test may be used in the selection of material for a particular purpose, or for deciding that the size or shape of a component, manufactured from a specific material, is sufficiently strong to fulfil its design purpose. As discussed in Section 4.3 above, hardness evaluation is a simple non-destructive alternative to tensile testing, that can achieve a reasonable estimate of tensile strength, *and* can be undertaken on small samples.

> *Advantages*: direct measurement of tensile strength in a known (and simple) stress state. Stressing of entire gauge section, both surface and volume. Therefore testing will encompass surface and volumetric flaws of the material under test.
>
> *Limitations*: the prime concern of any forensic investigation is to preserve evidence, not cut it up. Specimen size and geometry generally require access to volumes of material. This is very rarely the case during a failure investigation. Specimen machining costs and preparation time are high. Difficulty in machining a test sample from some materials (i.e. ceramics, composites, etc.). Influence of specimen dimensions on load/extension data. Care must be taken in identifying the position and direction from which the samples have been taken.

## 4.12.2 Flexure or Bending

Tension is also encountered in bending, and there are several kinds of test for bending simple shapes like beams of uniform section. Thus a beam can be loaded at three points and strained to failure. As the beam is deformed, one surface is deformed into a convex shape and the material here is in tension, while the opposite surface is deformed into a concave shape where it is under compression. It is often thought that forensic work, being scientific by nature, must always be accompanied by long mathematical arguments. However, this is not the case as simple logic matched against the evidence is the only way in which causation is determined. Nevertheless, there are occasions on which quantitative evaluation becomes valuable. A simple equation used for evaluating stresses and strains on beams in bending is the 'engineer's bending equation':

$$M/I = \sigma/y = E/R$$

where M is the moment on the beam, I the second moment of area of the beam section, $\sigma$ the stress at a distance $y$ across the beam section, E the tensile modulus of the beam material and R the radius of curvature of the bent beam.

Some of these variables become constant when a single type of beam is considered under two different loading conditions. Thus E is constant (same material), and I is constant (same section shape).

**Figure 4.25** Sections across 90° bends from cracked and uncracked aluminium sections taken from vehicle body panels. Note the two different bend radii. For scale, leg length = 50 mm (2 in.).

### 4.12.3  Case Study: Cracking of Aluminium Pressings

A particular case that required calculation of strain on the outside of a bend occurred when a manufacturer encountered a problem when forming 5-mm thick sheet aluminium in a vehicle body pressing plant. He supplied samples, of both intact and cracked press formed bends, stressing that all parameters (tooling, processing, etc.) were identical. However, visual observation of the two panels showed that forming processes had not been identical (Figure 4.25). Measurement of the un-cracked sample revealed an inside radius of 7 mm, whereas the cracked sample had been formed to an inside radius of 3 mm. A simple bending strain calculation (Figure 4.26) indicated that the strain of the outer skin was of the order of 26% in the satisfactory batch, compared with 45% in the unsatisfactory lot.

It was therefore considered that the bend radius was the major cause of the cracking, as the smaller radius was introducing 45% strain at the outer surface of the bend, well above the ductility of the aluminium sheet.[12]

### 4.12.4  Torsion Testing

Torsion testing provides a method for determining the modulus of elasticity in shear, shearing yield strength and the ultimate shear strength. It is useful for examining and testing parts such as axles, shafts, couplings and twist drills, and can be used to good advantage in the testing of brittle materials, as it has the advantage of no necking (as in a tensile test) or friction.

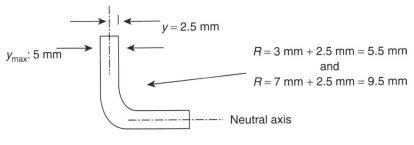

Situation (1) where inside radius of bend ≈ 3mm:
Strain = y/R = 2.5 × 10⁻⁶ / 5.5 × 10⁻⁶ = 0.45 or 45% strain

Situation (2) where inside radius of bend ≈ 7mm:
Strain = y/R = 2.5 × 10⁻⁶ / 9.5 × 10⁻⁶ = 0.26 or 26% strain

**Figure 4.26** Strain calculations made for the two different outside bend radii.

Results obtained from a torsion test will imply the shear properties of a material under a twisting action or opposing radial forces (ASTM E-143). Data obtained from a torsion test may be used to construct a stress–strain diagram, thus verifying an elastic limit, torsional modulus of elasticity, modulus of torsional rupture and torsional strength. Moreover, the test data are generally valid to larger values of strain than are tension test data. However, torsion testing as a tool for the failure investigator will be subject to the same limitations as tensile (and all destructive) testing above.

## 4.12.5 Shear Testing

Many engineering components have safety devices based on shear pins that limit the amount of torque or bending moment that may be transmitted. In this way, these cheap devices protect the rest of the more expensive mechanism from expensive overload. They function in a similar way to a fuse in an electrical circuit. Indeed, in some circumstances, shear pins are called 'fuse-pins' or 'fuse links'. Flywheels, storing energy in rotating machines, may cause severe damage if the machine is suddenly stopped for any reason. By incorporating a shear pin the damage is greatly reduced. Their designers need to know the applied force that will cause them to fail by shear overload. Many airliner engines are attached to the mounting pylons in this way and are designed to shear away if engine vibrations risk damage to the airframe. Figure 4.27A shows engine pylon mounting shear bolts ready to be installed.

As an engineering example, immense dynamic forces that can be generated by a large rotating flywheel mass would most certainly require the integration of fail-safe devices, just to (safely!) dissipate the flywheel kinetic energy in the event of an overload incident. One simple design solution is that of a keyway, incorporated between the flywheel and its shaft. Under any

 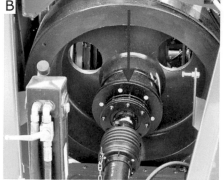

**Figure 4.27** (A) Airliner engine pylon mounting shear bolts ready to be installed. (Courtesy of MTU Aero Engines, Munich, Germany. (B) A large agricultural flywheel, weighing 600 kg, rotating at approximately 1,000 rpm, incorporated with a slip clutch protection system arrowed in red.

traumatic overload situation, the key would shear at a pre-determined load, at which point the flywheel would rotate freely, dissipating its kinetic energy and avoiding a potentially damaging torque transfer through the system load train.

When an unexpectedly higher than normal resistance is encountered through any power train, a torque-limiting clutch (slip clutch) is typical of a more current solution for the dissipation of kinetic energy in a rotating body. Figure 4.27B shows an agricultural round baler that incorporates a slip clutch to protect the baler, along with a freewheeling facility to protect the tractor (power take off) transmission from damaging torque load transmission.

> *Advantages*: for numerical simulations, material selections or a better knowledge of the material behaviour in general, it is often necessary to measure the shear-strength and shear-failure properties in a wide range of strain rates and temperatures.
> *Limitations*: very challenging to obtain reliable results, particularly at high strain rates.

### 4.12.6 Izod or Charpy Impact Test

The Izod or Charpy test has become the standard testing procedure for comparing the impact resistances of materials. Tough materials absorb a great deal of energy, whilst brittle materials tend to absorb very little energy prior to fracture. Furthermore, materials such as carbon steels undergo a 'ductile to brittle transition'. This behaviour becomes clear when plotting impact energy as a function of temperature. The resultant curve will show a rapid dropping off of impact energy as the temperature decreases. If the impact energy drops off very sharply, a transition temperature can be determined (Figure 4.28).

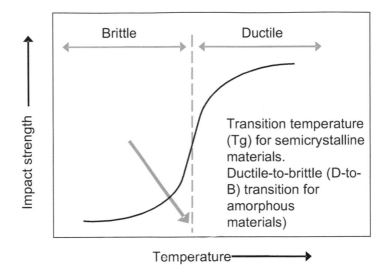

**Figure 4.28** The ductile-to-brittle transition as a function of temperature.

Therefore, with any equipment intended for low-temperature service operation, it is important to consider that low temperatures can (and do) adversely affect the tensile toughness – defined as the measure of the degree of ductility or brittleness – of many commonly used engineering materials. Tensile toughness can be determined by calculating the area beneath a conventional stress–strain curve. A ductile material will absorb significant amounts of impact energy before breaking, whereas a brittle material will tend to shatter on impact. In general, any material with high ductility and high strength properties will also possess a respectable tensile toughness. Nevertheless, tensile toughness can be exceptionally responsive to temperature changes, with many materials experiencing a shift from ductile to brittle behaviour if the ambient temperature falls below a particular level. The exact temperature (or temperature range), at which this ductile/brittle shift takes place, will vary from material to material.

The phenomenon of brittle fracture has been acknowledged since the early part of the last century but became recognised as a serious problem in ship construction during World War II.[13] As part of an American government project, the continuous block construction of all-welded cargo vessels (the 'Liberty Ships') was planned. Construction commenced at the outbreak of the Pacific war in 1942, with some 2,708 Liberty Ships constructed in a five-year period between 1939 and 1945. However, 1,031 incidents of damage or accidents due to brittle fracture were reported by April 1946. More than 200 Liberty Ships had sunk or were damaged beyond all hope of repair. Extremely low temperatures and stress raisers at hold corners were a feature in many of these cases, with onset of failure caused by occurrence and development of brittle cracking, due to a lack of fracture toughness of welded

**Figure 4.29** (A) The Liberty Ship SS Schenectady with complete hull fracture circled in red. (B) A closer view of the brittle hull fracture.

joints. Figure 4.29 shows the SS Schenectady, a T2-SE-A1 tanker built during World War II. On 16 January 1943, she was moored in calm water at the fitting dock at Swan Island, shortly after returning from her sea trials. Without warning, and with a noise audible for at least a mile, the hull cracked almost in half, just aft of the superstructure. Liberty Ship failures revealed the importance of fracture toughness, thus marking the birth of fracture mechanics.

> *Advantages*: is simple, straightforward and utilises small, simple, notched test specimen. Can be carried out over a range of sub-ambient temperatures. Can be used for comparing influence of alloy studies and heat treatment on notch toughness. Used for quality control and material acceptance purpose. The test may indicate the need for avoiding sharp corners in parts that are notch-sensitive and register relatively low values on the Izod impact test.
>
> *Limitations*: it is a destructive technique. Furthermore, factors that will affect the impact energy of a specimen will include yield strength and ductility, notches, temperature and strain rate and fracture mechanism. An impact test should not be considered a reliable indicator of overall toughness or impact strength as some materials are notch-sensitive and derive greater concentrations of stress from the notching operation. Large scatter inherent within the test may make it difficult to determine well-defined transition temperature curves.

### 4.12.7 Differential Scanning Calorimetry (DSC)

Differential scanning calorimetry is a technique used to examine the thermal properties of materials under carefully controlled conditions. Only milligram amounts of material are needed, so although the method is destructive, sampling can be considered minimal. The sample is heated at a constant rate (typically 10°C/min), and the heat flow into or from the sample is automatically recorded by the instrument. Heat is absorbed when a sample melts (an endothermic change), and the melting point $T_m$ is often characteristic of the chemical composition of the material under study, as the following examples indicate.

## 4.12.8 Case Study: Electronic Component Registration Difficulties

An electronic circuit board fabricator began to experience component registration difficulties on a batch of surface mount (SMT) boards that had been fabricated by wave soldering. The board profile was one that had up to that time never given rise to any production problem. However, for the batch in question, the incidence of chip miss-registration (along with associated board rework) climbed to an intolerable level. The board fabricator suspected the solder itself as the potential source of his problem. It was a eutectic tin–copper alloy, from which a sample of the parent stock was submitted for examination. Analyses were undertaken and compared to that of a known pure sample. DSC revealed (Figure 4.30) a liquid phase, existing at approximately 179 to 182°C. It was immediately apparent that such a liquid phase would allow time for surface mount components to 'float' out of registration, rather than freezing instantly at the eutectic temperature. The source of this unexpected phase was subsequently traced to cross-contamination from a previous melt, requiring the solder producer/supplier to review his quality control procedures with some urgency!

As electronic components and devices become smaller, the incidence of early failure is also on the increase. For domestic products, failure may be acceptable, but for avionics applications, a zero-failure rate is pursued.

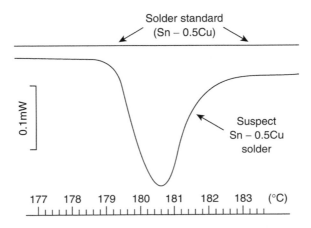

**Figure 4.30** Comparative differential scanning calorimetry (DSC) scan of suspect eutectic Sn-0.5Cu solder against that of a known standard sample. Melting temperature for the solder is 227°C. The suspect solder displayed a low-temperature (~186°C) melting phase.

Differential scanning calorimetry is even more useful for polymers, where additional information includes the glass transition temperature ($T_g$, the temperature at which the plastic becomes elastomeric) and the decomposition temperature (the temperature at which the material decomposes). While pure metals melt at sharply defined temperatures and alloys melt over a range, polymers melt over a wider range that is less sharply defined. That range is also sensitive to the heating rate and the molecular weight of the polymer as well as changes in chemical composition.

> *Advantages*: ease and speed for observation of transitions such as eutectic point. Observe phase changes or polymorphs. Determine the degree of purity in materials. Provides quantitative and qualitative data for endothermic (heat absorption) and exothermic (heat evolution) processes at play.[14]
>
> *Limitations*: interpretation of results is often difficult. 'The temperature range, for transitions in the different materials, often overlap. Measures only the sum or average value of the heat flow rate from overlapping processes – thus quantitative analysis of the individual' processes is difficult. 'Cannot optimize both sensitivity and resolution in a single experiment. Flat, straight baselines are required for detection of small transitions, [particularly glass transitions.] Absolute value of the DSC heat flow signal will be affected by instability of instrument electronics, its cooling systems, and environmental changes (such as temperature and humidity in the laboratory).'[14]

## 4.13 Novel Tools and Techniques

With increasing complexity and integration of many safety-critical systems, failure investigation has become increasingly more complex. Such technological advances are placing a considerable burden on the investigator attempting to trace the true failure path. However, advances in sensors, development of new materials and miniaturisation of devices are all paving the way for novel NDT tools and failure investigation techniques. It is of utmost importance that the forensic detective stays abreast with current developments in the field, simply to understand any limitations of use, along with the accuracy and reliability of any analysis gathered by such methods. Some of the up-and-coming analytical tools will now be briefly reviewed below.

### 4.13.1 The Contour Method[15]

This is a mechanical measurement technique for mapping residual stresses in engineering structures. First introduced in the year 2000, it 'relies on the surface deformation that results from stress relaxation over a newly cut plane to calculate the residual stresses that existed in the component before the cut. The main advantage of the contour method is its ability to produce a 2-D map of residual stress acting normal to the cut surface. It has also been adapted to measure residual stresses using the surface topography of plane-strain fractures[10,11] for forensic studies. The typical uncertainty for the Contour Method is around 30 MPa for steel and 10 MPa for aluminium.'[10]

## 4.13.2 Neutron Diffraction

This is another important residual stress measurement technique. The first article on record on measuring strains using neutron diffraction was published in 1985, so it is not as mature as XRD or hole drilling. One of the main advantages of neutron diffraction is that neutron beams can penetrate much deeper into materials than lab X-rays, enabling residual stress measurements deep inside polycrystalline materials. The International Organization for Standardization (ISO) has been publishing and revising a neutron diffraction standard since 2005 (ISO/TS 21432:2005) and, although the standard was last reviewed in 2010, it is again under review (May 2018).

## 4.13.3 Nano Test Systems

These offer flexible nano-mechanical property measurement. Techniques include a range of (nano) mechanical and tribological tests, including indentation, scratch and wear, impact and fatigue, elevated temperature indentation and indentation in fluids. Amongst other parameters, the systems are capable of measuring hardness, modulus, toughness, adhesion and many other properties of thin films and other surfaces or solids. They are currently in use for measurement of the mechanical properties of high-temperature and high-performance materials over a wide range of applications in the aerospace industry from airframe through avionics to engine development.[16,17]

## 4.13.4 Flash Thermography

In this method, 'a brief pulse of light is used to heat the surface of a sample, while an infrared camera records changes in the surface temperature. As the sample cools, the surface temperature is affected by internal flaws such as disbonds, voids or inclusions, which obstruct the flow of heat into the sample.'[18]

## 4.13.5 Thermographic Signal Reconstruction (TSR)

'Conventional approaches to analysis of thermographic NDT data are based on visual or computer analysis of the images from the infrared camera.

These methods are often difficult to interpret and limited in their ability to image deep or subtle features, or perform quantitative measurement.'[19] TSR addresses these limitations, and allows an unprecedented degree of sensitivity, depth range and resolution of subsurface defects. The introduction of TSR has started to play a significant role in the growth of thermography in real-world manufacturing, in-service inspection and failure applications.[19,20]

> *Advantages*: ideal for flaw detection. Material characterisation, including precise measurement of thickness, defect depth and thermal diffusivity. Use will encompass a wide range of composites, metals, polymers, ceramics and advanced materials.

### 4.13.6 Electromagnetically Induced Acoustic Emission (EMAE)

Used for damage assessment of thin walled conducting structures.[21]

### 4.13.7 Pulsed Eddy Current (PEC)

An emerging non-destructive examination technique used for detecting flaws or corrosion in ferrous materials or for measuring the thickness of objects. Being a non-destructive testing technique using a broadband pulse excitation with rich frequency information, it shows wide application potential. 'It can be used on materials as diverse as vessels, columns, storage tanks and spheres, piping systems, and structural applications with fireproofing.'[22] PEC shows excellent validation of both new and a combination of selected features in defect classification.[23]

> *Advantages*: non-contact operation. Can be used on both insulated and un-insulated materials. Concurrent information from a range of depths can be determined. A quick and cost-effective solution for corrosion detection.
> *Limitations*: only applicable to carbon steel and low-alloy steel. Variations in material properties within one object will result in spurious variations in PEC wall thickness readings, typically 10%. Integrates over a relatively large footprint. Therefore, isolated pitting defects, for example, may not be detected.

### 4.13.8 Microwave Technology

Concepts for the corrosion and fracture identification of tendons in pre-stressed concrete structures can be established by electromagnetic resonance methods (EMR) in combination with electromagnetic field strength measurements for the detection and localisation of pre-stressing steel fractures based upon microwave technology.[24,25]

## 4.14 Micrograph Acknowledgements

Micrographs 4.13 (B, C&D) and 4.16 (A,B,D) courtesy of: DoITPoMS Micrograph Library, University of Cambridge. www.doitpoms.ac.uk/miclib/browse.php

Micrograph 4.16 C courtesy of Vander Voort, G., Consultant in Metallography, Failure Analysis & Archeometallurgy. www.georgevandervoort.com

## References

1. Allgaier, M.W., P. MacIntire, S. Ness and P.O. Moore, *Nondestructive Testing Handbook, Visual and Optical Testing*. 2nd ed. 1993, Columbus: American Society for Nondestructive Testing.
2. Gagg, C.R., Failure of components and products by 'engineered-in' defects: Case studies. *Engineering Failure Analysis*, 2005. 12(6): pp. 1000–1026.
3. Lewis, P.R. and C.R. Gagg, *Forensic Polymer Engineering: Why Polymer Products Fail in Service*. 2010, Cambridge: Woodhead Publishing Limited.
4. Gagg, Colin R. and Peter R. Lewis, Environmentally assisted product failure – Synopsis and case study compendium. *Engineering Failure Analysis*, 2008. 15(5): pp. 505–520.
5. Semmens, J.E., L.W. Kessler and S.R. Martell, Evaluation of thin plastic packages using acoustic micro imaging. 1996, Bensenville, IL: Sonoscan, Inc, https://sonoscan.com/pdf/publications/53PEMThinSemiTest96.pdf.
6. Zwick/Roel, *The Brinell Hardness Test*. n.d., http://www.indentec.com/downloads/info_brinell_test.pdf.
7. Lewis, P., K. Reynolds and C. Gagg, *Forensic Materials Engineering Case Studies*. 2003, CRC Press, ISBN-10: 0849311829, ISBN-13: 978-0849311826.
8. Nadeau, Jay L., Michael W. Davidson and Richard G. Connell Jr., *Reflected Light Optical Microscopy*. 2012, Wiley, https://doi.org/10.1002/0471266965.com057.pub2.
9. Tompkins, H. and E. Irene, *Handbook of Ellipsometry*. 2005, William Andrew.
10. Araujo De Oliveira, J., J. Kowal, S. Gungor and M.E. Fitzpatrick, Determination of normal and shear residual stresses from fracture surface mismatch. *Journal of Materials & Design*, 2015. 83: pp. 176–184.
11. Hosseinzadeh, F., Y. Traore, P.J. Bouchard and O. Muransky, Mitigating cutting-induced plasticity in the contour method, part 1: Experimental. *International Journal of Solids and Structures*, 2016. 94–95: pp. 247–253.
12. Gagg, Colin, Domestic product failures: Case studies. *Engineering Failure Analysis*, 2005. 12(5): p. 784–807.
13. Hodgson, J. and G.M. Boyd, Brittle fracture in welded ships. *Transactions of the Institution of Naval Architects*, 1958. 100(3): pp. 141–180.
14. Thomas, L.C., Modulated DSC® paper #1: Why modulated DSC®? An overview and summary of advantages and disadvantages relative to traditional DSC. 2005, New Castle, DE: TA Instruments. http://www.tainstruments.com/pdf/literature/TP_006_MDSC_num_1_MDSC.pdf.

15. *Stress Map*. The Open University. 2015, https://www.stressmap.co.uk/contour -method/.
16. Wilson, G. and J. Sullivan, An investigation into the effect of film thickness on nanowear with amorphous carbon-based coatings. *Wear*, 2009. 266: pp. 1039–1043.
17. Korte, S. and W. Clegg, Micropillar compression of ceramics at elevated temperatures. *Scripta Materialia*, 2009. 60: pp. 807–810.
18. Shepard, S., Flash thermography the final frontier. *Quality Magazine: NDT Nondestructive Testing Including Materials Test*, April 2007: pp. 10–12, https:// www.qualitymag.com/articles/89334-flash-thermography-the-final-frontier
19. Shepard, S.M., J.R. Lhota, B.A. Rubadeux, D. Wang and T. Ahmed, Reconstruction and enhancement of active thermographic image sequences. *Optical Engineering*, 2003. 42(5): p. 1337–1343, doi: 10.1117/1.1566969
20. López, F., C. Ibarra-Castanedo, X. Maldague and V. Nicolau, Pulsed thermography signal processing techniques based on the 1D solution of the heat equation applied to the inspection of laminated composites. *Materials Evaluation, 72*, 2014. pp. 91–102.
21. Finkel, P., et al., Electromagnetically induced acoustic emission-novel NDT technique for damage evaluation, in *Review of Progress in Quantitative Nondestructive Evaluation*. AIP Conference Proceedings Volume 557. 2001.
22. Chen, T., G.Y. Tian, A. Sophian and P.W. Que, Feature extraction and selection for defect classification of pulsed eddy current NDT. *NDT & E International*, 2008. 41(6): pp. 467–476.
23. Beissner R.E., J.H. Rose and N. Nakagawa, Pulsed eddy current method: An overview. In: D.O. Thompson and D.E. Chimenti, eds., *Review of Progress in Quantitative Nondestructive Evaluation*. Boston, MA: Springer, ISBN: 978-1-4613-7170-0
24. Holst, A., H.-J. Wichmann and H. Budelmann, Novel NDT-techniques for corrosion monitoring and fracture detection of prestressed concrete structures, in *NDTCE'09, Non-Destructive Testing in Civil Engineering*. 30 June–3 July 2009, Nantes, France.
25. Budelmann, H., A. Holst and H.-J. Wichmann, Non-destructive measurement toolkit for corrosion monitoring and fracture detection of bridge tendons. *Structure and Infrastructure Engineering*, 2014. 10(4), https://doi.org/10.1080/1 5732479.2013.769009.

# Sources of Stress and Service Failure Mechanisms

# 5

## 5.0 Introduction

The way in which an engineered product or structure fails in service can often be directly related to operational conditions at play, both prior to, and at the instant of, demise. Operational circumstances will include both environmental influence as well as physical loading conditions, both of which must be identified by the forensic detective to accurately deduce a failure mechanism. Furthermore, component design features (such as stress raising notches, material type/defects, surface treatment, etc.) can, and will, influence any predominant failure mode or mechanism. Most failure modes manifest themselves as a physical breakdown of components due to adverse operational load conditions, design and/or operational environment. In addition, onset of failure can occur both within a component itself, at its interface with the environment or with adjacent components. Therefore, this chapter will consider a range of common failure modes found in service, along with characterisation of the mechanical response of a component or system subjected to the mode in question.

## 5.1 Fundamental Mechanical Background

Prior to reviewing a selected number of the most common failure mechanisms found in service, a brief background of the fundamentals of mechanical behaviour will be introduced. This review will not be presented at great depth, as there are a plethora of accomplished textbooks that can be consulted if required.[1–4] However, sections of this chapter will consider more complex, but highly relevant, behaviour that has been identified as a recurring root cause of service failure.

### 5.1.1 Stresses and Strains

The mechanical response or characteristics of a material are usually measured by a monotonic – the material's response to the single and continuously increasing force - laboratory test on a standard test piece. By far the most commonly quoted measurement is the tensile strength, and such values are widely used to make comparisons of different materials. However, the way the force is applied and the geometry of the test piece itself will affect the result.

For the forensic detective, it is important to recognise that there are other mechanical properties that are equally as important as that of absolute tensile strength in determining the ability of a component to withstand externally applied loads. Examples are toughness (which is a measure of the ability to absorb energy as a fracture develops) and ductility or malleability (the capacity to be deformed before fracture occurs). In many materials, events begin to happen that would destroy the integrity of an engineered component long before the tensile breaking force is reached. Among the most difficult types of material to use successfully and safely in highly stressed situations are those that behave in a brittle manner despite having very high tensile strength.

## 5.1.2 Ductile and Brittle Transition

Ferrous alloys can display a tendency to lose almost all of their inherent toughness when the ambient temperature drops below a ductile-to-brittle transition temperature for the material. This transition has been attributed as the cause for numerous dramatic and catastrophic past failures, e.g. the rupture of a 2.3-million-gallon molasses storage tank in the winter of 1911, World War II Liberty Ships breaking in half, etc. The notched-bar impact test is used to determine whether or not a material experiences a ductile-to-brittle transition as the temperature is decreased (Section 4.12.6). In such a transition, at higher temperatures the impact energy is relatively large, as the fracture is ductile. As the temperature is lowered, the impact energy drops over a narrow temperature range, as the fracture becomes more brittle. Over the ductile-to-brittle transition (Figure 5.1) features of both types will exist, and are found in body-centred cubic (bcc) metals, such as low-carbon steel, which become brittle at low temperatures or at very high strain rates. However, face-centred cubic (fcc) metals generally remain ductile at low temperatures. This behaviour becomes obvious when plotting impact energy as a function of temperature. The resultant curve is often a good indicator of the minimum recommended service temperature for a material.

Transition of a pure material may occur very suddenly at a particular temperature. However, for many materials, the transition will occur over a range of temperatures. This temperature range is a source of difficulty when trying to define a single transition temperature for service use. If a component (material) experiences a ductile-to-brittle transition, the temperature at which it occurs can be affected by the strain rate, the size and shape of the specimen and the relative dimensions of any inherent defect such as a notch. Therefore, critical applications will require analysis based on defect tolerance. The conventional engineering method used to determine structural reliability (as a direct function of applied stress, crack length and stress intensity at that crack tip) is that of linear elastic fracture mechanics (LEFM).

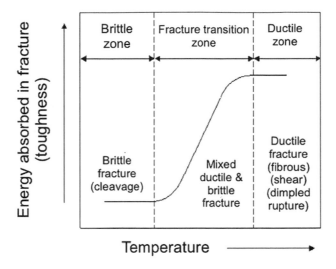

**Figure 5.1** A schematic of the ductile/brittle transition in metallic materials, showing the effect of temperature on notch toughness.

### 5.1.3 Fracture Toughness and Linear Elastic Fracture Mechanics (LEFM)

Fracture toughness of a material (or structure) denotes the intensity of stress required to grow (propagate) a pre-existing defect or flaw. The occurrence of flaws is almost unavoidable during the processing, fabrication or service life of a material or component. Therefore, characterisation of fracture toughness in a material or component that fails in a non-ductile manner may become paramount during the course of forensic analysis. A flaw can be described as a crack, void, weld defect, design discontinuity, metallurgical inclusion or a combination of any or all. As most engineers will appreciate, there are few (if any) materials that are flaw-free. It is common practice to assume that a defect of some size will be present in some components. With such an assumption the forensic or failure engineer can utilise an LEFM approach both to design and understand the failure of critical components. The approach incorporates flaw size and features, component geometry, loading conditions and fracture toughness, to assess the failure potential of a defect-bearing material or component.

Materials that fail in a non-ductile manner are extremely susceptible to the presence of defects or cracks. As there is little or no plastic deformation, stresses concentrate at the tips of pre-existing surface or interior defects. Such a flaw will serve as a crack initiator, and when stress levels attain a critical value, instantaneous crack propagation (growth to failure) will be the result. Failure is therefore caused by the propagation of a crack as distinct from yielding. When it contains a defect, it is toughness that is the measure

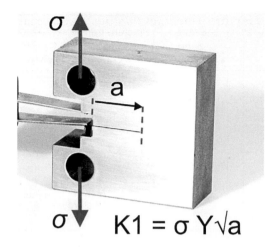

$$K1 = \sigma Y\sqrt{a}$$

**Figure 5.2** Schematic representation of stress intensity at a crack tip. The suffix I relates to the opening mode of the crack. Mode I is a tensile opening, whereas Modes II and III describe in-plane and out-of-plane shear opening respectively. (Courtesy of the Open University, Materials Engineering.)

of a material's ability to withstand stress. Plane strain fracture toughness symbolises this value as the critical stress intensity at fracture (Figure 5.2). So, stress intensity is defined as

$$K_1 = \sigma Y\sqrt{a} \tag{5.1}$$

where $Y$ is a geometric constant calculated using finite element methods and available in standard tables, $\sigma$ the applied stress and $a$ is the crack length. The units are MPa$\sqrt{m}$ (or MNm$^{-3/2}$) and very low values exist for ceramics and glassy polymers. Therefore, the strength of these materials will fall as the defect size increases.

There is a connection that exists between stress intensity (KI) and fracture toughness ($K^{IC}$), which is comparable to that of conventional stress in relation to tensile stress. $K^I$ is used to characterise the level of stress that will exist at the tip of any crack, whereas $K^{IC}$, is used to signify the peak value of stress intensity that any material (under plane-strain conditions) is able to withstand without rupture. If $K^I$ attains the $K^{IC}$ value, unstable fracture will occur. In a manner similar to other mechanical properties of materials, values of $K^{IC}$ can be found in a range of reference sources. For the purposes of this book, a selected range of values for $K^{IC}$ can be found in Table 5.1.

For any material, if fracture toughness ($K_{IC}$) and existing stress conditions are known, the length of crack that can be tolerated without further propagation (growth) can be calculated. A crack will grow if

$$\sigma \geq K_{IC}/\beta\sqrt{\pi a} \tag{5.2}$$

**Table 5.1 Fracture Toughness Values of Materials**

| Materials | E (GNm²) | σ (MNm²) | K$_{IC}$ (MNm⁻³/²) |
|---|---|---|---|
| *Steels* | | | |
| Medium carbon (AISI-1045) | 210 | 269 | 50 |
| Pressure vessel (ASTM-A5330-B) | 210 | 483 | 153 |
| High-strength alloy (AISI-4340) | 210 | 1,593 | 75 |
| Maraging steel (250-grade) | 210 | 1,786 | 74 |
| *Aluminium alloys* | | | |
| 2024-T4 | 72 | 330 | 34 |
| 7075-T651 | 72 | 503 | 27 |
| 7039-T651 | 72 | 338 | 32 |
| *Titanium alloys* | | | |
| Ti-6AL-4V | 108 | 1,020 | 50 |
| Ti-4Al-4Mo-2Sn-05 Si | 108 | 945 | 72 |
| Ti-6Al-2Sn-4Zr-6Mo | 108 | 1,150 | 23 |
| *Polymers* | | | |
| Polystyrene (PS) | 3.25 | | 0.6–2.3 |
| PMMA | 3.4 | | 1.2–1.7 |
| PC | 2.35 | | 2.5–3.8 |
| PVC | 2.5–3 | | 1.9–2.5 |
| PETP | 3 | | 3.8–6.1 |
| *Ceramics* | | | |
| Si$_3$N$_4$ | 310 | | 4–5 |
| SiC | 410 | | 3–4 |
| Al$_2$O$_3$ | 350 | | 3–5 |
| Soda-lime glass | 73 | | 0.7 |
| WC – 15 wt% Co (cermet) | 570 | | 16–18 |
| Electrical porcelain | – | | 1 |

where σ = the fracture stress, β = a dimensionless shape factor and *a* = the crack length for a surface opening crack (not an internal crack). There are a range of readily available books, and data sources for engineering calculations, that will include appropriate tables of values of beta for different geometries.

If the fracture toughness of a material is known, a fracture stress or critical crack size of a component can be calculated, as long as the stress intensity factor is also known. Such calculation will

- Establish permissible flaw size
- Ascertain the stress necessary to cause catastrophic failure
- Ascertain the load on a component at the time of failure

- Establish whether adequate materials were used in manufacturing
- Clarify whether a part design was adequate
- Elucidate any inherent potential for legal ramifications

Operational stresses can be calculated if the system that failed was efficiently documented. For example, the magnitude of load on a specific component at the instant of failure can then be determined. The service load history may also be known throughout the time that the part was used. Along with information on crack size at time of final failure, these data can then be used to calculate toughness. This will reveal whether or not the part performed according to its specification. Equally, if stresses are the unknown, toughness can be estimated from materials handbooks, again knowing the crack size and the area of the remaining sound metal at the time of failure. If both toughness and stresses are unknown, the toughness of a material can be found by undertaking a mechanical test. As discussed in Section 4.12.6, Charpy-impact testing of the parent material can be used to determine material toughness. Back-calculation can be used to determine stress in play at failure, thus providing a route to determine whether the material had – or had not – failed by an overload.

### 5.1.4 Limitations of a Fracture Mechanics Approach

Finding a pre-existing crack in any structure is an example of the weakest link for forensic or fracture mechanics analysis. In most practical cases, it is generally unknown whether or not the component in question actually contains a crack in it. Therefore, the options are as follows:

1. Estimate a size, based on microscopic examinations of representative samples of material. Alternatively, specify the largest flaw that can be tolerated, and manufacture with appropriately defect-free materials.
2. A popular route is to conduct a proof test (pressure vessel applications being a good example). Here, under controlled conditions, a structure (or component) will be subjected to a load significantly higher than that of its design service load. If the material fracture toughness is known, it then becomes possible to deduce the longest length of a crack that could be tolerated in the component or structure, without instigating failure during the course of proof testing.
3. Use a non-destructive test technique to attempt to detect cracks in the structure. Examples of such forensic engineering tools are described in Chapter 4: Analytical Methods, and include ultrasound, to detect echoes off crack surfaces; X-ray techniques; and inspection with optical microscopy. If a crack is detected, most of these tools will allow an estimate of crack length. If not, it must be assumed for design and (failure) analysis purposes, that the structure contains cracks that are just too short to be detected by the chosen method.

## 5.1.5 Stress Concentration $(K_t)$

One way in which the applied stress in a body can be amplified to a level above that of the ultimate strength of the material of construction is by the presence of stress concentrations. A stress concentrator is a local variation in geometric shape where stress lines through a component or structure are forced together (Figure 5.3A), and so magnified. The lines represent the tensile stress intensity, which is completely uniform across the whole section. However, if a notch is cut into the side, the load remains the same but the stress will increase, as it is being carried by a smaller cross-sectional area. It becomes highly concentrated at the root of the notch, as depicted by the stress distribution lines in Figure 5.3B. This stress at the root of the notch is determined by its depth and radius at the tip, according to the relationship:

$$S = 1 + 2(L/r) \qquad\qquad (5.3)$$

where S is the stress level, L is the notch depth and r the tip radius. A circular hole (L = r) would thus concentrate the stress by a factor of 3 (Figure 5.3C and D). Simple examples of stress concentrators include

- Cracks in or at the edges of bodies
- Holes in flat sheets
- Voids within products
- Corners and fillets
- Changes in profile of shafts
- Screw threads

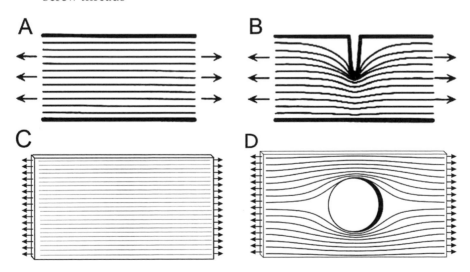

**Figure 5.3** (A and B) A notch as a stress raiser: local variations in shape where the stress lines through the product are forced together and so magnified. (C and D) Similar situation with stress concentration as a result of a through hole.

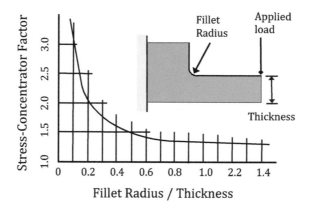

Fillet Radius / Thickness

**Figure 5.4** Stress concentration factor versus fillet-radius-to-thickness ratio.

Such design features are often inevitable in a product shape, but the magnitude of the stress concentration ($K_t$) can be minimised by forethought. Compilations of stress concentration factors allow designers to do just this.[5–7] However, stress raisers are often not given the consideration they warrant, until an application of a normal in-service load initiates a crack sited at a stress raiser. Sharp corners are a common factor, and have the potential to wreak havoc when the time is ripe. The importance of an adequate fillet radius cannot be overemphasised. Figure 5.4 illustrates how dramatically the stress concentration factor increases as the fillet-radius-to-thickness ratio decreases.

A simple formula for the stress at a notch tip is shown below. It relates the stress concentration $K_t$ to the length of the notch, D, and the radius of curvature at the tip, r (Figure 5.5):

$$Kt = \sigma/\sigma_0 = 1 + 2\left(D/r\right)^{\frac{1}{2}}  \tag{5.4}$$

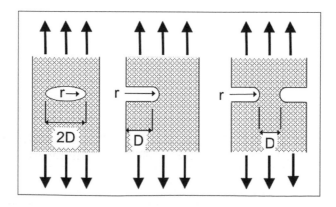

**Figure 5.5** Stress concentrations in flat bars.

So when r = D, Kt = 3, and Kt =1 when the feature disappears. The formula shows that a round circular hole, and a semi-circular notch in the edge of a sheet triples the applied load. Other simple examples include a spherical void, which doubles the stress. Sharp cracks represent a far more damaging scenario, with crack tip stress being multiplied many times. In a similar vein, abrupt geometric changes, such as sharp corners, will also exert a similar damaging influence on the strength of the product. As a failure mechanism, fatigue almost always starts at a design stress raiser (or sub-surface defect), so their presence and position are a crucial factor in examining failed products. These features usually determine the weakest link in a part or product, from which cracks can initiate and (often) propagate to failure.

## 5.1.6 Case Study: Stress Concentration – Cement Mill Power Input Shaft

The power input shaft from a cement plant mill failed after approximately 1.5 years in service. Failure was considered premature as the shaft supposedly had a design life of 25 years. Elemental analysis (Table 5.2) would suggest the shaft had been fabricated in a BS 970 817M40 (EN24) heat-treatable steel that had been through-hardened and tempered (Table 5.3). The shaft had an outside diameter of 30 mm and transmitted power with the aid of a keyway. Low-power magnification of the fracture surface revealed characteristic fingerprints of fatigue failure, initiating from a sharp corner of the shaft keyway (arrowed in red on Figure 5.6). Fatigue failures in shafts nearly always initiate at the surface, generally at points of mechanical or metallurgical stress concentration that locally increase stresses or reduce the material's fatigue resistance. Mechanical stress raisers include small fillets, sharp corners, grooves, press/shrink fits and in this case a keyway, the seat of which can create a stress concentration condition within the shaft.

Table 5.2 Elemental Analysis of the Shaft Material, Showing a Close Match to a BS 970 817M40 (EN24) Heat-Treatable Steel

| C% | Si% | Mn% | Ni% | Cr% | Mo% | Fe |
|------|------|------|------|------|------|------|
| 0.38 | 0.31 | 0.50 | 1.40 | 1.20 | 0.22 | Bal. |

Table 5.3 Hardness Evaluation, with Tensile Strength Inferred from Hardness Results and Yield Stress Taken from Literature

| Hardness (Hv) | Tensile Strength (MPa) | Yield Stress (MPa) | Shaft Dia, d (mm) | RPM | Power (hp) |
|------|------|------|------|------|------|
| 289–302 | 924–966 | 654 | 30 | 1,200 | 100 |

**Figure 5.6** Low-power magnification of fracture surface, showing classic fatigue beach markings. Crack initiation site arrowed in red.

## STRESS ANALYSIS

When any drive shaft is operating at constant speed, torque on a shaft is related to its rotational speed $\omega$ and the power W being transmitted:

$$W = T\omega$$

Therefore $T = W/\omega$

$$= \frac{100\text{hp} \times (1/0.001341)}{1200\text{rpm} \times 2\pi \times 1/60} = T = 528 \text{ Nm}$$

Shaft diameter $(d) = 30\text{mm} = 0.030\text{m}$

Shear stress $= Tr/J$, where $r = 0.015\text{m}$, and $J = \pi(0.033)^4/32 = 7.95 \text{ e}^{-8}$

Therefore shear stress $= (528 \times 0.015)/7.95 \text{ e}^{-8}$

$$= 99.6 \text{ MPa}$$

From ASME, the permissible shear stress, $\tau_{max} = 0.3 \times \sigma_{yield}$ which = 196 MPa. It can be seen that service shear stress of 99.6 MPa, developed on the shaft, was far less than the permissible value. However, fatigue failure in shafts nearly always initiates at points of mechanical or metallurgical stress concentration that locally increase stresses or reduce the material's fatigue resistance. There are different stress concentration (Kt) factors for bending and torsional loads. Peterson (R.E. Peterson, *Stress Concentration Factors*,

Wiley, New York, 1974) gives Kt = 2.14 for bending and Kt = 2.62 for torsion. As the fracture surface clearly shows a torsional fatigue pattern, a Kt of 2.62 was utilised. Although analysed at the worst possible operating conditions, as a direct result of stress intensification, $\tau_{max}$ increased to 260 MPa, suggesting fatigue crack could well have initiated almost immediately after entering service.

The stress-raising influence of keyways can be reduced by using key seats profiled with bull-nose end mills. The stress concentration factor for a sled runner key seat will be significantly lower than for a profile key seat, whereas a circular key seat (woodruff) will have a lower stress concentration factor than any other key geometry.

## 5.1.7 Residual Stresses

Any process transforming the shape or properties (i.e. thermal treatment) of a material will introduce residual stresses. These stresses are universally present and are a multifaceted function of the material or component processing (Section 5.12 below) and service history. As such, residual stresses cannot readily be determined by calculation or predicted from first principles with any great accuracy. However, for complete design assessment of integrity, knowledge of the total stress field is required. Nevertheless, in many circumstances, designers limit their assessments simply to consideration of stresses generated by externally applied loads that the component will be exposed to during its service lifetime. Although an essential requirement in the design process, adoption of such a narrow focus holds the potential to miss a key integrity factor. Consideration of residual stresses may well provide an indicator as to how close a material is to its ultimate limit. Such reflection may also define performance under fatigue loading and can often influence response to environmental conditions. Development of residual stresses in engineered components is expanded in the section 'Manufacturing Fabrication as a Source of Stress' later in this chapter (Section 5.12).

## 5.1.8 Summary of Fundamental Mechanical Background

The way in which engineered products fail in service can often be directly related to how materials of manufacture respond to operational conditions at play, both prior to, and at the instant of, demise. The above review of material behaviour under mechanical loading conditions presented a limited and basic appraisal of initial fundamental mechanical reactions, simply as there are a plethora of perfectly adequate textbooks available for expanding knowledge in this arena. However, a review of more complex mechanical behaviour, being a recurring root cause of service failure, was considered a necessity for any forensic engineering textbook. When investigating failure and/or attempting to quantitatively evaluate service loading and stresses, an understanding of material response to both manufacturing and service loading stress is a paramount pre-requisite for the failure investigator.

It is of importance to appreciate that there rarely is a fixed value of a mechanical property of a material. This is of particular relevance when describing or modelling the mechanical response of materials or structures with constitutive equations. 'Constant' values may change significantly with temperature, strain rate and, from the material perspective, with microstructure. Although this variability has gone largely unrecognised in many modelling activities, it should be fully appreciated by the failure detective. The incorporation of realistic service property values, which take account of these factors, represents a major challenge to both reliable design of any engineered artefact, and for any subsequent forensic failure analysis. Attention will now be focussed on a range of typical failure mechanisms that are found to continually materialise in practice.

## 5.2 Service Failure Mechanism

A failure mechanism is the sequence of steps that, together, constitute a failure mode. Initially, knowledge of a failure mechanism is important for the design integrity process as it may allow inspection, monitoring or other integrity management measures to identify and monitor stages of a failure process prior to any final demise. The evaluation of failure modes is designed to focus on those modes to which the component is considered to be most susceptible in practice. This evaluation is based on the practical knowledge and experience of the investigator, who will have an intuitive feel for failure modes specific to the application in question.

There are numerous failure mechanisms that can overtake any engineered system, even when working within its specified service loading and operating conditions. In addition, specific failure mechanisms appear more frequently than others, playing a repeated role in the demise of engineered systems, structures and components. Consequently, a brief review of the most common recurring failure mechanisms will be reviewed and their salient features highlighted. However, there are many publications that specifically focus on the subject of failure and associated failure mechanisms. Therefore, the reader is referred to them for additional detail.[8–12]

A failure mechanism is simply the way in which a system or component might fail whilst in service use. In general, stresses imposed on any product during the course of its lifetime can be grouped as mechanical (yielding, ductile rupture, bending, torsion, shear, fatigue, wear, impact), thermal, chemical, environmental, biological, electrical and misuse or abuse. Other failure mechanisms can result in technical shortcomings without being destructive, such as an unacceptable increase of friction in a bearing. In addition, there are failure modes that will result from interaction with other systems or users.

### 5.2.1 Case Study: Hubble Space Telescope Mirror and the Team Philips Catamaran

'The initial failure of the Hubble Space Telescope is an example of problems caused by relying on computer simulations. In 1990, when the orbiting telescope sent its first photographs back to Earth, the images were unexpectedly fuzzy and out of focus. NASA determined that the problem was the result of a human error made years before the launch: the telescope's mirror had been ground into the wrong shape[13] (Figure 5.7A). The mirror, tested prior to launch like the telescope's other separate components, functioned properly on its own. However, the manufacturers did not actually test the mirror in conjunction with the other components. The manufacturers relied on computer simulations to determine that the separate components would work together. The simulation didn't take into account the possibility of a misshapen mirror. As a direct result of the Hubble problems, NASA revised its opinion regarding 'the merits of actually testing a system rather than depending upon theory and simulation' Implicit reliance on computers, to predict and counteract failure', could be considered the weak link of this story.[13]

However this is not a 'one-off' incident; another classic example is the demise of the Team Philips catamaran (Figure 5.7B), where failure to correctly model loading conditions of rough seas led to structural failure and ultimate loss of the catamaran.

The Hubble Telescope mirror failure and the Team Philips catamaran loss represent two typical examples of an interaction failure.

Any failure mechanism is, therefore, influenced by a range of factors that must be known (or deduced) in order to predict the presence and severity of the actual process at play. So, for the forensic investigator, gaining a clear understanding of the circumstances in operation at the instant of failure will be paramount. As a simple example, if evidence revealed that a certain type

**Figure 5.7** (A) The mirror from the Hubble Space Telescope. (Courtesy of Wikimedia Commons, public domain.) (B) The Team Philips catamaran. (Courtesy of Pete Goss at http://www.petegoss.com/gallery.php?t=team-philips&gID=26.)

of chemical had been in use with a failed product or device, the chemical compatibility of that product can be readily checked. Therefore, a range of common failure modes, likely to be encountered by the forensic investigator, will be reviewed.

## 5.3 Mechanical Failure

Mechanical failure will arise 'from application of external forces that cause a product or component to deform, crack, or break when the yield strength of the material is exceeded.'[14] The applied force mode may be tensile in nature, compressive, torsion, 'or impact – with the force being applied over short or long time spans, and at varying temperatures and/or environmental conditions'.[14] Failure analysis can model failure by laboratory testing under monotonic loading conditions, with application of a gradually increasing force – in any mode – until failure occurs. Most laboratory testing will be carried out at a constant isothermal temperature. However, most service applications experience fluctuating temperatures, and simulation will often demand and thus encompass thermal transients. As a mode of loading, creep may be regarded as a special case (Section 5.8 below), as a fixed force is applied and held constant. Resultant strain is monitored over time, i.e. creep response is a time-dependent phenomenon, whereas monotonic tensile loading, for example, is considered time-independent.

Any material subjected to monotonic uniaxial force will eventually fail. Features of failure will depend on several factors that include

- The material in question and its microstructure
- The applied stress state (uniaxial, biaxial, triaxial, hydrostatic, etc.)
- Loading rate
- Temperature
- Ambient environment (corrosive, water, etc.)

Individual loading modes will produce characteristic fracture morphologies (shape), a range of which are discussed in Chapter 6. It is these fracture morphologies that provide a fingerprint of failure for the forensic detective.

'Any engineered system or individual component can fail from application of a single overload force. Traumatic overload can produce either a ductile or a brittle fracture mode. Ductile and Brittle failures are terms that simply describe the amount of macroscopic plastic deformation that preceded fracture.'[14] Direct observation of the fracture surface will reveal which of the two failure modes were at play at the instant of failure (Section 6.3).

## 5.4 Hydrogen Embrittlement

There is an insidious mode of brittle failure that will often be encountered in service – that of hydrogen embrittlement. When atomic hydrogen enters a metallic material, it can result in a loss of load-carrying ability, loss of ductility or sub-microscopic cracking. Hydrogen embrittlement is an insidious failure mechanism that can cause unexpected and sometimes catastrophic brittle failure at applied stresses well below the yield strength of the material. Hydrogen embrittlement is therefore considered to be the Achilles' heel of high-strength ferrous steels and alloys. Hydrogen embrittlement does not affect all metallic materials equally; the most vulnerable are high-strength steels, titanium alloys and aluminium alloys. Although the phenomenon is well-known, extensive research has failed to pinpoint the precise mechanism at play. However, hydrogen embrittlement mechanisms are thought to be diffusion controlled, with current thinking suggesting the susceptibility of any material is directly related to the characteristics of its trap population. In turn, the trap population of any material is related to its microstructure, dislocations, carbides and other elements present in the structure. Therefore, diffusion is controlled by the rate of escape of hydrogen from the traps. It follows that the nature and the density of traps will control the diffusion coefficients.

### 5.4.1 Sources of Hydrogen

Hydrogen entry is a pre-requisite for embrittlement, with available atomic hydrogen being derived from a range of possible sources. (Defence Standard 03-30, October 2000 - Treatments for the Protection for Metal Parts of Service Stores and Equipment against Corrosion, Ministry of Defence Defence Standard 03-30, Issue 3, Publication: 4 February 2005).

During the primary steelmaking process, as part of the manufacturing cycle, hydrogen can be introduced during component heat treatment, carbonising, cleaning, pickling, phosphating and electroplating. There can be particular concern in the embrittlement of 'as carburised' parts which can have a surface hardness in the range of HRC 45–50. The problem is magnified further if the steel parts are coated with sacrificial coatings (e.g. Zn, Cd) that accelerate hydrogen charging. Fabrication processes such as roll forming, machining and drilling can introduce hydrogen as a result of breakdown of any unsuitable lubricants. Welding or brazing operations also have potential to introduce hydrogen. A special case is arc welding, in which the hydrogen is released from moisture in the coating of the welding electrodes; to minimise this, special low-hydrogen electrodes are used for welding high-strength steels. During the service lifetime of a component, hydrogen can be introduced as a result of cathodic protection reactions or corrosion reactions.

### 5.4.2 Hydrogen-Induced Brittle Failure

As a mechanism, failure by hydrogen embrittlement requires the presence of tensile stresses, susceptible material, and of course the presence of hydrogen. It is an insidious type of failure as it can occur without the material being subjected to an externally applied load (residual stresses). It can also occur at levels of applied loading that are considerably lower than the material yield stress. Therefore, catastrophic failure can occur with little or no obvious sign of any significant structural/material deformation, or noticeable weakening of the component (Figure 5.8). Although it is high-strength steels that are the most vulnerable to hydrogen embrittlement, all materials exhibit varying degrees of susceptibility to hydrogen embrittlement.

### 5.4.3 Reducing the Probability of Hydrogen Embrittlement

In the electroplating industry, the accepted method to alleviate hydrogen embrittlement is with a 'relief-baking' process, with the hardness and strength criteria of Table 5.4 used to determine the temperature/time for hydrogen embrittlement relief (baking).

For steels of actual tensile strength below 1000 MPa (HRc-31), heat treatment relief baking after plating is not essential.

Most specifications require relief baking within four hours of plating, while some require baking within one hour of plating. The process serves to remove or redistribute the hydrogen, but may further reduce the fatigue limit of high strength steels. This fact was a phenomenon described by R.A.F.

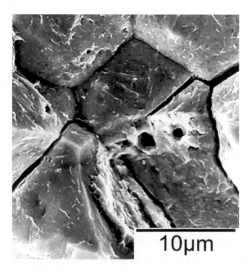

**Figure 5.8** Scanning electron micrograph showing a brittle fracture through an alloy steel socket head cap screw. The fracture features are characteristic of hydrogen embrittlement. (1,300×).

**Table 5.4 From: ASTM B 850-94 – Specific Bake Times for Various Strength Steels**

| Tensile Strength MPa | Rockwell Hardness $H_R$ | Post-Plate Bake @190–220°C. Hour |
|---|---|---|
| 1,700–1,800 | 49–51 | 22+ |
| 1,600–1,700 | 47–49 | 20+ |
| 1,500–1,600 | 45–47 | 18+ |
| 1,400–1,500 | 43–45 | 16+ |
| 1,300–1,400 | 39–43 | 14+ |
| 1,200–1,300 | 36–39 | 12+ |
| 1,100–1,200 | 33–36 | 10+ |
| 1,000–1,100 | 31–33 | 8+ |

Hammond and C. Williams ('Metallurgical Reviews', 5, 165, 1960) and again by J.K. Dennis and T.E. Such ('Nickel and Chromium Plating -2nd Edition', 1986, p. 72). For the production electro-plater, having to remove parts from the production line to bake – followed by a separate chromating process – is a laborious business. There is no visible difference between a baked and un-baked component, and it is at this point of production that components can (and *do*) miss the final bake process. Any component not subjected to baking will thus enter service with a clear potential for brittle failure.

## 5.5 Impact

In general, structures, components and equipment are not intentionally designed to be exposed to instantaneous loading conditions. However, they may be exposed to such conditions by accident when, for example, a vehicle crashes. Once such an event is encountered, it is those materials which are already vulnerable to brittle fracture, such as ceramics, polymers below their glass transition temperatures and metallic materials sensitised by ductile/brittle transition, that are most likely to fracture. Toughness values and the quality of the material, in terms of defect size and distribution, are the key parameters here. Ductile materials can usually accommodate this type of instantaneous loading, although the permanent deformation induced in them may cause subsequent problems.

As reviewed in Section 4.12.6, an impact test would be the preferred method used to gauge the resistance of a material to breaking under a suddenly applied force. The aim of any impact test is to gauge the relative toughness or impact toughness of a material. The Charpy test method is recognised as being most commonly used, often in quality control applications where it is a fast and economical comparative test, rather than a definitive test. Figure 5.9

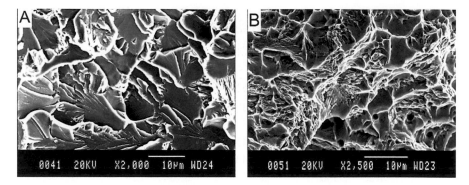

**Figure 5.9** (A) Charpy low temperature impact fracture surface of plain medium carbon steel (0.4%C) showing cleavage, with twist and tilt grain boundaries being evident. (B) Charpy room temperature impact fracture surface of plain medium carbon steel (0.4%C), showing micro-voidal coalescence (MVC) along with regions of brittle inter-pearlitic fracture.

provides a comparative illustration between the fracture surface, of a plain medium carbon steel (0.4%C), impacted at low temperature and room temperature respectively. The resultant fractograph provides a (visual) fingerprint of the process (impact at low or room temperature) at play.

As reviewed in Section 4.12.6, features that will influence Charpy-impact energy results, gathered from test samples, will include the following: yield strength and ductility, notch geometry, temperature and strain rate and fracture mechanism.

An in-service failure feature that can sometimes be found in the microstructures of particles, displaced by impact from hardened steels, is a white etching band that will follow the line of maximum shear displacement as the particle was expelled. Two such bands are illustrated in Figure 5.10A and B, which show etched micro-sections of two fragments broken from hardened tools: (A) the rim of a hammer and (B) a highly alloyed high-speed tool. Such bands can only be observed if the particle is sectioned at right angles to the direction of the shearing displacement. To examine the structure at such an orientation will usually require sectioning the particle while it is standing on one point. White etching bands are only found when the particle has been sheared by an impact or explosion, not when sheared slowly, as in a compression-testing machine for example. They are believed to arise from adiabatic heating from the energy of deformation causing a phase transformation along the shear plane.

Identification of a white etching band proved useful in proving that a Stilson wrench had been used as a hammer. One corner of the end had chipped and flown into a man's eye. He was removing the timing gear on a car engine and claims he released the nut using the wrench and then tapped the gear with a soft hammer to release it from the shaft. Figure 5.10C shows

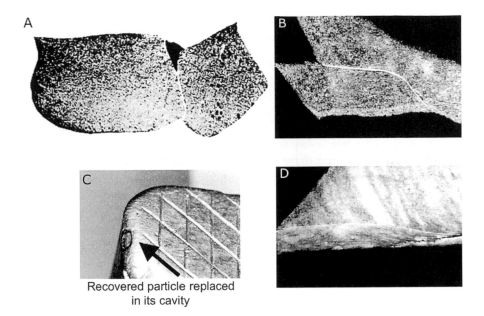

**Figure 5.10** White etching layers following shear planes in particles of steel broken from hardened tools by impact. (A) From rim of engineer's hand hammer, etched, ×25. (B) From perimeter of alloy steel punch, etched, ×75. (C) Recovered fragment replaced in cavity at a corner on the Stilson wrench. (D) Lightly etched micro-section across fracture of fragment, showing white layer in zone of shear deformation.

the particle recovered from his eye replaced in a cavity on the wrench, confirming that that was where it originated. Figure 5.10D is an etched section at the sheared particle edge, which quite clearly reveals shearing deformation and a white layer, establishing the particle was dislodged by an impact.

## 5.6 Bending

When in service, bending is the single most damaging mode of load application that any component can be subjected to. A simple equation used for evaluating stresses and strains on beams in bending is the 'engineer's bending equation':

$$M/I = \sigma/y = E/R$$

where $M$ is the moment on the beam, $I$ the second moment of area of the beam section, $\sigma$ the stress at a distance $y$ across the beam section, $E$ the tensile modulus of the beam material and $R$ the radius of curvature of the bent beam. Some of these variables become constant when a single type of beam is considered under two different loading conditions. Thus $E$ is constant (same material) and $I$ is constant (same section shape).

## 5.7 Torsion

Torsion will occur when any shaft is subjected to a torque. This holds true for both rotating (engine drive shafts, motors, turbines, etc.) situations and for stationary (bolts, screws, etc.) situations. Applied torque will induce a twist in the shaft, and one end will rotate relative to the other, thus inducing shear stresses on any cross section. Failure can occur as a result of shear stresses on their own, or in combination with tensile and/or bending stresses. As with bending, a simple engineering equation can be used to evaluate torsion stresses and strains:

$$T/J = G\theta/L = \tau/r$$

where $T$ = applied torque; $J$ = polar second moment of area; $G$ = elastic constant of the material; radius of the shaft = $r$; its length = $L$; $\theta$ = angle of twist; $\tau$ = shear stress.

Torsion *testing* will therefore provide a method for determining the modulus of elasticity in shear, shearing yield strength and the ultimate shear strength. It is useful for examining and testing parts such as axles, shafts, couplings and twist drills, and can be used to good advantage in the testing of brittle materials. Torsion testing has the advantage that there is no necking (as in a tensile test) or friction. Furthermore, the test data are generally valid to larger values of strain than are tension test data. However, torsion testing, as a tool for the failure investigator, will be subject to the same limitations as tensile testing above. Many engineering components have safety devices based on shear pins that limit the amount of torque or bending moment that may be transmitted. In this way these cheap devices protect the rest of the more expensive mechanism from expensive overload. They function in a similar way to a fuse in an electrical circuit. Indeed in some circumstances shear pins are called 'fuse-pins' or 'fuse links'. As discussed in Section 4.12.5 (Shear Testing), flywheels, storing energy in rotating machines may cause severe damage if the machine is suddenly stopped for any reason. By incorporating a shear pin the damage is greatly reduced. Designers of such equipment must calculate the applied force that will cause such fuse pins to fail by shear overload. Many airliner engines are attached to mounting pylons by shear pin bolts, that are designed to shear away if engine vibrations risk damage to the airframe.

## 5.8 Creep

'At ambient and elevated temperatures, most materials can fail at a stress which is much lower than its ultimate strength. This group of failure modes are time-dependent, and termed creep deformation and creep rupture. More generally, materials or components undergoing continuous deformation over time under a constant load or stress' is said to be creeping.[14,15] Therefore, creep of materials and structures is generally considered a high-temperature progressive deformation at constant stress. Creep is commonly found in boiler applications, gas turbine engines and ovens. These types of service application have components that will experience creep under normal use. Generally, failure by the process of creep is readily identifiable by the extensive deformation that will occur. However, failures may appear ductile or brittle, and cracking may be either trans-granular or inter-granular. Elastic, plastic, visco-elastic and visco-plastic deformations will also be encompassed within the process of creep, being dependent on the material, service temperature and time of creep deformation. The designer encounters most problems from the creep of metals or polymers above their glass transition temperature. Again, temperature is a critical parameter, and for metals, creep must be considered when values of the homologous temperature ($T_h$) exceed about 0.4, where

$$T_h\left(^0K\right) = T\left(^0K\right)/T_m\left(^0K\right) \tag{5.5}$$

Here, homologous temperature = $T_h$; service or operating temperature = T and melting temperature of the material in question = $T_m$.

However, 'high' temperature is a relative term that is dependent on the materials being evaluated. Take the case of eutectic lead/tin solder:

$T_m = 183°C = 456K$; $T_h = 0.4$ (for creep onset)
Therefore, $0.4 = T^0K/456K$
So, $T^0K = 182.4K = -90°C$

It is clear that eutectic solder *will* creep at temperatures as low as −90°C. Therefore, although creep is considered to be a high-temperature process, 'high' is totally dependent on the material system. This can be readily demonstrated in the graphic of Figure 5.11. The figure demonstrates that solders at room temperature are subjected to homologous temperatures equal to those of superalloys operating in an aircraft jet engine. It is the high homologous temperatures of solders in normal service that are the cause of microstructural instability and therefore their stress, strain and strain rate sensitivity.[16]

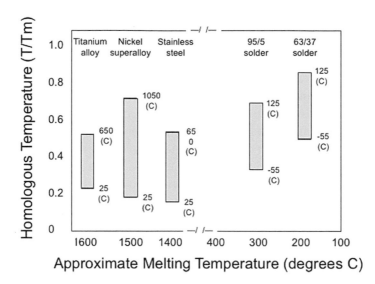

**Figure 5.11** Homologous temperatures for a range of engineering materials. The numerical values on each of the bars represent approximate service operating range in degrees Celsius.

## 5.9 Fatigue

There is one particularly insidious mechanical failure mode that is reported to be responsible for the majority of in-service failure – it is that of 'fatigue'. Fatigue is the progressive, localised and permanent structural damage that occurs when a material is subjected to cyclic or fluctuating stresses or strains under nominal stresses that have maximum values lower than the yield strength of the bulk material. When designing a system or component destined for cyclic service loading applications, the engineer will be interested in the number of cycles of stress that cause failure – this being principally determined by the stress range $\Delta\sigma$ ($\sigma_{max} - \sigma_{min}$) – or the equivalent strain range $\Delta\varepsilon$ ($\varepsilon_{max} - \varepsilon_{min}$) if service conditions dictate cycling between limits of strain rather than stress. Under conditions of stress control, the traditional means of conveniently presenting data to compare performance is with an S–N graph, i.e. a plot of stress range versus the number of cycles to failure (Figure 5.12). As might be expected, the smaller the stress range, the larger the number of cycles required for failure. A few materials, notably steels, exhibit a fatigue limit, which is a value of the stress range below which fatigue failure does not occur. The majority of materials display increasing fatigue endurance as the stress range diminishes. Data in this form are difficult to analyse or to use in expressions for life prediction. A property more fundamental to materials behaviour is strain, as it is a measure of the extent to which the material is deformed. Plots of strain range against numbers of

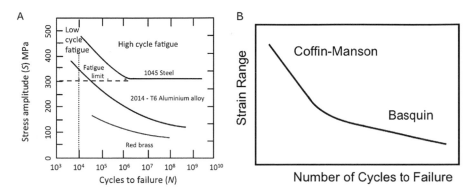

**Figure 5.12** (A) Traditional S-N (stress range versus number of cycles to failure) plots for materials with and without a fatigue limit. (B) Stress range versus number of cycles to failure showing the domains of elastic (Basquin) and plastic (Coffin–Manson) dominance.

cycles to failure are considerably more useful to the engineer or designer, particularly when plotted on logarithmic scales, when linear relationships often exist (Figure 5.12A). Elastic strains usually only extend to a fraction of 1%, after which permanent deformation (plasticity) occurs. The total strain in this case, $\varepsilon_t$, is made up of an elastic component, $\varepsilon_e$, and a plastic component, $\varepsilon_p$. Similarly, in terms of strain ranges,

$$\Delta\varepsilon_t = \Delta\varepsilon_e + \Delta\varepsilon_p \tag{5.6}$$

A plot of log $\Delta\varepsilon_t$ versus log$N_f$ will derive a curve as shown in Figure 5.12B. However, under conditions where either $\Delta\varepsilon_e$ or $\Delta\varepsilon_p$ dominate, plotting these terms against $N_f$ gives a linear relationship. Hence, either

$$\Delta\varepsilon_e N_f{}^\alpha = C_1 \text{ (Basquin) or} \tag{5.7}$$

$$\Delta\varepsilon_p N_f{}^\beta = C_2 \text{ (Coffin-Manson)} \tag{5.8}$$

'Metal fatigue cracks will initiate and propagate in regions where the strain is most severe. The fatigue process entails three stages – crack initiation, crack propagation (growth) through the load bearing cross-section, culminating with final fracture of the remaining load bearing section. Fatigue failures are frequently caused by poor detailing and material inhomogeneities such as; notches, grooves, surface discontinuities and flaws. It is therefore of importance that the significance of sharp section changes, surface scratches, and/ or structural irregularities are both recognised and, if at all possible, avoided by both designer and manufacturer of any structure destined to be exposed to cyclic loading conditions during its service lifetime. Recognition must be given that specific features such as holes, screw threads, keyways, inclusions, existing cracks etc, may locally concentrate stresses – thereby sensitising the component or structure to failure by fatigue.'[15]

### 5.9.1 Minimising Fatigue

Investigations sometimes reveal minor procedural changes that eventually have disastrous consequences. It is important to recognise these changes and comprehend the consequences that may result from such change. A lucid report emphasising your findings – backed up with a clear and common-sense argument – will highlight your opinion of any potential problem area. There are some additional design features that enhance fatigue life, and if they are omitted they can cause premature fatigue failure.

- Improve surface finish and texture.
- Select materials and heat treatments to give high yield stresses and high post-yield ductility. This reduces the accumulated plastic strain and delays the onset of crack embryo formation.
- Introduce residual compressive stresses that reduce the effective stress amplitude. This is particularly effective in the propagation stage of cracking. The cracks do not suffer stress concentration effects if they are held shut for more of the stress cycle by the residual compression. This is achieved by the following means.
  - (a) Surface plastic strain by shot peening. The surface flows, but is pulled back into compression by the underlying elastic material. This effect is also utilised when threads are rolled rather than machined.
  - (b) Ion bombardment into interstitial sites, for example, nitrogen.
  - (c) Case hardening, which is doubly effective in steels. The surface martensite formed is of high yield stress and under compression.

Fatigue as a failure mechanism was brought sharply to the attention of aircraft engineers in 1954, where investigators from the Royal Aircraft Establishment at Farnborough, England were the first to undertake full-scale testing on a brand-new fuselage (Figure 5.13A). This testing was undertaken after three de Havilland Comet passenger jets broke up in mid-air and crashed within a single year. It was found that sharp corners around the plane's window openings (actually the forward ADF antenna window in the roof) acted as initiation sites for cracks (Figure 5.13B). From that point onwards, all aircraft windows were immediately redesigned with rounded corners.

However, any tight radius can act as a stress raiser, 'severely curtailing service lifetime of a product or device. This can be illustrated by considering a failure that overtook an upmarket, therefore expensive, vacuum cleaner. It had been in use for some eighteen months when a spring catch', that retained

**Figure 5.13** (A) Full scale testing on full-size Comet fuselage. (B) Fatigue failure initiating at tight window radius.

extension pipes, unexpectedly failed. Furthermore, it came to light that a large number of this particular model of vacuum 'were failing prematurely and at an identical position. The catch in question had been injection moulded in ABS, and flexed about a tight radius each time the pipe was removed and replaced, inflicting two cycles per use.'[14]

'Observation of the fracture surface revealed a multi-start low-cycle fatigue fracture that had initiated from a tight radius moulded as part of the catch profile.' The presence of classic clamshell beach markings (Figure 5.14) exposed low-cycle fatigue as the culprit of failure, exacerbated by the intensified stress at the initiation site. 'The obvious solution to this failure was a simple design change – an increase in the bend radius was all that was required to alleviate the problem.'[14]

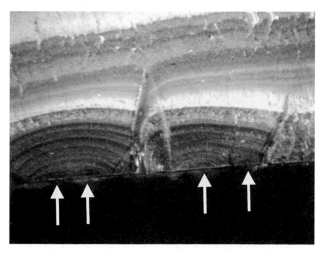

**Figure 5.14** ABS spring catch from an 18-month-old vacuum cleaner, showing classic multi-start bending fatigue. Note: clamshell beach markings that are typical of low cycle fatigue crack growth. Individual initiation sites are arrowed in red.

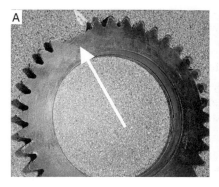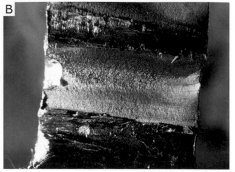

**Figure 5.15** (A) A power train gear wheel with two missing teeth. (B) Under low-power magnification, the presence of beach markings on both gear tooth fracture surfaces is clearly seen. The presence of beach markings presents a fingerprint of fatigue as the failure mode at play.

## 5.9.2  Case Study: Failure of a Power Transmission Gear Wheel

A ring gear wheel failed in service, with overload of the power transmission train being the likely suspect. On investigation, two teeth were found to have detached (Figure 5.15A). In essence, a gear tooth is a cantilever plate with tensile stresses on the contact side of the tooth and compressive stresses on the opposite side. Three common modes of failure of gear teeth are

   (i)  Surface contact fatigue (appearing at the tooth flank)
   (ii)  Tooth root bending fatigue
   (iii)  Traumatic overload

Observation under low-power magnification revealed that the fracture surface of two separate gear teeth contained classic beach-marking features, providing a fingerprint of a fatigue fracture mechanism having been at play (Figure 5.15B). Therefore it could be stated that failure of the gear teeth in question was as a direct result of a classic tooth root bending fatigue mechanism. By definition, fatigue is a mechanical process that progresses over time – it is *not* an instantaneous or traumatic failure mode. This failure was therefore not the result of a traumatic overload caused by some unusual impact or shock loading incident, but was as a direct result of mechanical breakdown following a lengthy period of wear and tear, over its service lifetime. Once fatigue cracking had been initiated at some prior point in its working history, failure of the gear was inevitable.

### 5.9.3  Fatigue-Related Failure Modes

Uniaxial tensile, bending, rotating bending and torsion are conventional modes of fatigue failure experienced in practice. However, there is a range of additional fatigue-related failure modes that must be recognised by the forensic failure analyst. Fatigue as a failure mode can be enhanced by the interaction of temperature, pressure and environment as follows:

- Thermal fatigue cracking of a material or structure can be initiated by repeated temperature cycling of a mechanically constrained structure or component, e.g. soldered joints of components such as ceramic chip carriers on a fine pitch technology (FPT) electronic circuit board. Thermal fatigue could also be introduced by temperature gradients in a component or part. Thermal fatigue can be used to 'prove' components prior to market release, i.e. thermal cycling (burn-in) of micro-electronics as a method of weeding out parts susceptible to infant mortality, thus eliminating premature service failure.
- Contact fatigue – contacting elements that articulate (roll, or roll and slide) against each other, whilst under high contact pressure will, over time, be prone to the onset of surface pitting or spalling of their contacting surface, e.g. bearing elements rolling in their raceways are being subjected to the action of rolling contact fatigue.
- 'Corrosion fatigue can be initiated by the combined action of repeated or variable stress whilst operating or working in a corrosive environment. The process frequently initiates at a surface corrosion pit, with the prevailing environment holding influence over the crack growth rate, and/or the probability of fatigue crack initiation. Test data show that for high strength steels, the fatigue strength at 10 million cycles in salt water can be reduced to as little as 10% of that in dry air.'[17] Carbon steels exhibit trans-granular fracture. Copper and its alloys fail by inter-granular fracture.

### 5.9.4 Approximating the History of a Fatigue Failure

When examining a fracture surface, the first and most vital undertaking for the forensic engineer is to identify the site where the fracture initiated and determine what relevance (if any) the initiation site has to the normal loading or service stressing. This having been accomplished, attention may then be focused on the number of fatigue cycles that may have been experienced during the course of its service lifetime. The minimum number of stress cycles a fractured component has been subjected to can be approximated by estimating the number of striations on the critical part of the fracture surface (Miner's Law, Section 1.5.1). However, as discussed, a blind application of this law may lead to a gross underestimate, simply as a result of 'non-striating' stress cycles. Nevertheless, even an approximation can provide a guideline or useful supporting evidence in casework.

An alternative approach is to make use of fracture mechanics, as reviewed in Section 5.1.3 above. However, as fracture mechanics is the mechanics of cracked bodies, it follows that it can be applied only to cases where crack initiation has occurred. As discussed in Section 5.1.3, this technique is based on the stress at the end of a sharp crack and develops a stress intensity factor along with a strain energy release force.

### 5.9.5 Summation of Fatigue as a Service Failure Mechanism

Metals and polymers are typically susceptible to fatigue failure, while ceramics tend to be more resistant. There are a number of different modes of fatigue that include high-cycle fatigue, low-cycle fatigue, thermal fatigue, surface fatigue, impact fatigue, corrosion fatigue and fretting fatigue, some of which are discussed above. Although fatigue of a specific material type can be effectively characterised in a laboratory environment, dependable design against fatigue failure can be challenging to safely account for 'in-service'. It transpires that fatigue is found to be a common failure mode by the failure or forensic engineer, attributed as playing a role in approximately 90% of all material structural failures. Failure investigations often reveal minor design or procedural changes that eventually have disastrous consequences. It is therefore important for the design and manufacturing engineer to recognise these changes and comprehend any consequences that may result from such change. Taking a retrospective view of the Dee Bridge failure of 1847, an artistic flourish to the casting introduced a stress-raising situation at one particular corner. It is likely that a crack initiated on first loading by a train. One could question how little has been assimilated as, even today, fatigue failures are frequently caused simply by poor detailing and material inhomogeneities such as notches, grooves, surface discontinuities and flaws. It is therefore of importance that the significance of sharp section changes, surface scratches or structural irregularities are both recognised and, if at all possible, avoided by both designer and manufacturer of any structure destined to be exposed to cyclic loading conditions during its service lifetime. Recognition must be given to specific features, such as holes, screw threads, keyways, inclusions, existing cracks, etc., that may locally concentrate stresses, thereby sensitising the component or structure to failure by the onset of fatigue.[15]

## 5.10 Wear

There is a further mechanical mode that can result in premature service failure – wear. Friction is the resistance to motion between two surfaces that are forced to slide relative to each other. Frictional properties of materials in intimate contact will result in wear of surfaces when such contacts side, impinge or oscillate relative to each other. Friction and wear are of considerable importance when considering the efficiency and/or operating lifetime of a product or component; friction will result in wasted power and generate heat, whereas any ensuing wear will lead to poor working tolerances, loss of efficiency and may ultimately lead to premature failure. The wear process has been defined as 'the progressive loss of substance from the operating surface of a body occurring as a result of relative motion at that surface'. Wear is relatively gradual, with the exception of galling. Here, excessive friction between high spots will result in localised welding. Subsequent splitting creates further roughening of rubbing surfaces, accelerating the breakdown process. In contrast to outright breakage, product or machine performance may degrade slowly rather than cease suddenly, so the point of failure may not always be obvious. Products or components that had ostensibly failed by 'wear' are often encountered, and it is often necessary to establish if the *rate* of wear was acceptable and reflected good engineering practice or not.

The study of the processes of wear is part of the discipline of tribology. The complex nature of wear has delayed its investigations and resulted in isolated studies towards specific wear mechanisms or processes.[18] A range of commonly found wear processes or mechanisms include[19]

- Adhesive wear
- Abrasive wear
- Surface fatigue
- Fretting wear
- Erosive wear

A number of different wear phenomena are also commonly encountered and represented in literature.[20] Impact wear, cavitation wear, diffusive wear and corrosive wear are all such examples. However, these wear mechanisms do not necessarily act independently in many applications. Take fretting as an example – it is simply a type of wear caused by repeated movement of two surfaces against one another, so there are elements of a fatigue-like process occurring in this failure mode. As such, this is typical of all real failure modes, when different modes combine with one another to produce a failed product. So although each distinct mode may be easy to recognise when occurring alone, it becomes more difficult when in combination with others. This is where the forensic engineer or failure analyst will be required to utilise and demonstrate his/her investigative ability. Therefore, as an *'aide-memoire'*, a limited atlas of wear morphologies will be presented in Section 6.8.5, and can be considered a further tool in the forensic toolbox.

## 5.11 Corrosion

Even if fatigue cracking or excessive wear is avoided, the life of an engineering component is not infinite. If the size of the section of a component or structure is reduced to cause general overloading, or poor fitting with neighbouring components, equally disastrous events can result. There are mechanisms that will remove material from the component throughout its service lifetime and a major cause of bulk loss (metal removal or wastage) in components is *corrosion*. In general terms, corrosion is simply the surface decay that will occur when metals are exposed to reactive environments. There are two major divisions found in the study of corrosion: dry corrosion and wet corrosion. Dry corrosion is dominated by high-temperature oxidation of metals and is controlled by the rates at which oxygen can diffuse through the oxide skin, along with the stability of oxide adhesion to its base metal. Wet corrosion involves electrolytic interactions that require the presence of an aqueous solution (electrolyte). The majority of cases that will be of concern to the forensic or failure engineer will involve aqueous solutions, with salt (sea) water turning out to be a particularly troublesome liquid to steels.[21]

Only metals and alloys corrode, as electrons associated with each atom are not tied up in chemical bonding to other atoms. Instead, electrons exist like a cloud or gas surrounding the regular array of atoms in the crystal. If placed in a chemical solution, the outermost metal atoms can lose electrons and become positively charged ions, in which form they are able to enter solutions. In this state they are able to take part in chemical reactions as positively charged ions ($Fe^{2+}$ or $Fe^{3+}$ for iron, depending whether two or three electrons have been given up), and the electrons so freed are able to form an electric circuit. This process is called 'corrosion', and results in a wasting away of the surface of the metal and the formation of a chemical corrosion product, universally termed 'rust'.

The formation of rust can occur at some distance away from the actual pitting or erosion of iron. Factors such as the presence of salt greatly enhance the rusting of metals as dissolved salt will increase the conductivity of aqueous solutions formed at the surface of a metal, therefore enhancing the rate of electrochemical corrosion.[21] For example, iron or steel tend to corrode much more quickly when exposed to salt used to melt snow or ice on roads – hence the great attention to corrosion protection that is shown by both the car manufacturer and bridge deck designer. The consequent loss of cross-sectional area thus reduces the capacity of a component to support a service load. The role of the forensic engineer in this arena is often to assess if basic design and application, along with appropriate material choice, reflect professional engineering practice.[22] However, it should be remembered that other forms of corrosion may actually have aesthetic appeal, such as the green patina on copper roofs or bronze sculptures.

## 5.11.1 Case Study: The Statue of Liberty

On 4 July 1986, the American nation held centenary celebrations for 'Miss Liberty', a gift from France to the USA. The Statue of Liberty (Figure 5.16A) had been designed by the sculptor August Bartholdi, assisted by an engineer named Gustave Eiffel. Although many people think the statue is made of stone, it actually consists of a copper skin over a steel skeleton. From a central steel pylon there extends a secondary structure of horizontal steel bars which supports a 'crinoline' of wrought iron. This defines the external shape of the statue as shown in Figure 5.16B.

The surface of the statue consists of sheets of copper (2.4 mm thick) attached with copper rivets to the bars of the crinoline by copper saddles; the arrangement is shown in Figure 5.17A. However, the arrangement of crinoline and saddles puts two dissimilar metals (copper and iron) into close proximity, with the risk that in the presence of water a galvanic cell might be formed. Iron, the metal lower in the electrochemical series, would form the anode in such a cell, and corrosion would occur at a rate dependent on current flowing.

Bartholdi and Eiffel had foreseen this risk. Wads of asbestos felt, impregnated with shellac, were used within the saddles to separate the two metals. Unfortunately, this system did not prevent electrolytic cells from being set up when water accumulated inside the statue, from condensation and from leaks

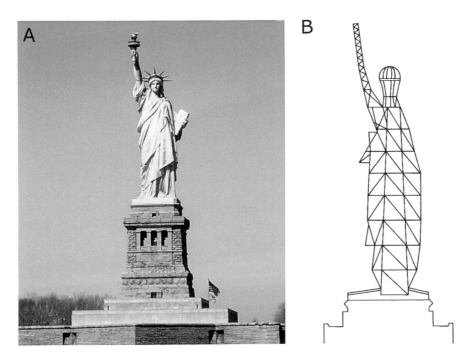

**Figure 5.16** (A) The Statue of Liberty. (B) The 'crinoline' (or armature) that forms a skeleton for the Statue of Liberty.

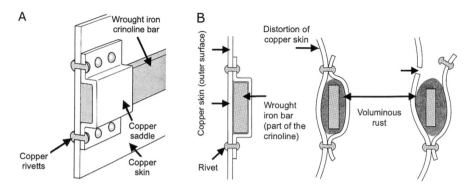

**Figure 5.17** (A) Method of attaching skin to 'crinoline'. (B) The effect of corrosion between copper skin and wrought iron crinoline.

of rain water, particularly via the torch at the top of the statue. Over the 100 years of its lifetime, about one-third of the 1,500 saddles were affected by corrosion of the iron within. The anodic reaction is the same as that described by bulk wet corrosion, Vis:

$$Fe \rightarrow Fe^{2+} + 2e^-$$

The cathodic reaction might be expected to be the deposition of copper ions from solution onto the copper cathode

$$Cu^{2+} + 2e^- \rightarrow Cu$$

However there are insufficient copper ions in solution for this to be significant, and the dominant cathodic reaction is

$$\frac{1}{2}O_2 + H_2O + 2e^- \rightarrow 2OH^-$$

When a volume of iron is converted to rust it will occupy a much larger volume. When this swelling occurred in the statue's crinoline, the saddles became distorted and their rivets pulled out of the copper skin (Figure 5.17B). The skin became deformed and disfigured, forming additional holes for rainwater to enter. In some places, over half the section of the wrought iron had been lost by corrosion, so the ability of the crinoline to carry the load of the skin was increasingly in doubt.

In the two years prior to the centenary, $75 million dollars was spent on restoration. The remedy used was two-fold – to replace the wrought-iron crinoline with one of stainless steel, with an electrode potential almost the same as that of copper. There would then be little driving force for galvanic corrosion. As a belt to these braces, a layer of the noble plastic polytetrafluoroethylene (PTFE), carried on a self-adhesive tape of woven glass fibre, was used to cover the steel and thereby separate it from the copper. Together with the repair of the leaking torch and air conditioning of the interior, the restoration is considered to have cured the problem. Just to be on the safe side accessible saddles will be inspected periodically!

## 5.11.2 Stress Corrosion Cracking (SCC)

Stress corrosion cracking (SCC) is a phenomenon related to a combination of static tensile stress along with a corrosive environment.[23] Failure will often occur in a mild environment and under tensile stresses well below the macroscopic yield strength of the material. The origin of tensile stresses may be external forces, thermal stresses or residual stresses. In general, as these stresses increase, the time required to initiate SCC decreases. Trace amounts of powerful reagents can induce micro-cracks which then grow slowly under applied loads or through another problem known as frozen-in strain.[24] The classic example comes from the 1920s in India when cartridges exploded in the rifles rather than firing a bullet. The cause was traced to hairline cracks in the brass cases, which in turn were caused by small amounts of ammonia gas emitted by dung heaps. The ammonia reacts with copper in the brass to form a complex (cupro-ammonium) ion, and cracking is encouraged by inbuilt stresses produced by forming the cartridge shape.

## 5.11.3 Case Study: SCC of Brass Impeller Blades

A service situation of SCC occurred when a manufacturer experienced a problem with brass impellor blades (Figure 5.18A). The blades were cracking and disintegrating after a few hours in operation, stirring a pitch mixture for use in ceramic crucible manufacture. The forensic investigator noticed an ammonia smell from the pitch, and identified SCC as the most probable cause of early demise. The brass unit was replaced by bronze, and coated with PTFE to resist wear (Figure 5.18B). The blade had not failed after several months usage.[25]

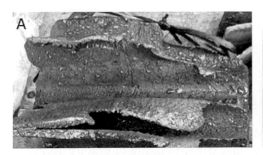

**Figure 5.18** (A) Severely damaged brass impeller. (B) Changing the material of manufacture to bronze, and applying a PTFE impeller blade coating provided a cure for the service environment use.

## 5.12  Manufacturing Fabrication as a Source of Stress

The forensic investigator must recognise that all manufacturing and fabrication processes offer the possibility of introducing detrimental contaminant and/or stresses into a material or structure. However, casting and welding are considered to be the two most prevalent sources of manufactured-in defects. Therefore, these two processes will be singled out for specific attention.

### 5.12.1  Casting

Casting is a process used throughout the engineering and manufacturing industries to produce a wide variety of products, for example steel castings of several hundred tons that are used in power generation plants; vehicle engine blocks and cylinders; through small, precision castings, such as orthodontic components, to single crystal aerospace engine turbine blades. The reasons for such diverse use include

(a)  Casting can produce parts with a complex geometry, internal cavities and hollow sections.
(b)  It can be used to make small (few hundred grams) to very large size parts (thousands of kilograms).
(c)  It is economical, with very little wastage: the extra metal in each casting is re-melted and re-used.
(d)  Cast metal is isotropic – it has the same physical/mechanical properties along any direction.

As a production process, casting involves introducing liquid metal into a mould where it then solidifies into the required shape. It is then removed from the mould to reveal the cast object or component.

There are a variety of casting processes, including

• Non-permanent moulds, e.g. greensand, chemically bonded sand systems
• Permanent moulds/dies, e.g. die casting – gravity, low and high pressure
• Other processes, e.g. ingot, continuous, investment (lost wax), lost foam, shell, centrifugal and squeeze casting

Major production systems are based on metal mould processes, such as die, or permanent mould casting, which use either pressure or gravity feed in the casting process. Another metal mould process is centrifugal casting, which takes advantage of radial forces on the solidifying product to reduce casting porosity; it is, for example, used to produce pipes. More recently, technologies that encompass squeeze casting have taken advantage of mechanical forces on the solidifying metal. The semi-solid metal casting (SSM) process routes of thixo-casting, rheo-casting, thixo-forming or thixo-molding are near net shape processes utilised for the production of non-ferrous metal parts, where the process combines the advantages of both casting and forging. The process is named after the fluid property thixotropy, which is the phenomenon that allows this process to work. Simply, thixotropic fluids shear when the material flows, but thicken when standing.

Many of the casting processes shown in Table 5.5 can lead to a range of commonly encountered discontinuities. These defects are a problem if the

**Table 5.5  Advantages and Disadvantages Associated with a Range of Casting Routes.**

| Process | Advantages | Disadvantages | Examples |
|---|---|---|---|
| Sand | Wide range of metals, sizes, shapes, low cost | Poor finish, wide tolerance | Engine blocks, cylinder heads |
| Shell mould | Better accuracy, finish, higher production rate | Limited part size | Connecting rods, gear housings |
| Expendable pattern | Wide range of metals, sizes, shapes | Patterns have low strength | Cylinder heads, brake components |
| Plaster mould | Complex shapes, good surface finish | Non-ferrous metals, low production rate | Prototypes of mechanical parts |
| Ceramic mould | Complex shapes, high accuracy, good finish | Small sizes | Impellers, injection mould tooling |
| Investment | Complex shapes, excellent finish | Small parts, expensive | Jewellery |
| Permanent mould | Good finish, low porosity, high production rate | Costly mould, simpler shapes only | Gears, gear housings |
| Die | Excellent dimensional accuracy, high production rate | Costly dies, small parts, non-ferrous metals | Precision gears, camera bodies, car wheels |
| Centrifugal | Large cylindrical parts, good quality | Expensive, limited shapes | Pipes, boilers, flywheels |
| Semi-solid metal casting (SSM) | Complex parts produced net shape, porosity-free, pressure tightness, tight tolerances, thin walls | High cost of raw material, higher die development costs, operators require a higher level of training | MMC components, wheels, pistons, rocker arms, pump housings, engine brackets |

*Note:* Casting of Single Crystal Materials Is a New Technology That Has Become of Extreme Importance. Moulding and Casting Methods Used Include Crystal Growing by Techniques Such as Crystal Pulling, Zone Melting and Vapour Deposition. Applications Range from Gas Turbine Blades to Semiconductor Materials part with such a defect is subjected to cyclic loads during use. Under such conditions, the defects will act as stress raisers, initiating cracks, which then propagate under repeated stress, causing fatigue failure. In addition to acting as a stress concentrator, internal holes also reduce the load-bearing section, thus reducing the actual strength of the part below that of its expected strength of design. Typically common casting discontinuities that may be encountered include

- *Inclusions*: non-metallic particles in the metal matrix/lighter impurities appearing the casting surface are dross.
- *Porosity*: (blow holes, pinholes) and micro-porosity: a network of small voids distributed throughout the casting caused by localised solidification shrinkage of the final molten metal in a dendritic structure.
- *Distortion*: uneven shrinkage of casting. Also distortion can occur during annealing, stress relieving and high-temperature service.
- *Shrinkage/Piping*: a depression in the surface or an internal void in the casting caused by solidification shrinkage that restricts the amount of the molten metal available in the last region to freeze. It often occurs near the top of the casting in which case it is referred to as a pipe.
- *Cold Cracking*: cracks in cold or nearly cold metal due to excessive internal stress caused by contraction. Onset of cold cracking is often brought about when the mould is too hard or casting is of unsuitable design.
- *Hot Cracking (or hot tearing)*: occurs when the casting is restrained or in the early stages of cooling after solidification. The defect manifests as a separation of the metal (hence the terms tearing or cracking) at a point of high tensile stress caused by metal's inability to shrink naturally.
- *Cold Shuts*: occurs when two portions of the metal flow together, but there is lack of fusion between them due to premature freezing. Its causes are similar to those of miss-runs.
- *Miss-run*: is a casting that has solidified before completely filling the mould cavity.
    Typical causes include
  (1)  Fluidity of the molten metal is insufficient.
  (2)  Pouring temperature is too low.
  (3)  Pouring is done too slowly.
  (4)  Cross section of the mould cavity is too thin.
- *Surface Irregularities*: such as a blow hole, scar, blister, penetration, buckle.
- *Improper Composition*: alloy composition is often modified to enhance fluidity. However, any compositional change will also influence mechanical response of the material.
- *Microstructure*: cast structures – columnar, dendritic or equiaxed grains, eutectic phases; segregation; grain size.

Some of the above casting discontinuities or imperfections may have little or no effect on the function or service life of the cast component. Many imperfections are easily corrected by blast cleaning or grinding, with other imperfections often acceptable in some locations. For just this reason, it is not uncommon for the design engineer to segment a casting drawing. Depending on the criticality of the location, the same imperfection would be judged acceptable in one location while being totally unacceptable in another – a germane point that the failure detective should be both familiar and comfortable with.

## 5.12.2 Casting Failure Analysis

Failure of a casting can be due to a large range of various causes, as discussed above. Some castings may fail due to design shortcomings, while others fail due to casting imperfections. Improper loading or environment may also contribute to the cause of failure. The role of the failure investigator is primarily to determine the cause of casting failure and establish whether a casting imperfection was the primary or contributing cause of failure.

## 5.12.3 Case Study: Extrusion Press

In the following case study, casting defects were both extensive and apparent. However, the problem was not that the defects were unseen during manufacture, but more to the point the manufacturer appeared determined to simply ignore them. The machine in question was an extrusion press, where the first indication of a problem was the detection of cracks as shown in Figure 5.19.

Note the apparent excellent surface finish of the casting. It is the practice to use body filler on the surface of a casting to hide imperfections, in the same way that a car workshop will repair a body panel. Dismantling of the press revealed unpainted machined surfaces with defects that would have

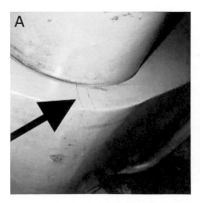

**Figure 5.19** (A) Cracking arrowed in the body casting of an extrusion press. (B) The body cracked open for inspection.

 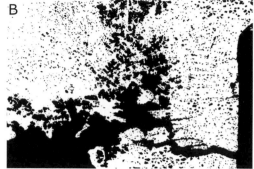

**Figure 5.20** (A) The casting contained a number of visual surface defects. (B) A mass of internal voiding with surface breaking cracks, as seen to the bottom right.

been very apparent to the manufacturing machinist. However, there had been no attempt at repair. When the casting was fractured open, as shown in Figure 5.20, it became apparent that fatigue cracks had initiated at a number of internal casting defects, as seen in Figure 5.20.

This case classically illustrates how human nature remains the same: not wanting to accept what they see in front of them, despite overwhelming evidence to the contrary.

## 5.12.4 Welding

Welding is a joining process in which joint production can be achieved with the use of high temperatures, high pressures or both (Table 5.6). It is often seen as the universal panacea of joining technology. However, welding is essentially a casting process and as such is vulnerable to all defects associated with casting (Figure 5.21; a key to this image may be found in Table 5.7). The

**Table 5.6  Classification of Welding Processes**

Fusion welding:
*Gas (Chemical)*
• Oxy-acetylene welding
*Electrical*
• Arc welding
Shielded metal arc welding (SMA);  Submerged arc welding (SA);  Metal inert gas welding (MIG); Tungsten inert gas welding (TIG)
• Resistance welding
Spot welding; Seam welding

Solid state welding:
Friction welding (FRW); Friction stir welding (FSW); Explosion welding (EXW); Forge Welding (FOW); Ultrasonic welding (USW); Cold Welding (CW); Pressure Welding (PW); Roll Welding (ROW); Diffusion Welding (DFW)

Other types of welding:
Electron beam welding (EBW); Laser beam welding (LBW); Percussion welding (PW); Thermite welding (TW)

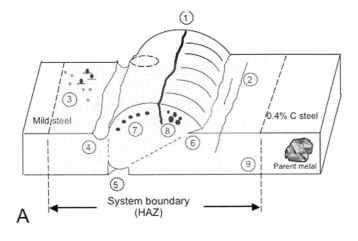

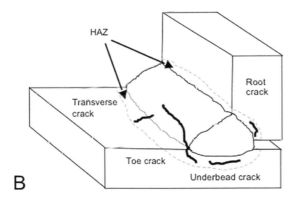

**Figure 5.21** (A) Schematic collage of possible welding defects from manual metal arc (stick) welding of mild steel to plain carbon steel. (B) Possible hydrogen cracking in weldment.

**Table 5.7   Key to Figure 5.21**

| | |
|---|---|
| 1. | Centreline cracking – too high tensile stresses on cooling |
| 2. | HAZ cracking – formation of (brittle) martensite on cooling of austenite |
| 3. | Spatter and arc strikes – surface defects that are potential fatigue initiation sites |
| 4. | Undercutting and root concavity – welding power too high |
| 5. | Lack of penetration – welding power too low or welding speed too slow |
| 6. | Overlap – welding speed too high or welding power too low |
| 7. | Slag entrapment – ineffective de-slagging between weld runs |
| 8. | Gas porosity – damp electrodes or other contamination |
| 9. | Problematic microstructure in HAZ – grain growth, possible martensitic formation |
| 10. | Hydrogen (embrittlement) cracking – originate in HAZ (Figure 5.21B), may extend into weld metal – hydrogen from damp electrodes, cleaning fluids, rust, etc. |

intense heat involved in the welding process influences the microstructure of both the weld metal and the parent metal close to the fusion boundary (the boundary between solid and liquid metal – Figure 5.22A). As such, the welding cycle influences the mechanical properties of the joint.[26]

The molten weld pool is rapidly cooled, as the metals being joined act as an efficient heat sink. This cooling results in the weld metal having a chill cast microstructure. In the welding of structural steels, the weld filler metal does not usually have the same composition as the parent metal. If the compositions were the same, the rapid cooling could result in hard and brittle phases, such as martensite, forming in the weld metal microstructure. This problem is avoided by using weld filler metals with a lower carbon content than the parent metal.

The parent metal close to the molten weld pool is heated rapidly to a temperature that will be dependent on the distance from the fusion boundary. Close to the fusion boundary, peak temperatures near the melting point are reached, whilst material only a few millimetres away may only reach a few hundred degrees Celsius. The parent material close to the fusion boundary is heated into the austenite phase field. On cooling, this region will transform into a microstructure that is different from the rest of the parent material. In this region, the cooling rate is usually rapid, and hence there is a tendency towards the formation of low-temperature transformation structures, such as bainite and martensite, which are harder and more brittle than the bulk

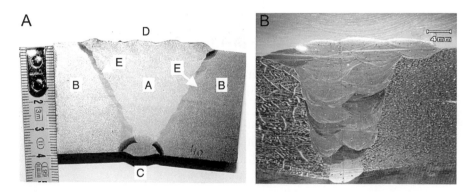

**Figure 5.22** (A) Low magnification view of a weld cross section. Key: (a) weld metal; (b) patent metal; (c) root weld; (d) free surface; (e) heat-affected zone. (Courtesy of Wikimedia Commons, public domain.) (B) Microhardness survey performed across a weld, running through the HAZ and into the base metal. In this particular instance, hardness profiles indicated a brittle HAZ (above 50 HRC). Note: the micro indenter pattern. (Courtesy of University of Cambridge DoITPoMS Micrograph Library, licensed under Creative Commons.)

of the parent metal (Figure 5.22B). This is the region that is termed the heat-affected zone (HAZ).

The microstructure of the HAZ is influenced by three factors:

(1) The chemical composition of the parent metal
(2) The heat input rate during welding
(3) The cooling rate in the HAZ after welding

The chemical composition of the parent metal is important as it determines the hardenability of the HAZ. The heat input rate is significant since it directly affects the grain size in the HAZ. The longer the time spent above the grain-coarsening temperature of the parent metal during welding, the coarser the structure in the HAZ. Generally, a high heat input rate leads to a longer thermal cycle and thus a coarser HAZ microstructure. It should be noted that the heat input rate also affects the cooling rate in the HAZ. As a general rule, the higher the heat input rate the lower the cooling rate. The value of heat input rate is a function of the welding process parameters: arc voltage, arc current and welding speed. In addition to heat input rate, the cooling rate in the HAZ is influenced by two other factors. First, the joint design and thickness are important since they determine the rate of heat flow away from the weld during cooling. Second, the temperature of the parts being joined, i.e. any pre-heat, is significant, as it will determine the temperature gradient which will exist between the weld and parent metal.

In conclusion, although welding is often considered a prime fabrication route, it is not the universal panacea of joining technology. Welding is essentially a casting process, carrying with it a distinct possibility of incorporating any defect associated with the casting process. Therefore, as a joining method, welding has a clear potential to introduce a vast range of defects into a structure – each of which will carry the potential to initiate premature demise.[26]

## 5.12.5  Quench Cracking

An additional recurring phenomenon that is encountered time and again by the forensic or failure investigator is that of quench cracking. A prime reason for the popularity of ferrous structural materials (steels) is the ability to tailor their mechanical properties by simple thermal treatment. This operation consists of heating the metal to a high temperature (austenitising) for a specified time to complete transformation to austenite and diffusion of constituents and then cooling in a quenching medium to produce the desired microstructure and as-quenched hardness. The objective of quenching the steel from the austenite phase field to room temperature is to produce martensite, a hard, exceedingly brittle and highly stressed material structure. However, this hardening treatment is most often followed by a lower-temperature heating process (tempering) for stress relieving the microstructure to achieve the desired physical property of toughness.

The transformation process of austenite to martensite produces a microstructure in which the Fe atoms have re-arranged themselves into a bcc crystal structure, but the C atoms are trapped in interstitial sites. The crystal structure transformation occurs so that the C atoms stretch the crystal lattice preferentially in one direction, producing a body-centred tetragonal crystal structure. Therefore, the cooling rate must be fast, but not too fast. If the cooling rate is not fast enough, the C atoms have time to diffuse. Given this opportunity, they will agglomerate to form $Fe_3C$ particles, and the steel will transform to ferrite + cementite, instead of to martensite. On the other hand, the transformation to martensite produces high internal stresses; the stretching of the crystal lattice is one manifestation of these stresses. If the transformation occurs too quickly, the internal stresses may build to the point where the steel cracks. This phenomenon is particularly insidious if the cracks are completely contained on the inside of a component. From the outside, the component may look absolutely fine. However, when a load is applied to the component, the internal crack can grow and cause catastrophic failure.

To avoid such quench cracking, the rate of quench must be controlled to allow the material time to adjust to the new crystal structure. This explanation is somewhat simplistic, but it conveys the idea of how quench cracking is avoided by slowing the quench rate. The quench rate can be adjusted by changing the viscosity of the coolant: higher viscosity reduces the rate at which heat is removed from the steel, with oil as a commonly used quenching medium. However, to date, environmental concerns have dictated a movement toward a solution of polymer molecules in water as a more environmentally acceptable quench medium.

As an example of quench cracking, a pin punch was being used by a DIY enthusiast.[26] It was one of a set of punches that he had purchased from a local market. He was driving out corroded bolts on some equipment that he was attempting to repair. As he was hitting the striking end of the punch with a ball peen hammer, the end of the punch shattered, causing splinters to embed in his cheek. Although not seriously hurt, he was shocked that the punch had failed in such a way when being (correctly) hit on its striking end. A polished transverse section through the striking end revealed large quench cracks (Figure 5.23). The presence of quench cracks was indicative of poor thermal control, with the punches being quenched at a rate that was far too severe for the material. Furthermore, microstructural observation showed that the punches had not received a tempering treatment, having excessive hardness at the striking face and not having had 'toughness' imparted by the tempering treatment.

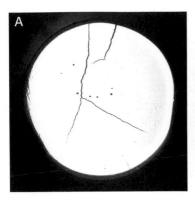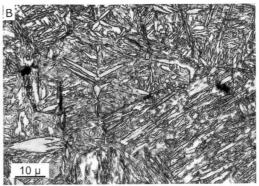

**Figure 5.23** (A) Polished transverse section showing a number of quench cracks through the striking end of the pin punch. (B) Martensitic microstructure of the punch striking face.

## 5.12.6 Tempering and Toughening

As discussed above, martensite in a quenched tool steel is exceedingly brittle and highly stressed. Consequently cracking and distortion of the object are liable to occur after quenching. Furthermore any retained austenite is unstable and, as it changes, physical product dimensions may alter, e.g. die dimensions may change in excess of 0.012 mm. Therefore, it becomes a necessity to re-heat the steel below the critical range, in order to relieve stresses and to allow the arrested reaction of cementite precipitation to take place. This is the process of tempering, and the steps involved are as follows:

- 150–250°C. The object is heated in an oil bath, immediately after quenching, to prevent related cracking, to relieve internal stress and to decompose austenite without much softening.
- 200–450°C. Used to toughen the steel at the expense of hardness. Brinell hardness is 350–450.
- 450–700°C. The precipitated cementite coalesces into larger masses and the steel becomes softer. The structure is known as sorbite, which at the higher temperatures becomes coarsely spheroidised. It etches more slowly than troostite and has a Brinell hardness of 220–350. Sorbite is commonly found in heat-treated constructional steels, such as axles, shafts and crankshafts subjected to dynamic stresses. A treatment of quenching and tempering in this temperature range is frequently referred to as toughening, and it produces an increase in the ratio of the elastic limit to the ultimate tensile strength.

Tempering reactions occur slowly, with reaction time as well as the temperature of heating being of importance. Increasingly, the process of tempering is performed under pyrometric control in oil, salt (e.g. equal parts sodium

**Table 5.8    (A) Tempering Colour Series. (B) Tempering Temperatures for Some Engineering Artefacts**

(A)

| English Colour Text | (°F) | Colour | (°C) |
|---|---|---|---|
| Clear – as fully hardened | 100 | | 38 |
| Pale yellow | 420 | | 216 |
| Very pale yellow | 430 | | 221 |
| Light yellow, straw | 440 | | 227 |
| Pale straw-yellow, straw | 450 | | 232 |
| Straw-yellow | 460 | | 238 |
| Deep straw-yellow | 470 | | 243 |
| Dark yellow, light orange | 480 | | 249 |
| Yellow-brown, orange | 490 | | 254 |
| Brown-yellow, bronze | 500 | | 260 |
| Spotted red-brown, dark brown | 510 | | 266 |
| Brown with purple spots | 520 | | 271 |
| Light purple, purple | 530 | | 277 |
| Full purple, purple | 540 | | 282 |
| Dark purple | 550 | | 288 |
| Full blue | 560 | | 293 |
| Dark blue, blue | 570 | | 299 |
| Dark blue | 590 | | 310 |
| Pale blue | 600 | | 315 |
| Light blue | 610 | | 321 |

*(Continued)*

**Table 5.8 (Continued)   (A) Tempering Colour Series. (B) Tempering Temperatures for Some Engineering Artefacts**

(B)

| Temperature °C | Engineering Artefacts |
| --- | --- |
| 230 | Planing and slotting tools |
| 240 | Milling cutters, drills |
| 250 | Taps, shear blades for metals |
| 260 | Punches, cups, snaps, twist drills, reamers |
| 270 | Press tools, axes |
| 280 | Cold chisels, setts for steel |
| 300 | Saws for wood, springs |
| 450–650 | Toughening for constructional steels |

and potassium nitrates for 200–600°C) or lead baths and also in furnaces in which the air is circulated by fans. After tempering, the objects may be cooled either rapidly or slowly, apart from steels susceptible to temper brittleness.

Temper colours formed on a cleaned surface are still used occasionally as a guide to temperature. They exist due to the interference effects of thin films of oxide formed during tempering, and they act similarly to oil films on water. Alloys such as stainless steel form thinner films than do carbon steels for a given temperature and hence produce a colour lower in the series. For example, pale straw corresponds to 300°C, instead of 230°C (Table 5.8).

For turning, planing, shaping tools and chisels, only the cutting parts are required to be hardened. This is often undertaken in the engineering workshop by heating the tool to 730°C, followed by quenching the cutting end vertically. When the cutting end cools, it is cleaned with the stone and the heat from the shank of the tool is allowed to temper the cutting edge to the correct colour. At this point, the whole tool will be finally quenched. As a matter of normal practice, surface oxidation will be reduced to a minimum, by coating the tool with charcoal and oil.

# References

1. Dowling, N., *Mechanical Behavior of Materials*. 4th ed. 1990, Pearson, ISBN-10: 0131395068, ISBN-13: 978-0131395060.
2. Hibbeler, R.C., *Mechanics of Engineering Materials*. SI ed. 2013, Pearson, ISBN-10: 9810694369, ISBN-13: 978-9810694364.
3. Ashby, M. and D. Jones, *Engineering Materials 1: An Introduction to their Properties & Application*. 2011, Butterworth-Heinemann.
4. Ashby, M. and D. Jones, *Engineering Materials 2, Third Edition: An Introduction to Microstructures, Processing and Design*. International Series on Materials Science and Technology. 2005, Butterworth-Heinemann.

5. Murakami, Y., *Stress Intensity Factors Handbook*. 1986, Pergamon Press.
6. Rooke, D. and D.J. Cartwright, *Compendium of Stress Intensity Factors*. 1976, London: HMSO.
7. Roark, R., *Formulas for Stress and Strain*. 8th ed. 2012, McGraw-Hill Book Company.
8. Jones, D., *Failure Analysis Case Studies II*. 1st ed. 2013, Elsevier, ISBN0080545556, 9780080545554.
9. Jones, D., *Failure Analysis Case Studies III*. 1st ed. 5 May 2004, Elsevier Science, ISBN: 0080444474, 9780080444475.
10. Moalli, J., *Plastics Failure Analysis and Prevention*. 2002, William Andrew Publishers.
11. Becker, W.T. and R. Shipley, *ASM Handbook: Volume 11: Failure Analysis and Prevention*. 10th ed. 1 July 2002, ASM International.
12. Pantelakis, S. and C. Rodopoulos, Engineering against fracture, in *1st Conference on Engineering Against Fracture*. 2008, Patras, Greece: Springer Science and Business Media.
13. Hendley, T. Vicky, *The Importance of Failure*. http://www.prism-magazine.org/october/html/the_importance_of_failure.htm.
14. Lewis, P. and C. Gagg, *Forensic Polymer Engineering: Why Polymer Products Fail in Service*. 1st ed. 2010, Woodhead Publishing Limited, ISBN: 9781845691851.
15. Gagg, C.R. and P.R. Lewis, In-service fatigue failure of engineered products and structures. *Engineering Failure Analysis*, 2009. 16(6): pp. 1775–1793.
16. Plumbridge, W., Solders as high temperature engineering materials. *Materials at High Temperatures*, 2000. 17(3): pp. 381–387.
17. https://corrosion-doctors.org/Failure-Analysis/10-physical-testing.htm.
18. Jones, M.H. and D. Scott, *Industrial Tribology: The Practical Aspects of Friction, Lubrication, and Wear*. 1983, New York: Elsevier Scientific Publishing Company.
19. Gagg, C.R. and P.R. Lewis, Wear as a product failure mechanism – Overview and case studies. *Engineering Failure Analysis*, 2007. 14(8): pp. 1618–1640.
20. Hutchings, I. and P. Shipway, *Tribology*. 2nd ed. 2017, Butterworth-Heinemann, ISBN: 9780081009109.
21. Talbot, D.E.J. and J.D.R. Talbot, *Corrosion Science and Technology*. 2nd ed. 2018, CRC Press, ISBN 9781498752411.
22. Gagg, C.R. and P.R. Lewis, Environmentally assisted product failure – Synopsis and case study compendium. *Engineering Failure Analysis*, 2008. 15(5): pp. 505–520.
23. Jones, D., *Principles and Prevention of Corrosion*. 2nd ed. 2013, Pearson, ISBN13: 9781292042558, ISBN10: 1292042559.
24. Raja, V.S. and T. Shoji, *Stress-Corrosion Cracking*. 2011, Woodhead, ISBN: 9780857093769.
25. Gagg, C.R., P.R. Lewis and C. Tsang, Premature failure of a vacuum pump impeller rotor recovered from a pitch impregnation plant. *Engineering Failure Analysis*, 2008. 15(5): pp. 606–615.
26. Gagg, C.R., Failure of components and products by 'engineered-in' defects: Case studies. *Engineering Failure Analysis*, 2005. 12(6): pp. 1000–1026.

# Fracture Morphology and Topography

# 6

## 6.0 Introduction

No police detective or criminologist would consider leaving the scene of a crime without a careful inspection for fingerprints. Their discovery and subsequent identification will often lead the investigator to the culprit and/ or to a solution of the crime. Similarly, the detailed markings on a fracture surface will often provide the forensic metallurgist or failure analyst with important clues as to why a system, structure or component failed. Each individual failure mechanism described in Chapter 5 will give rise to unique fracture morphology and fracture surface topography that can be considered a fingerprint of failure. Therefore, a clear understanding of the relationship between failure mechanism and fracture morphology/ topography will allow characterisation of the range of probable conditions at play at the instant of failure. Fractography is the tool that will provide an insight into the inter-relationship between failure mechanism, fracture morphology and mechanical characteristics. Fractography contains a basic premise that the topography and morphology of a growing crack are characteristic of the material of construction (microstructure), its service loading/operating conditions and environment. So, by observing, measuring and interpreting the shape of a failure and its fracture surface topography, it is possible to determine material features and the mechanics of crack growth. Consequently, it can be considered a principal tool to support the development of a probable chain of events leading to failure, and ultimately to the root cause of failure. Therefore, this chapter will consider a range of common failure mechanisms found in service (introduced and reviewed in Chapter 5), along with identification and interrelationship of associated fracture morphologies and their correlation to mechanical response of the material, component or system.

## 6.1 What Is Meant by the Term 'Failure'?

Failure can be defined by one of two categories, either: (1) *physical fracture* or (2) *failure to meet design criteria*.[1] Considering each in turn:

(1) *Physical fracture*: when a component, machine or a process fails and grinds to a halt or, more worryingly, breaks suddenly and without warning (catastrophic failure). Failure of a component, such as a bearing, crankshaft or bracket, is usually obvious. However, the concept of machine or system failure can range over something as complex as an aircraft or as simple as a ladder. The difficulty in determining the root cause of failure will increase with the complexity of the component or machine, although in most situations the root cause can usually be determined. The failure of processes such as heat treatment, coating, plating or welding can be more complex, and it naturally follows that root cause determination will also be more complex. Therefore, it also follows that any cause-and-effect relationship for processes (Section 3.2.1) may, also, not be readily apparent.

(2) *Failure to meet design criteria*: when a component, machine or a process fails to achieve performance criteria such as life, operating limits and specification requirements. Consumers have an expectation for the life of certain items, such as car engines and tyres, household appliances, biotechnical implants, etc. These lifetime criteria may not be given in writing by the manufacturer but, if they fail prematurely, they are often deliberated in the courtroom. More specific are operating limits, such as fuel consumption, production scrap rate, computer chip speed or acquisition frequency. These items are usually well-defined, and in most cases, mutually agreed upon by all parties involved. Finally, specification requirements, such as mechanical properties, plating thickness, weight and coating optical properties, will be well-defined in written documentation (product specification). Specific definitions or limits are usually demanded/defined by the customer, with identifiable testing being required to ensure compliance.

It is recognised that all materials have limits as to what they are able to withstand in terms of forces that can be applied and environmental conditions under which they can safely operate. Designers and manufacturers strive to ensure their products meet any appropriate materials and engineering specifications, aiming for an acceptably long, trouble-free service lifetime in a particular application. That is to say, *no* product designer or manufacturer would deliberately produce a product that would fail prematurely in service. However, no material is universally suited for every type of engineered product, and no product is unbreakable or will not fail in some way if the service conditions are hostile. When a product does not achieve the expected level of performance it is said to have 'failed'. That does not necessarily mean that it has broken but simply that it may have worn at an unusually high rate, suffered debilitating environmental attack or degradation, or succumbed to an issue that would prejudice its future performance if it had been allowed to remain in service.

When an engineering system or component is exposed to service loading, it will distort under that load. Within a sound design, service loading should not be excessive, therefore the working stress will not exceed the material yield point, i.e. the part will deform elastically so that when the load is removed the part will return to its original shape. However, if a working load takes the material past its yield strength, the part will permanently deform. An even greater increase in that working load will cause the component or system to crack or fail. The type of applied loading can be tensile in nature, compressive, torsion or impact, with the force being applied over short or long time spans, and at varying temperatures and/or environmental conditions.[2] Lange[3] introduced a systematic classification of cracks and fracture, as shown in Table 6.1. Just a cursory glance will give an impression of the many different aspects of fracture and of the difficulties in identifying a fracture mode. As a matter of fact, the table itself is somewhat limited as a fracture mode identification tool (in-depth failure mode identification charts are presented later within this chapter as Tables 6.4 and 6.5). However, aspects covered in the table most certainly suggest that it is the metallurgist

**Table 6.1   Classification of Cracks and Fracture**

| Mechanically Caused Cracks and Fracture | Cracks and Fracture Caused by Corrosion | Thermally Induced Cracks and Fracture |
|---|---|---|
| 1. Ductile fracture | Intercrystalline corrosion | Creep cracks |
| Dimples | Intercrystalline stress corrosion cracking | Weld cracks |
| Transcrystalline dimple fracture | | |
| Intercrystalline dimple fracture | Anodic stress corrosion cracking | Hot cracks |
| | Hydrogen-induced stress corrosion cracking | Heat treatment cracks |
| 2. Cleavage fracture | Hydrogen-induced cracks and fracture | Grinding cracks |
| Transcrystalline cleavage fracture | Corrosion fatigue | Thermal shock cracks |
| Intercrystalline cleavage fracture | Soldering fracture | |
| 3. Mixed mode fracture | | |
| 4. Fatigue fracture | | |

*Source:* after Lange[3].

*Note:* Lange does not distinguish between brittle and ductile fracture. Moreover, his classification chart presents a somewhat limited picture of fracture mode identification. To appreciate the degree of limitation, this table should be compared with Tables 6.4 and 6.5.

who can provide important clues to the cause of failure, as he will be able to determine a failure mechanism by fractographic and metallographic investigation. Developed in the 16th century as a quality control practice employed for ferrous and nonferrous metal working,[4] fractography is the tool that will provide an insight to the inter-relationship between failure mechanism, fracture morphology and mechanical characteristics. Awareness of material structure, inherent defects and mechanical or thermal history can be gained by metallographic application. Therefore, the appearance of the fracture and the classification of cracks and fractures hold the potential to yield significant information for accurate deduction of both the failure mode at play, and the cause/s of failure initiation.

## 6.2 Fracture Path and Fracture Morphology

Any engineered system or individual component can fail from the application of a single instantaneous overload force or by a progressive mode such as fatigue, corrosion, wear, etc. Instantaneous loading will be considered initially, with progressive failure modes being considered in the latter stages of this chapter.

Instantaneous overloading of either ductile or brittle materials will generate specific fracture surface characteristics. Furthermore, particular load application will produce characteristic fracture path morphologies (shape), a range of which are shown in the schematic of Figure 6.1 and pictorially in Figure 6.2. As clearly illustrated, ductile and brittle materials behave differently when subjected to the same type of load application (stress), namely the following:

*Tension*: ductile metals in pure tension will undergo lateral contraction, necking down as the material of construction passes its yield strength and

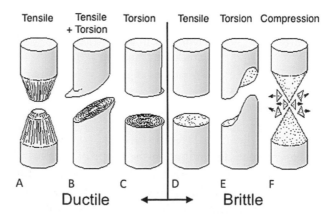

**Figure 6.1** Schematic of failure morphologies produced by individual loading application, and for ductile or brittle material.

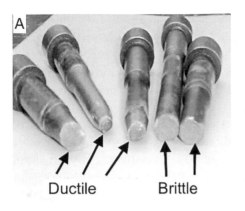

**Figure 6.2** Actual specimens or components having failed under different load conditions: (A) tensile and compressive. Cap head bolts failed by tensile and compressive overload. (B) Buckling. There are a multitude of issues that influence track buckling, but the prime factors at play are the track longitudinal and lateral resistance. (Courtesy of Wikimedia Commons, licensed under Creative Commons.) (C) Bending. A telescopic hydraulic lift ram failed by bending when its load shifted sideways whilst the ram was fully extended. (D) Torsion. Failure of an F-18 engine shaft. (A, D courtesy of Dudley Knox Library, Naval Postgraduate School, Monterey, CA.

deforms. Microscopic voids between the grain structure link up and enlarge as the metal begins to tear apart. Finally, the edges tear away at a 45° angle to the load. This sharp edge or shear lip is a classic indicator of where the part ultimately failed. The two fracture pieces mate together in a cup-and-cone form. By contrast, brittle metals under tension will fracture perpendicular to the applied load. There will be little or no obvious deformation, and the part will have a clean, smooth break.

*Compression*: ductile metals overloaded in compression deform in a fairly predictable manner, depending on the shape of the part. A cylinder, for instance, will 'barrel out' in compression, while a thin-walled tube or strut will buckle. It is safe to say that a ductile metal will not generally crack in compression; it will simply continue to deform to a point where the ductility is exhausted. At this point, brittle cracking can be expected. On the other hand, brittle materials will fail in a pattern perpendicular to the load. Using the same example of a cylinder, a compression load will create force on the outside surface of the part. As the material will not easily deform, it will maintain or carry the load up to failure, whereupon it will split explosively along its length.

*Torsion*: ductile metals yield to severe twisting in the direction of force. The part continues to twist until it finally wrings off across the axis of torsion. Compared to the cup-and-cone fracture seen in tension, this torsional break will appear relatively fat, with concentric rings and a central pip. Brittle metals in torsion will exhibit the same resistance to deformation. However, the part will withstand the twisting energy until it finally fractures in a rapid manner, following a helical path that appears to be 45° to the axis of torsion. However, although the material fails perpendicular to the applied loading, the fracture path twists in a direction opposite to that which it was loaded.

The macro- and micro-scale appearances of fracture surface features will provide an additional narrative of how and (often) why fracture occurred. Content of this 'failure' story can be collated and presented in terms of the following fracture features and information present on a fracture surface:

- Crack initiation site
- Crack propagation direction and path (Figure 6.1)
- Crack mechanism
- Load conditions at play (tensile, compressive, buckling, torsion, shear and monotonic or cyclic load application)
- Environmental influence
- Influence of geometric constraints on crack initiation and/or crack propagation
- Influence of engineered-in defects (fabrication imperfections) on crack initiation and/or propagation

From the forensic detective viewpoint, it cannot be emphasised strongly enough that any fracture surface will possess features that are a fingerprint of failure. In practice, any classification of fracture is made on the basis of fracture surface appearance, the nature of fracture path and the amount of plastic deformation prior to failure (with any environmental considerations also playing their part). Therefore, the forms of various typical fracture surfaces will be presented and discussed, to illustrate a range of general fracture characteristics.

## 6.3 Ductile or Brittle Failure

In terms of macroscopic behaviour, metallic failure can be ductile, brittle or a combination of the two, as shown in the schematic of Figure 6.1. Ductile fracture will occur as a result of nucleation, growth and coalescence of voids in the material. This mode of failure is controlled by the rate of nucleation of the voids, their rate of growth and the mechanism of coalescence. Therefore, a ductile failure is one where there is substantial distortion or plastic deformation of the failed part. Normally, a component will fail in a ductile manner when it plastically deforms, and the steadily reducing cross section can no longer carry the applied service load.

Ductile failure can be identified from the following physical features:

- The high degree of deformation and distortion that will be present around the fracture zone
- Tearing of metal accompanied by appreciable gross plastic deformation
- Grey or fibrous appearance on fracture surface
- Exhibits necking – cup-and-cone formation
- Micro-void formation and its coalescence forming a dimpled fracture surface structure (Figure 6.3A)

Materials that fail in a non-ductile manner are extremely susceptible to the presence of defects or cracks. As there will be no plastic deformation, stresses concentrate at the tips of pre-existing surface or interior flaws. These defects act as crack initiators (nuclei), and when stress levels attain a critical value, crack propagation will result. Failure is therefore caused by the propagation of a crack as distinct from ductile yielding.

With any brittle fracture, a crack will form:

- With little or no necking.
- Without gross plastic deformation.
- Giving a bright and granular appearance on fracture surface.
- And will take either an intergranular or trans-granular path through the material structure (Figure 6.3b).

## 6.4 Brittle Overload (Fast) Fracture Features

With any brittle overload failure, separation of the two fracture halves is not quite instantaneous. However, the crack front proceeds through the load-bearing section at nearly the speed of sound in the material. A crack will initiate at a point of maximum stress, and will then propagate across the section by cleavage (Figure 6.4A) of individual material grains, or along grain boundaries (Figure 6.4B). One result of this failure mechanism is that the direction of the fracture path will frequently be indicated by chevron marks (sometimes called river markings) that point toward the origin of the failure. Figure 6.5A shows a photograph of a section of tube drawing die that exploded when it

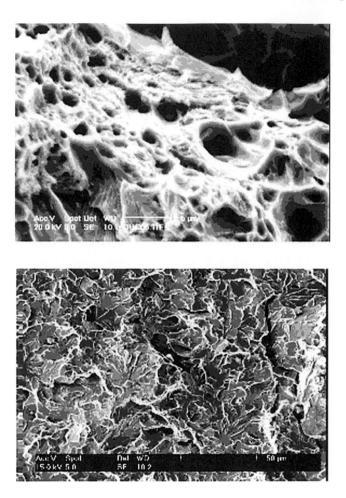

**Figure 6.3** Scanning electron micrographs showing fracture surface features of (A) a ductile failure. (B) A brittle failure.

wandered out of its bolster confines during a drawing operation (bolts holding a retaining ring had failed, thus initiating the train of events). Rapidly accelerating hoop stresses almost immediately exceeded the strength of the die material, and explosive failure instantly ensued. As a result of tracking the chevron marks back to the fracture origin, it was possible to determine that there was no inherent metallurgical problem associated with the actual die material. Further examples of chevron (or river) markings are shown in Figure 6.5.

## 6.5 Ductile/Brittle Transition

The possibility of ductile/brittle material transition has been discussed previously in Section 5.1.2. However, a word of caution has to be raised at this

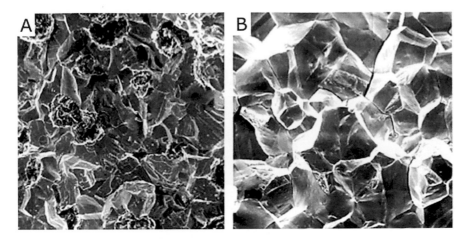

**Figure 6.4** Crack propagation in brittle fracture can be either: (A) trans-granular (through the grains) also termed cleavage, or (B) intergranular: along the grain boundaries. In both cases, the surface usually appears shiny, as the facets reflect light.

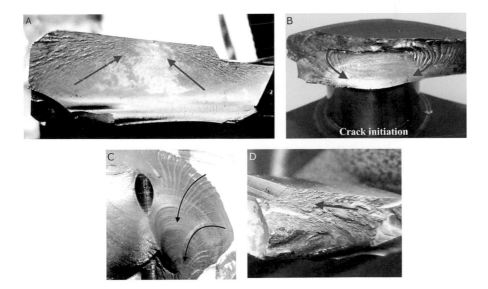

**Figure 6.5** (A) The presence of chevron (or river) markings on the fracture surface of a tube drawing die where a crack had initiated and propagated in a series of fan-shaped ridges (termed 'chevron' or 'river' markings) that are typical of fast (brittle) fracture. (B–D) Chevron markings on fracture surfaces of three different components, each of which failed in a brittle (fast) manner. Note: chevron markings point back towards the crack origin, therefore indicating the point of crack initiation.

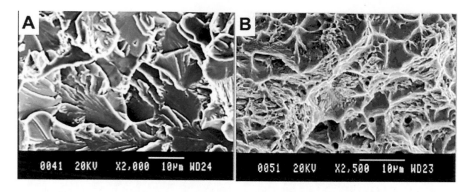

**Figure 6.6** (A) Charpy-impact fracture surfaces of plain medium carbon steel (0.4%C). Low-temperature fracture showing cleavage – twist and tilt grain boundaries evident. (B) Room-temperature fracture showing micro-voidal coalescence (MVC) and regions of brittle fracture.

point: the type of fracture – ductile or brittle – should be compared with the expected nature of the material. There are circumstances where brittle fractures can and do appear in normally ductile materials. This ductile/brittle transition situation can indicate either

- A rapid loading event, where a dynamic (severe shock) load on the most ductile material or component can cause it to fracture like glass.
- Or that a change has occurred in the material – such as low-temperature embrittlement – so that the material will no longer act in a ductile manner.

This ductile/brittle transition has been attributed as the cause for numerous dramatic and catastrophic past failures, e.g. the rupture of a 2.3-million-gallon molasses storage tank in the winter of 1911, World War II Liberty Ships breaking in half, etc. Figure 6.6 presents side-by-side micrographs of brittle and combined ductile/brittle features, for comparison purposes.

In summary, ductile and brittle fractures can often be determined by visual observation for evidence of the degree of deformation at and around the fracture site. However, macro-scale observations alone may not convey the full story surrounding fracture. Deducing both the macro- and micro-scale appearance and mechanism of fracture is often a prime requirement for a more complete understanding of failure. Furthermore, a complete macro- and micro-scale understanding of fracture will allow development of a robust failure hypothesis.

## 6.6 Fracture Surface Features

As previously suggested, simple observation of the face of any fracture can and will provide a plethora of information regarding the cause or causes of that failure. Information that can be gleaned from a fracture face 'per se' will include

- Type and direction of force(s) acting on the component
- Magnitude and fluctuations of the imposed forces

Furthermore, fracture observation may also provide a general indication of the length of time from crack initiation to final failure (Section 5.9.4).

No police detective or criminologist would consider leaving the scene of a crime without a careful inspection for fingerprints. Their discovery and subsequent identification will often lead the investigator to the culprit and/ or to a solution of the crime. Similarly, the detailed markings on a fracture surface will often provide the forensic metallurgist or failure analyst with important clues as to why a system, structure or component failed as demonstrated by the following example:

Fatigue failure of a shaft will almost always start at a stress concentration on the periphery (outside) of the shaft, as the local stress will be maximised at that point (Figure 6.7). However, a (hatched) instantaneous fracture zone, present on the fracture face, will be carrying the load at the instant before the part breaks. Therefore, by observing the size of the instantaneous zone,

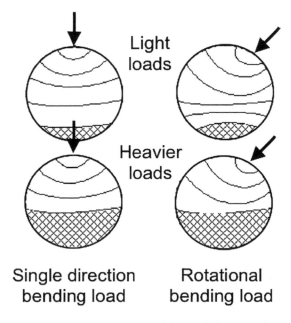

**Figure 6.7** Schematic diagram of fracture surfaces of shafting, showing features associated with light or heavy loading, with load application either unidirectional or rotational. Note: uniform progression marks in the uni-directionally loaded shaft, whereas progression marks are distorted by rotational loading.

it is possible to gauge the magnitude of the load on the part. Furthermore, the fracture path taken through the cross section will provide an indication of operational loading at failure. An example of fatigue failure in a shaft is shown graphically in Figure 6.7. Here a comparison between a lightly and a heavily loaded shaft for both plain bending and rotational bending is given, with black arrows indicating the initiation point on each shaft. This example clearly demonstrates the similarity between fracture features and fingerprints in a criminal investigation.

## 6.7 Fatigue Fracture

As discussed in Section 5.9, fatigue is a progressive localised permanent structural change that occurs in a material subjected to repeated or fluctuating stresses well below the ultimate tensile strength (UTS).[5] Fatigue fractures are caused by the simultaneous action of cyclically applied stress, tensile stress, and plastic strain, all three of which must be present to initiate a fatigue crack in the first instant. The cyclic nature of loading can instigate initiation and growth of a crack, and, ultimately, when crack growth is significant enough such that the remaining cross-sectional area can no longer support service loading, the material fractures. Final sudden failure of the remaining cross section will occur either by ductile overload, shear or brittle fracture.[5] What must be emphasised is that the actual circumstances of use when the final break takes place are of no consequence. All that is needed to cause final separation is one last application of a normal service load. The presence of beach markings or striations on the fracture surface is the classic sign of fatigue fracture. Beach markings are macroscopic fatigue features marking an interruption (a dwell) of some form in the fatigue cracking process. However, they do not delineate individual cycles of crack growth, and the absence of such markings does not necessarily mean that fatigue was not at play.[5] Conversely, striations are semi-elliptical (conchoidal) lines on a fatigue fracture surface that emanate outward from the origin and delineate the crack-front position with each consecutive stress cycle, as shown by the photographs of Figure 6.8 and by the schematic of Figure 6.9. Striations are discriminated from striation-like features on the fracture surface in that true fatigue striations never cross or intersect one another. Beach marks (also sometimes clamshell or tide marks) are macroscopic fatigue features marking an interruption (of some form) in the fatigue crack propagation (growth) progress. However, both features are used to identify fatigue as a fracture mechanism. The spacing of fatigue striations is often uniform and, if the cyclic stress frequency is known, it can be used to calculate the fatigue crack growth rate. Striations can be discriminated from striation-like artefacts on the fracture surface, in that true fatigue striations never cross or intersect one another.[5]

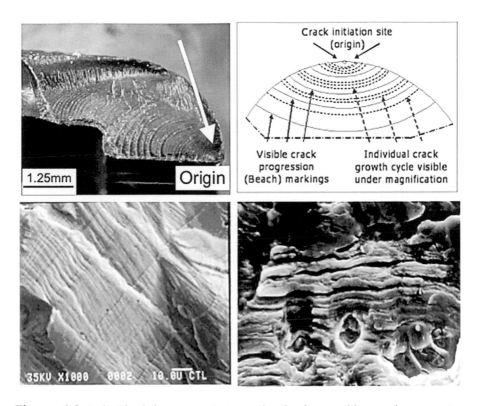

Crack initiation site
(origin)

Visible crack                    Individual crack
progression                      growth cycle visible
(Beach) markings                 under magnification

1.25mm                                      Origin

35KU X1000    0002    10.0U CTL

**Figure 6.8** Individual fatigue striations clearly discernible on the scanning electron micrographs.

## 6.7.1 Low- and High-Cycle Fatigue

There are several different modes of fatigue that include both high-cycle fatigue and low-cycle fatigue. However, while of utmost importance to the forensic investigator, other modes such as thermal fatigue, surface fatigue, impact fatigue, corrosion fatigue and fretting fatigue, will not be considered at this point. Low-cycle fatigue cracking will initiate when subjected to high-strain amplitude loading conditions, with final fracture occurring in less than $10^4$ cycles. Conversely, it is low-strain amplitudes that will drive the process of high-cycle fatigue, with final fracture occurring after a minimum of $10^6$ (millions) of load cycles, or greater. When considering high-cycle fatigue, striations will rarely be visible to the naked eye. Observation would be undertaken under magnification, generally on a scanning electron microscope. A point to be made here is that, if not fully aware, the investigator may confuse beach markings with fatigue striations. Under low-cycle fatigue, striations may well be visible, being wide pitched (as a result of low cycles to failure) and often intermittent and/or irregular (Figure 6.10). Apart from the generally coarse region of final fracture, there

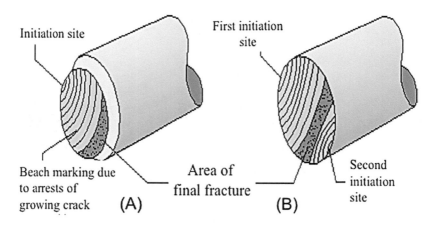

**Figure 6.9** Schematic diagrams of bending fatigue. (A) uni-directional bending (for example, a leaf spring or a diving board). (B) reverse bending (such as a motor shaft with a heavy belt load).

may be striation-free zones having a polished or plane featureless appearance. Again, however, the absence of such fatigue markings does not mean that fatigue had not been at play.

As an '*aide-memoire*', an illustration of fatigue striations on the fracture surface of a lead-free solder is shown in Figure 6.10. In contrast, beach markings are shown in Figure 6.11. Finally, a range of commonly encountered fatigue fracture surface features can be represented by the schematic illustration of Figure 6.12. Each feature represents a unique fingerprint that will describe both mode and loading characteristics in play prior to, and at, the point of failure.[6]

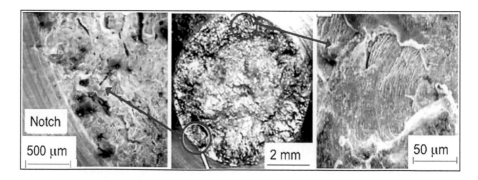

**Figure 6.10** Fracture surface of Sn-3.5Ag lead-free solder. The dark outer ring seen on the fracture surface is where a crack propagated by low-cycle fatigue loading. The magnified left-hand photograph shows partial intergranular cracking whereas the right-hand photograph shows clear striation marks, the characteristics of low-cycle fatigue in this instant.

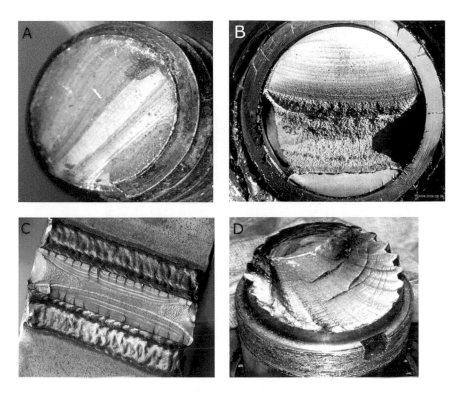

**Figure 6.11** (A–C) Three macro photographs showing reverse bending fatigue failures. (D) A torsional fatigue failure as a comparator.

## 6.7.2 Chevron and Striation Combination

As previously discussed, macroscopically visible fractographic features will serve to identify the fracture origin and direction of crack propagation. Chevron (herringbone) patterns are radial marks resembling nested letters 'V' and pointing towards the origin. Chevrons are indicative of a brittle (fast) fracture mode, whereas beach marks are visible semi-elliptical lines on a fracture surface that radiate outward from the origin, running perpendicular to the overall direction of fatigue crack propagation. Beach markings indicate successive positions of the advancing crack front and are indicative of fatigue as the failure mode. However, it is not uncommon to see both chevron and beach markings on one fracture surface, being the result of a rapidly increasing fatigue fracture. As a general rule, chevron lines and beach markings (or striations) will only intersect one another at right angles, as demonstrated by the following example.

### 6.7.3 Case Study: Internal Combustion
Engine Crank-Shaft Failure

An internal combustion engine crank-shaft failed after only 7,000 miles in service (Figure 6.13). The garage from which the vehicle had been purchased

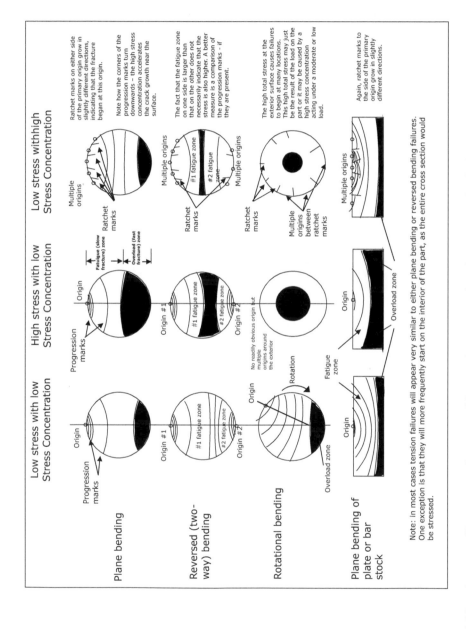

**Figure 6.12** Schematic illustration of a range of fatigue fracture surface features, each of which are a unique set of fingerprints that describe both the mode and loading type at failure. After NW Sachs[6].

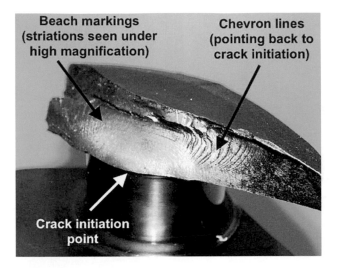

**Figure 6.13** Fracture surface on the web of an engine crank, displaying both beach markings and chevron lines, in addition to crack initiation, sited at a tight ground internal radius on the bearing journal.

was insistent that the failure had been due to the 'boy-racer' nature of the vehicle owner. Furthermore, they refuted the validity of the vehicle warranty, due to their insistence that the vehicle had been abused. However, the fracture surface clearly shows classic fatigue failure of the first crank throw. The crack initiation site is arrowed on the figure, along with clear fatigue beach marks (striations visible under high magnification), and chevrons (river marks) that point back to the crack initiation site. A fatigue crack had initiated at a sharp grinding corner, and may well have been exacerbated by fine grinding chatter marks and grinding burn on the ground radius and bearing surface. This failure was, without doubt, an engineered-in manufacturing fault, and had no connection with the owner's driving habits. When presented with this simple evidence, the vehicle warranty was honoured in full, with no penalty to the vehicle owner.

## 6.7.4 Splined Shafting

There is a classic example of a fingerprint of fatigue as a failure mode, that is not found under any other situation: that of fatigue failure of a splined shaft. The fracture surface will always resemble that of a 'star burst', as illustrated by the following hydraulic pump shaft failure. A broken section of hydraulic pump shaft is shown entering its female transmission sleeve (Figure 6.14A), in a normal working position at which it failed. It can be clearly seen that the resultant fracture surface resembles a 'star burst'. This classic 'fingerprint' feature is not observed on any other kind of fracture. Fatigue failure of splined shafting is a complex failure mode caused by cyclic torsional loads,

**Figure 6.14** (A) A broken splined shaft entering its transmission sleeve. (B) A closer view of its fracture face, showing a star-burst morphology. (C and D) Fatigue crack initiation and propagation at each and every spline.

with fatigue cracking initiating in each of the individual splines (Figures 6.14C and D). Cracks propagate by following a helical path across each spline and then continue towards the central region, eventually leaving a small area (or pip, as seen in the centre of Figure 6.14B) to sustain service loading. This finally failed in a ductile mode by torsional overstress – it should be considered that the stress on a circular shaft, subjected to torsion, will vary as the cube of the radius – so when there is only a small area of sound metal near the centre, the stress level under normal loads is much greater than when the cracking is still at the outer, splined, region of the shaft.

### 6.7.5 Ratchet Marks

The presence of ratchet marks will suggest multiple crack initiation points (origins) along with relatively high total stresses. They indicate multiple crack initiation sites on different planes in the metal, which have then joined up via tear ridges, or shear, as the crack increases in size. Ratchet marks can form as a result of either high stress on the part or from intrinsic high-stress concentrations (see Figure 6.12). However, by observation of both ratchet marks and the size of the instantaneous fracture zone, it can be determined whether load or stress concentration had been the major cause of fracture.

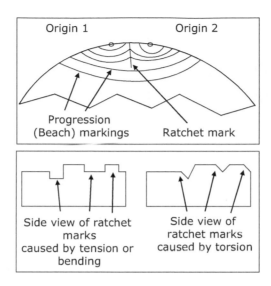

**Figure 6.15** Schematic of ratchet profile, showing either tensile, bending or torsion as the failure mechanism.

As an example, the combination of many ratchet marks and a small overload zone will indicate a light load, with high-stress concentrations. In addition, by observing the edges of the ratchet marks, it is possible to determine whether torsional forces were involved with the failure. If plane bending or tension has instigated failure, the sides of the ratchet marks will be essentially perpendicular to the fracture face. If the primary load causing the failure was torsional, the sides will be tapered[6] (Figure 6.15).

## 6.7.6 Thermal Fatigue

As a failure mechanism, thermal fatigue will entail a gradual deterioration and eventual cracking of a material or component, simply by alternate heating and cooling cycles. During the course of thermal cycling, free thermal expansion (TE) of the component material may be partially or completely constrained. However, any constraint of TE will introduce thermal stresses, the level of which may eventually initiate and propagate a fatigue crack. As thermal fatigue cracking usually initiates below 50,000 cycles, it can be classified as a low-cycle fatigue mechanism. The following example will illustrate thermal fatigue cracking in action, where a lorry driver was seriously injured when the vehicle clutch exploded.

Due to a combination of an incorrectly adjusted clutch plate, and an inexperienced driver, the surface of a lorry clutch pressure plate became subjected to a series of sudden frictional heating cycles followed by quenching. Quenching, at the surface of the clutch plate, arose from a combination of a large thermal

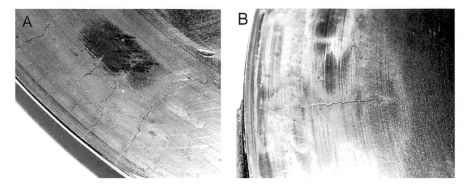

**Figure 6.16** (A) Photo-macrograph of a thermally fatigued cast iron clutch plate from a truck engine. (B) Closer view of thermal fatigue crack. Dimensions: outside diameter, 42 cm × 2.5 cm thick.

mass of the clutch, and the cooling effect of its rapid movement through surrounding air. The net result of such thermal cycling was the development of a large number of radial cracks that are shown in the macrograph of Figure 6.16. When the clutch eventually disintegrated, pieces of the plate flew into the cab, seriously injuring the driver.

## 6.8  Additional Modes of Progressive Failure

The remainder of this chapter will focus on (briefly) introducing several additional modes of progressive failure that will be encountered at some point by the forensic or failure investigator. A better understanding of these failure mechanisms will enable more appropriate decisions when undertaking failure investigations. Even basic knowledge and awareness will help the engineer to be better equipped in understanding and ultimately preventing the premature failure of a material or component. The progressive failure modes to be reviewed will cover corrosion, environmentally assisted cracking (EAC) and wear and creep, each of which will be referenced at appropriate points within the narrative.

### 6.8.1  Corrosion

The role of the forensic or failure engineer in the corrosion arena is often to assess if basic design, manufacture and material choice reflect professional engineering practice. Corrosion science exhibits a major division between dry and wet corrosion. Dry corrosion is dominated by high-temperature oxidation of metals and is controlled by the rate at which oxygen can diffuse through the oxide skin and by stable oxide adhesion to the metal substrate

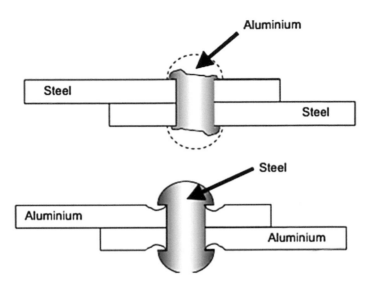

**Figure 6.17** Schematic of galvanic corrosion in action on riveted plates.

involved. Wet corrosion involves electrolytic interactions that require the presence of an aqueous solution. Galvanic corrosion is an electrochemical process where one metal preferentially corrodes when in electrical contact with a different metal type, when immersed in an electrolyte (Figure 6.17). Good design can prevent the onset of in-service (wet) corrosion by limiting the contact of the electrolyte. However, changes in operational procedures and manufacture can reverse any good design practice.[7] Both dry and wet corrosion result in the wasting away of a component, reducing the thickness of section and thereby the load-carrying capacity of the affected structure until failure takes place by mechanical overstress or fatigue. A review paper[7] covers salient points relating to corrosion, and environmentally enhanced break down, as a product failure mechanism and is illustrated by numerous case studies.

Dry corrosion is seldom a problem for structures and components operating in normal environmental conditions, though it clearly has serious consequences for parts operating at elevated temperatures. It is not generally realised that stainless steels rely on a thin surface film of chromium oxide, which forms virtually instantaneously in an oxidising atmosphere. If the (oxide) film is broken by scratching or abrasion it reforms and effectively becomes self-healing. However, under chemically reducing conditions stainless steel is no better than plain unalloyed varieties.

The forensic or failure engineer must be able to recognise the main forms of corrosion as, again, each will have its own unique fingerprint. As an aid to recognition, a range of corrosion types are shown in sketch format by Figure 6.18 and pictorially in Figure 6.19.

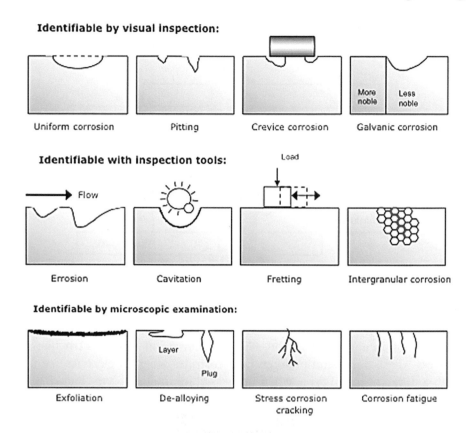

**Figure 6.18** A range of schematics depicting the main forms of corrosion that may be encountered in the course of a failure investigation.

## 6.8.2 Stress Corrosion Cracking (SCC)

For certain metallic alloys, the effects of brittle fracture, stress and corrosion combine to produce the phenomenon termed stress corrosion cracking.[7] This particular form of corrosion mechanism requires the following to act together:

(a) A susceptible microstructure
(b) Externally applied service stress levels or residual internal tensile stress from a manufacturing operation
(c) A mildly aggressive chemical environment that, on its own, would cause no significant corrosive attack

If the above three features are present, then crack-like defects (that start as corrosion crevices) can grow rapidly and proliferate throughout the micro-structure. In simple terms, SCC is the cracking induced from the collective effect of a susceptible microstructure, a tensile stress and a corrosive environment.

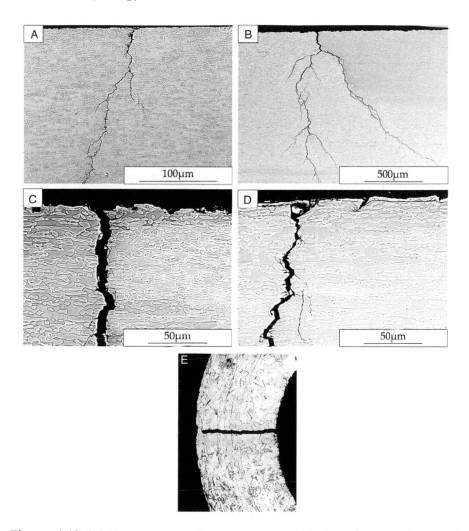

**Figure 6.19** (A) Cross-section microstructures of 2304 Duplex Stainless Steel showing stress corrosion cracking after exposure under four-point bend loading: (A) 700 MPa, sectioned parallel to loading direction. (B, C) 700 MPa, sectioned parallel to loading direction. (D) 500 MPa, sectioned parallel to loading direction. (A–D courtesy of Zhou et al., SCC of 2304 Duplex Stainless Steel— Microstructure, Residual Stress and Surface Grinding Effects, *Materials* 2017, 10, 221.) (E) A corrosion fatigue failure of a stainless-steel high-pressure pipe, shown as a direct comparison between SCC (highly branched) and corrosion fatigue failure (little or no branching). (Courtesy of Wikimedia Commons, public domain.)

Observation of the crack propagation path will provide a fingerprint of SCC in action.[7] Usually, the majority of the component surface will remain intact (unaffected), but fine cracks, penetrating into the material body, will be readily seen. Furthermore, observation of the fracture surface microstructure will reveal either an inter-granular or trans-granular morphology, as

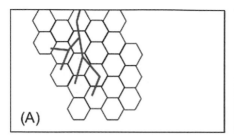 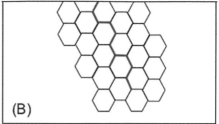

**Figure 6.20** Propagation path for either: (A) trans-granular or (B) intergranular stress corrosion cracking (SCC).

shown by Figure 6.20. Macroscopically, SCC fractures have a brittle appearance as shown by Figure 6.21. As a failure mode, SCC can be classified as a catastrophic form of corrosion, as identification of fine SCC cracking is generally difficult and damage not readily predictable. Therefore, a disastrous failure may occur unexpectedly, with minimal overall material loss.

As discussed above and in Section 5.11.2, there are three criteria for onset of stress corrosion cracking: (1) susceptible material, (2) presence of stress, and (3) a corrosive environment. If just one of the criteria is missing, then failure by SCC will not occur.

### 6.8.3 Crevice Corrosion

Crevice corrosion will occur in shielded areas of structures or assemblies. Shielded areas will include washers, lap joints, insulating materials, clamps, threads, fastener heads, gaskets, de-bonded coatings, surface deposits, etc. There is again a wide range of pit shapes that will identify a crevice corrosion mechanism at play, as shown in the sketch of Figure 6.22 and the pictures of Figure 6.23.

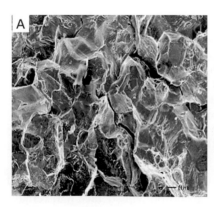 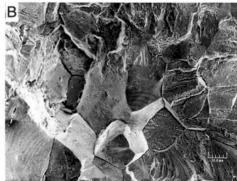

**Figure 6.21** (A) Fracture surface image of inter-granular Stress Corrosion Cracking (SCC) failure. (B) Fracture surface image of trans-granular stress corrosion cracking (SCC). (Courtesy of LinkedIn Slideshare, licensed under Creative Commons.)

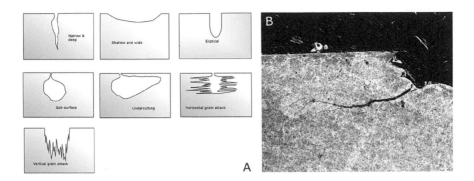

**Figure 6.22** (A) Crevice corrosion pit morphology. (B) A transverse SEM image showing continued formation of a sub-surface pit.

## 6.8.4 Environmentally Assisted Cracking (EAC)

When considering environmental influence in the failure investigation process, unequivocal statements regarding the characteristics of any of the forms of environmentally assisted cracking (EAC) cannot be made, as exceptions are not unusual. However, two environmentally assisted failure processes – hydrogen embrittlement (HE) and stress corrosion cracking (SCC) – may be distinguished from each other by consideration of the background information; the extent, nature and distribution of corrosion; the sites of crack

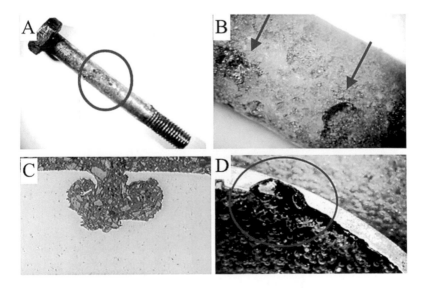

**Figure 6.23** (A) Ferritic stainless-steel rigging bolt, with expected failure location circled. (B) Closer view of circled area showing deep crevice corrosion pits in the bolt shank after four months in service. (C) A transverse section through a pit. (D) In contrast, a transverse section of inner brass cooling tube showing extent of pitting attack penetrating through the wall thickness.

initiation and other factors. Factors which favour SCC as a failure mechanism include[8]

(1) An in-service environment and residual manufacturing stresses capable of producing SCC (Figure 6.24), and conversely, an absence of pre-treatments which could introduce hydrogen into the material prior to service
(2) Multiple cracks initiated from corrosion pits
(3) Corrosion products on the fracture surface, which become thicker closer to the origin

Factors which favour HE include

(1) Inadequate baking following coating, or the use of other treatments that could introduce hydrogen into the steel
(2) Exposure of a component to low-humidity air in service[8]

Fractures produced by HE could become corroded after cracking, but such corrosion is generally manifested by patches of rust on both IG and dimpled overload areas rather than by the more uniform, often black, corrosion film found on SCC fractures. HE failures sometimes initiate from subsurface inclusions, but it is often not possible to distinguish between surface initiation sites and initiation sites at small inclusions close to the surface. Moreover, SCC often exhibits more crack branching and less pronounced, partially formed, dimples than cracking produced by HE. However, the extent of crack branching and dimples also depends on the stress intensity at the crack tip, on hydrogen concentration or environment and on the composition and thermal condition of the steel. Because some of this information is often not known for in-service failures, these features cannot be reliably used to distinguish between HE and SCC.[9]

**Figure 6.24** Two identically formed deep drawn brass cups. The deep drawn brass cup on the right shows stress corrosion cracking (SCC) under the influence of the residual manufacturing stresses and a mildly corrosive environment. The cup on the left had been annealed before putting it into an identical environment. Annealing has removed any SCC susceptibility.

## 6.8.5 Wear

Wear is a general term used to describe the deterioration of the surface of a material or component caused by frictional forces generated as a result of contact between two surfaces that are moving in relation to one another. Temperature will influence the wear rate (the rate at which a material deteriorates under frictional forces) as friction generates heat. In turn, the heat generated can affect the material microstructure, rendering it more susceptible to deterioration. However, the attention of the forensic (or failure) investigator will often focus on one area, that of any resultant wear processes at play during the service lifetime of a system, device or component.[10] When engaged on a product failure investigation, it will be implicit in any instruction to establish whether or not the rate of wear was acceptable, reflected good engineering practice or that the device had simply reached the end of its useful service lifetime. A review paper[10] covers salient points relating to wear as a product failure mechanism and is illustrated by numerous studies. However, common types of wear that exist are abrasive, erosive adhesive and surface fatigue, as shown in the flow-chart of Figure 6.25.

With regard to the common types of wear mechanisms, they generally do not occur in isolation. It is more common to find the mechanisms acting in parallel, or in succession. As an example, with any bearing failure, it is generally found that fretting corrosion and rolling contact fatigue will occur together, or the formation of adhesion particles may then lead to, or be the cause of, abrasion. Such hard particles suspended in lubricating oil will 'dent' rolling element bearings, leading to contact fatigue. In addition to those listed on Figure 6.25, there are other wear processes, e.g. Peterson has included impact, gouging, wire drawing, metal/metal, fluid erosion, rain erosion impingement, erosion-corrosion and deformation.[11] Archard's list includes surface fracture, tearing and melting.[12] The following are examples of the four prime wear mechanisms of abrasion, erosion, adhesion and surface fatigue (Figures 6.26 and 6.27). Again, the illustrations are general and include some schematics. However, they are intended simply as a visual atlas, or reference, for the forensic failure engineer.

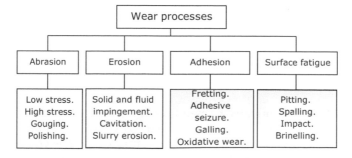

**Figure 6.25** A flow-chart depicting a range of various wear mechanisms.

## A Visual Atlas of Abrasion, Erosion, Adhesion and Surface Fatigue

**Abrasion**:

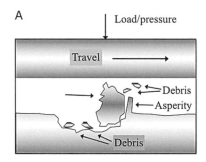

**Figure 6.26** (A) Illustration of abrasive wear mechanism.

**Erosion:**

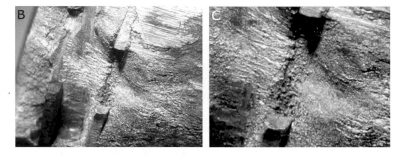

**Figure 6.26** (B) General view of erosion around the rings of a piston. (C) Closer view of erosion channelling, along with a view of unsupported end of a piston ring.

**Adhesion:**

**Figure 6.26** (D) Adhesion followed by abrasion under reciprocating sliding wear of an overheated (and, therefore, un-lubricated) piston/cylinder contact.

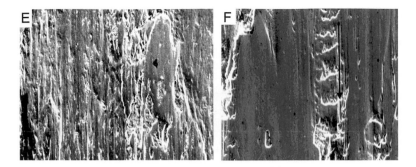

**Figure 6.26** (E) Appearances of adhesion on the cylinder lining wall of the overheated piston shown in Figure 6.26D. (F) A closer view of the cylinder lining which was manufactured from a hardened and tempered steel.

**Surface Fatigue:**

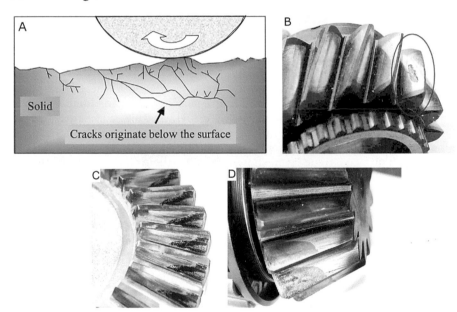

**Figure 6.27** (A) Illustration of the mechanism of surface fatigue. (B) Rolling contact fatigue on the flank of a gear wheel tooth. (C and D) Mis-alignment of a crown wheel and pinion gear, leading to onset of surface contact fatigue. Note the matching contact footprint on both crown wheel and spur gear tooth flanks.

## 6.8.6 Wear of Rolling Contact Elements

Rolling contact bearings are used in almost every type of rotating machinery, and can be a critical factor in service lifetime and performance.[13,14] Reliability is dependent on both the type of bearing selected and the geometrical accuracy of associated components. Previously, design engineers

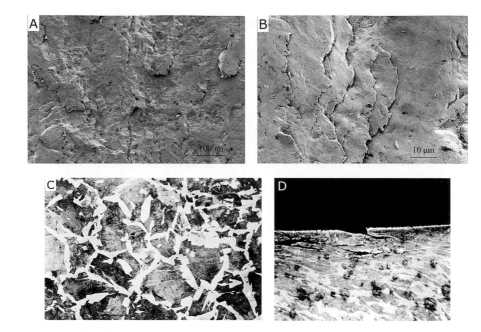

**Figure 6.28** (A and B) Appearances of surface fatigue on a wheel of a harbour crane. (C) Bulk microstructure consisting of pearlite (grey, fine lamella) grains, within a network of grain boundary ferrite (white) and a few side plates of ferrite. This structure is typical of carbon steel such as EN8 in a normalised (austenised) condition (heated to approximately 850°C, and air cooled). (D) Transverse section through the surface fatigue layer, showing microstructural flow and sub-surface cracking.

have established lifetime estimations based on fatigue as the normal service failure mode.[15-17] Such calculations depend on an explicit assumption that bearings are properly installed, operated and maintained. However, as a result of improvements in manufacturing technology and materials, bearing fatigue is now not a limiting life factor.[18,19] Fatigue, which is related to subsurface stresses, now accounts for less than a few percent of failures experienced in service.

There is an 'infinite life' theory for rolling bearings;[20] 'under good operating conditions and provided its fatigue load limit is not exceeded, bearing life will not be limited by fatigue and should therefore exceed the life of the machine'. However, in real-world operations, countless differences will exist in bearing life expectancy as a direct result of a variety of mechanical and environmental influences,[21] some of which are presented in Table 6.2. Each of these factors will give rise to unique morphology of primary and secondary damage.[22] In turn, each factor will be initiated by a range of causes (Table 6.3)

**Table 6.2 Primary, Secondary and Cage Damage Mechanisms Linked to Bearing Failure in Service**

| Primary Damage | Primary Damage | Secondary Damage | Cage Damage |
| --- | --- | --- | --- |
| Wear | Surface distress | Flaking | Vibration |
| Indentations | Corrosion | Cracking | Excessive speed |
| Smearing | Electrical current | | Wear |

**Table 6.3 Seven Common Bearing Failure Mechanisms with Associated Causes of Each**

| Wear | Indentations | Cracks | Flaking |
| --- | --- | --- | --- |
| Abrasive particles | Faulty mounting | Rough treatment | Preloading |
| Inadequate lubrication | Overloading | Excessive drive-up | Oval compression |
| Vibration | Foreign particles | Smearing | Axial compression |
| | | Fretting corrosion | Misalignment |
| Smearing | Corrosion | Cage Damage | Indentations |
| At roller ends | Deep-seated rust | Vibration | Smearing |
| Guide flanges | Fretting | Excessive speed | Deep-seated rust |
| Rollers, raceways | | Wear | Fretting corrosion |
| External surfaces | | Blockage | Fluting |

that are a fingerprint of breakdown, thus allowing the failure engineer to determine (and eliminate) the root cause of failure.

In contrast to traumatic failure, bearing (and ultimately machine) performance may degrade slowly rather than cease suddenly, so the point of failure may not always be obvious. Once bearing break down is detected, finding and correcting the root cause of premature failure is not even an issue – the failed bearing is simply replaced. However, *any* premature bearing failure is symptomatic of additional problems that, if left untreated, will cause an identical failure to occur again. Instruction to the forensic engineer or investigator will often focus on unearthing a root cause of bearing demise, simply to prevent the reoccurrence of failure.

It could well be part of the investigator's instruction to establish whether or not the service lifetime was acceptable, choice and maintenance of the bearing reflected good engineering practice or that the bearing or device had simply reached the end of its useful service lifetime. Again as an aid to bearing failure recognition, a range of primary and secondary damage mechanisms taken from both Tables 6.2 and 6.3 are pictorially illustrated as Figures 6.29 to 6.33 inclusive.

**Abrasive wear**

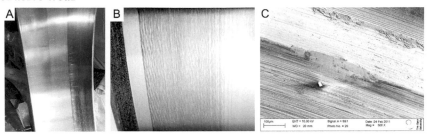

**Figure 6.29** (A) The outer race of a spherical roller bearing with raceways that have been worn by abrasive particles. (B) Note: the dividing line between worn and unworn sections can generally be felt by finger touch alone. (C) SEM image of element rolling track surface smearing.

**Surface distress**

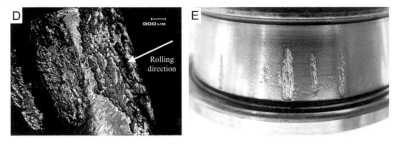

**Figure 6.29** (D) Appearances of surface fatigue under rolling wear on the cage of a roller bearing. (E) Surface flaking.

**Indentations**

**Figure 6.30** When pressing bearings onto a shaft or into a housing, appropriate presses or hydraulic tools must be used. Hammers and punches must never be used if premature spalling failure of a new bearing is to be avoided. Here, clear evidence of hammer blows to the outer ring during installation can be seen.

## Lubrication

**Figure 6.31** The frosted appearance of this bearing race illustrates the result of oil viscosity being too low (thin), allowing metal-to-metal contact to occur. This type of premature failure will occur during initial start-up of heavily loaded bearings. In this particular case, damage occurred after less than one minute of operation.

## Corrosion

**Figure 6.32** Water contamination is clearly evident on this outer ball bearing ring. Excessive water contamination will cause severe corrosion, as can be seen on the rolling track, while small amounts of water will stain the surfaces with a brown discoloration.

## Fluting (electrical flow)

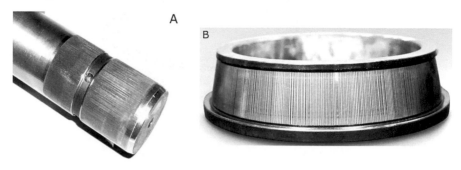

**Figure 6.33** Electrical damage (fluting) generated when electric currents pass through a bearing. There is arcing and burning at the points between the races and the rolling elements where the current jumps the air gap, leaving a classic linear 'washboard' fingerprint as seen.

## Rolling element retaining cage failure

Bearing cage damage includes deformation, fracture and wear to the cage, cage pillar, side face, pocket surface or guide surface. These problems can also be caused by large moment load, sudden acceleration or deceleration, excessive rotation speed, poor lubrication or rises in temperature. However, it is important to appreciate that rolling bearings do have a finite life, with repeated high stressing at the contacts between rolling elements and tracks eventually resulting in damage to the contacting surfaces.

There are a large number of reasons why rolling bearings fail prematurely and it is possible to postulate a train of events that led to the failure shown in Figure 6.34. There is a likelihood that the bearing had been faulty right from the point of manufacture. However after completing 10,000 hours of service prior to its failure, this is not a likely scenario. Based on observations of damage sustained by the ring, it is safe to say that on the balance of probability, failure of the bearing in question would have initiated through lack of (or contaminated) lubricant, leading to initial breakdown of the bearing cage, thus allowing the rolling elements freedom to move out of registration. Wear damage sustained by the raceway of the inner bearing ring bears witness to the observation that the failure process had been in play for a good period of time prior to final demise. With the rolling elements free to move, damage would accumulate rapidly, with debris from both the cage and bearing rings accelerating the attrition process. Heat generated from the failure process would not only soften the contact surfaces (thereby accelerating failure) but would also provide vibration forces that would instigate fatigue crack initiation of the output shaft. The actual sequence of events leading to failure may be up for debate, but what is clear is that the bearings failed through a wear process, initiating at the bearing rolling element cage.

**Figure 6.34** An inner bearing raceway showing clear evidence of surface blue-ing, and no evidence of rolling element indents expected in an overload situation. The heavily worn rolling track showing extensive surface damage accumulated over time, with markings from free roaming rolling elements arrowed in red.

**Figure 6.35** (A and B) These agricultural tractor bearings have failed as a result of continual welding contact between asperities on the metal surfaces. Such contact will result in metal 'pull-out' as the surfaces adhere to each other during rotation. This condition can be caused by use of incorrect oil viscosity, excessive loads or speeds, incorrect internal clearances, or a combination of these problems. An increase of as little as 4 or 5°C in temperature may have contributed to this failure, due to unacceptable thinning of the lubricant's viscosity. When analysing the root cause of a failure, all possible contributing causes must be considered.

## 6.8.7 Case Study: An Agricultural Bearing Failure

As an example of bearing failure investigation, the bearing detritus shown in Figures 6.35 and 6.36 was recovered from an agricultural tractor that had completed 5,419 recorded hours of service. At the time of failure, the policyholder had stated 'his driver was pulling a trailer down the road with a full sand load in it when he had to move over on the road for a passing car. The tractor went up a bank and into a drainage gully, but continued its journey after this incident. Within metres of the incident the tractor broke down with no drive.'

**Figure 6.36** The tractor driveshaft bearing detritus in question. Note: only the inner ring of bearing had been recovered, and the extensive wear and plastic deformation of the bearing cage.

Clearly, the inference here was that failure was as a result of a traumatic incident/event (dropping into the drainage gully).

In addition to damaged bearing rings shown in Figure 6.35, a broken driveshaft bearing was recovered from the tractor (detritus seen in Figure 6.36). Visual inspection determined that failure had initiated as a result of cage breakdown. Cage breakdown had allowed the rolling elements to move out of their captive positioning, thus contributing to cracking and subsequent splintering damage of the rolling elements. Any bearing cage failure would suggest inadequate or contaminated bearing lubricant. Although the actual scenario of events that led to failure may be open to differing opinion, it was clear that the bearing had most certainly succumbed to failure mechanisms normally associated with wear of raceways, retaining cage and rolling elements. Damage inflicted to the bearing inner ring and associated raceway was mechanical in nature, rather than the result of a traumatic loading incident. There was no evidence on the raceway that would substantiate initiation of failure, or actual failure, from a single traumatic loading event. (Figure 6.37.)

## 6.9 Summary

A central aim of this chapter was to compare and contrast a typical range of failure morphologies and fracture surface topographies that may be encountered during the course of a forensic or failure analysis. Detailed markings on a fracture surface or failure morphology will often provide the forensic metallurgist or failure analyst with important clues as to why a system, structure or component failed, a fingerprint of that particular failure. Each

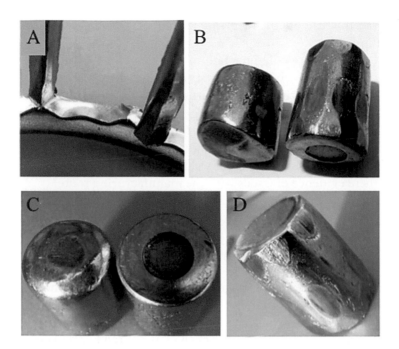

**Figure 6.37** (A) Bearing cage displayed extensive plastic deformation and wear damage from rolling element contact. (B) Rolling element cracking had clearly occurred. (C) Extensive element end wear resulting in polished and domed the ends of individual rolling elements. (D) Wear and grinding attrition noted on some rolling elements.

individual failure mechanism described in Chapter 5 will give rise to unique fracture morphology and fracture surface topography that can be considered a fingerprint of failure. Therefore, a clear understanding of the relationship between failure mechanism and fracture morphology/topography will allow characterisation of the range of probable conditions at play at the instant of failure. It is again worth reiterating that the cause of many failures can be detected simply by careful examination of both fracture morphology and fracture surface. This is accomplished by the use of low-power magnification coupled with some basic chemical or physical testing and analysis.[23] The way products are made is important for the features that can arise from the particular way a component is shaped, and when those features turn into defects. Visual inspection of failure can determine the forces involved, whether the load was applied cyclically or was a single overload, the direction of the critical load, and the influence of outside forces such as residual stresses or strains.

In order to correctly interpret any failure mode from visual inspection, fractographs and higher-magnification micrographs it is important for the forensic engineer to bear in mind some key questions:

- Is the fracture ductile or brittle (Figure 6.3), is it a simple overload failure?
- Is there evidence of fatigue failure? (Section 6.7)
  - Are there beach marks? Or striations? (Figure 6.11)
- Is there evidence of any environmental influences (Section 5.4; Section 6.8.1)
- Is there evidence of creep failure? (Section 5.8)
- Is the microstructure what it should be?
- If welded, does it exhibit poor weld quality? (Section 5.12.4)
- Is there evidence of issues with the manufacturing route?
  - Are there any stress-raising features?
  - Is porosity present (particularly in castings)? (Figure 5.22 and 5.23)
  - Is there evidence of flow (directionality in microstructure)?
  - Does the manufacturing route match the specification?

These parameters or failure fingerprints, combined with other additional (and often necessary) assessment of the components and their loading history, will generally provide a good starting guide to probable service failure mechanisms.[20] Knowing the roots of the failure, the investigator can both pursue the cause and/or causes of failure and postulate a scenario or train of events leading to failure. As a final aid for deducing the correct morphology and topography of failure, two tables are presented (Tables 6.4 and 6.5). They can be used as a check-list for failure mode identification, with traumatic (instantaneous) and mechanical (progressive) failure being differentiated on each table. The reasons for discriminating between traumatic and mechanical failure lie with the insurance industry, where a traumatic failure would be considered a valid claim, whereas a mechanical event would be repudiated as an insurance claim.

**Table 6.4   A Check-List for Failure Mode Identification**

| Monotonic Overload (Time-Independent) | Fatigue Failure (Time-Dependent) | Progressive Failure (Time-Dependent) |
| --- | --- | --- |
| Tensile | Tension-tension | Creep |
| Tensile tear | Tension-compression | Corrosion |
| Buckling | Unidirectional bending | Stress corrosion cracking |
| Shear | Reverse bending | Corrosion fatigue |
| Bending | Rotational bending | Hydrogen embrittlement |
| torsion | Torsion | Radiation damage |

*Note:* single-cycle overload failure is classed as a traumatic event, whereas time-dependent failure is classed as mechanical in nature. This classification has serious ramifications in any litigation or insurance claim.

**Table 6.5 Failure Mode Identification Chart with Traumatic (Instantaneous) and Mechanical (Progressive) Failure Modes Differentiated**

| Method | Instantaneous / Failure Modes | | Fatigue | Progressive Failure Modes | | Creep |
|---|---|---|---|---|---|---|
| | Ductile Overload | Brittle Overload | | Corrosion | Wear | |
| Visual: 1 to 50× (fracture surface) | • Necking or distortion in direction consistent with applied loads<br>• Dull fibrous fracture<br>• Shear lips | • Little or no distortion<br>• Flat fracture<br>• Bright or coarse texture, crystalline, grainy<br>• Chevrons or river markings point to origin. | • Flat progressive zone with beach marks<br>• Overload zone consistent with applied loading direction<br>• Ratchet marks where origins meet | • General wastage, roughening, pitting or trenching<br>• Stress-corrosion and hydrogen damage may create multiple cracks that appear brittle | • Gouging, abrasion, polishing or erosion<br>• Galling or storing in direction of motion<br>• Roughened areas with compacted powdered debris (fretting)<br>• Smooth gradual transitions in wastage | • Multiple brittle appearing fissures<br>• External surface and internal fissures contain reaction scale coatings<br>• Fracture after limited dimensional change |
| Scanning electron microscopy: 20 to 20,000× (fracture surface) | • Microvoids (dimples) elongated in direction of loading<br>• Single crack with no branching<br>• Surface slip band emergence | • Cleavage or inter-granular fracture<br>• Origin area may contain an imperfection or stress concentrator | • Progressive zone: worn appearance, flat, may show striations at magnifications above 500×<br>• Overload zone: may be either ductile or brittle | • Path of penetration may be irregular, intergranular or a selective phase attacked<br>• EDS may help to identify corrodent (chemical species) | • Wear debris and/or abrasive can be characterised as to morphology and composition<br>• Rolling contact fatigue has the appearance of wear in early stages | • Multiple intergranular fissures covered with reaction scale<br>• Grain faces may show porosity |
| Metallographic inspection, 50 to 1,000× (cross section) | • Grain distortion and flow near fracture<br>• Irregular, trans-granular fracture | • Little distortion evident<br>• Inter-granular or trans-granular<br>• May relate to notches at surface or brittle phases internally | • Progressive zone: usually trans-granular with little apparent distortion<br>• Overload zone: may be either ductile or brittle | • General or localised surface attack (pitting, cracking)<br>• Selective phase attack<br>• Thickness and morphology of corrosion scales | • May show localised distortion at surface consistent with direction of motion<br>• Identify embedded particles | • Microstructural change typical of over-heating<br>• Multiple intergranular cracks<br>• Voids formed on grain boundaries or wedge-shaped cracks at grain triple points<br>• Reaction scale or internal precipitation<br>• Some cold flow in final stages of failure |

*(Continued)*

**Table 6.5 (Continued)　Failure Mode Identification Chart with Traumatic (Instantaneous) and Mechanical (Progressive) Failure Modes Differentiated**

| Method | Instantaneous Failure Modes | | Progressive Failure Modes | | | |
|---|---|---|---|---|---|---|
| | Ductile Overload | Brittle Overload | Fatigue | Corrosion | Wear | Creep |
| Contributing factors | • Load exceeded the strength of the part<br>• Check for correct alloy and processing by hardness or destructive testing, chemical analysis<br>• Loading direction may show failure was secondary<br>• Short-term high-temperature rupture has ductile appearance (see creep) | • Load exceeded the dynamic strength of the part<br>• Check for correct alloy and processing as well as correct toughness, grain size<br>• Loading direction may show failure was secondary, or impact induced<br>• Low temperatures (embrittlement) | • Cyclic stress exceeded the endurance limit of the material<br>• Check for correct strength, surface finish, assembly and operation<br>• Prior damage by mechanical or corrosion modes may have initiated cracking<br>• Alignment, vibration, balance<br>• High cycle low stress: large fatigue zone; low cycle high stress: small fatigue zone | • Attack morphology and alloy type must be evaluated<br>• Severity of exposure conditions may be excessive; check pH, temperature, flow rate, dissolved oxidants, electrical current, metal coupling, aggressive agents<br>• Check bulk composition and contaminants | • For gouging or abrasive wear, check source of abrasives<br>• Evaluate effectiveness of lubricants.<br>• Seals or filters may have failed<br>• Fretting induced by slight looseness in clamped joints subject to vibration<br>• Bearing or materials engineering design may reduce or eliminate problem<br>• Water contamination<br>• High velocities or uneven flow distribution, cavitation | • Mild overheating and/or mild over stressing at elevated temperature<br>• Unstable microstructures and small grain size<br>• Ruptures occur after long exposure times<br>• Verify correct alloy |

*Note:*　EDS = energy dispersive spectroscopy.

# References

1. Dennies, Daniel P., The organization of a failure investigation. *Practical Failure Analysis*, June 2002. 2(3): pp. 33–41.
2. Lewis, Peter and Colin Gagg, *Forensic Polymer Engineering: Why Polymer Products Fail in Service*. 2010, Cambridge: Woodhead Publishing Limited, ISBN-10: 1439831149, ISBN-13: 978-1439831144.
3. Lange, G., *Systematic Analysis of Technical Failure Cases*. 1992, Oberursel: Informationsgesellschaft-Verlag.
4. Biringuccio, V., De La Pirotechnia, C. Smith and M. Gnudi, ed., *Venice 1540*. 1942, New York: AIME.
5. Gagg, C. and P. Lewis, In-service fatigue failure of engineered products and structures - Case study review. *Journal of Engineering Failure Analysis*, 2009. 16(6): pp. 1775–1793.
6. Sachs, N.W., Understanding the surface features of fatigue fractures: How they describe the failure cause and the failure history. *Journal of Failure Analysis and Prevention*, April 2005. 5(2), https://doi.org/10.1361/15477020522924.
7. Gagg, C. and P. Lewis, Environmentally assisted product failure - Synopsis and case study compendium. *Engineering Failure Analysis*, 2008. 15(5): pp. 505–520.
8. Lynch, S., Failures of structures and components by environmentally assisted cracking. *Engineering Failure Analysis*, 1994. 1(2): pp. 77–90.
9. Eliaz, N., A. Shachar, B. Tal and D. Eliezer, Characteristics of hydrogen embrittlement, stress corrosion cracking and tempered martensite embrittlement in high-strength steels. *Engineering Failure Analysis*, 2002. 9(2): pp. 167–184.
10. Gagg, C. and P. Lewis, Wear as a product failure mechanism - Overview and case studies. *Engineering Failure Analysis*, 2007. 14(8): pp. 1618–1640.
11. Peterson, M.B., *Classification of Wear Processes*. Wear Control Handbook. 1980, New York: ASME.
12. Stachowiak, G.W., ed., *Wear – Materials, Mechanisms and Practices*. 2014, John Wiley & Sons, ISBN:9780470016282.
13. Palmgren, A., The service life of ball bearings. *Zeitschrift des Vereins Deutscher Ingenieure*, 1924. 68(14): pp. 339–341.
14. Gustaffson, O. and T. Tallian, Detection of damage of assembled rolling element bearings. *ASME Transactions*, 1962. 5: pp. 197–209.
15. Asme Tribology Division Technical Committee, *Life Ratings for Modern Rolling Bearings: A Design Guide for the Application of International Standard ISO 281/2*. 2003, American Society of Mechanical Engineers, ISBN13: 9780791836637.
16. Institute, A.n.s., Load rating and fatigue life for roller bearings, in *ANSI/AFBMAStandard 11-1990*. 1990. Washington, DC: American Bearing Manufacturers Association, https://webstore.ansi.org/Standards/ABMA/ANSIABMA112014
17. Kurfess, T.R., S. Billington and S.Y. Liang, *Advanced Diagnostic and Prognostic Techniques for Rolling Element Bearings*. Springer Series in Advanced Manufacturing. 2006, Cham, Switzerland: Springer, ISBN 978-1-84628-268-3.

18. Zaretsky, E.V., *Rolling Bearing Life Prediction, Theory, and Application National Aeronautics and Space Administration*. 2013, NASA Glenn Research Center, NASA/TP—2013-215305.
19. Johnson, K., The strength of surfaces in rolling contact. *Proceedings of the Institution of Mechanical Engineers (IMechE)*, 2000. 203(3): pp. 151–163.
20. Harris, T. and R. Barnsby, Life ratings for ball and roller bearings. *Journal of Engineering Tribology. Proceedings of the Institution of Mechanical Engineers Part J*, 2001. 215(6): pp. 577–595.
21. Loewenthal, S. and D. Moyer, Filtration effects on ball bearing life and condition in a contaminated lubricant. *Journal of Lubrication Technology, Transactionsof the ASME*, April 1979. 101(2): pp. 171–179.
22. Tallian, T., *Failure Atlas for Hertz Contact Machine Elements*. 2nd ed. January 2000, American Society of Mechanical Engineers, ISBN: 0791800849.
23. Lewis, P.R., K. Reynolds and C. Gagg, *Forensic Materials Engineering: Case Studies*. 2003, Boca Raton, FL: CRC Press, ISBN-10: 0849311829, ISBN-13: 978-0849311826.

# The Engineer in the Courtroom

# 7

## 7.0 Introduction

Much depends on the way in which an investigator expresses him or herself in a report, whether interim in nature, a final version for the instructing party or whether it entails litigation or not. Available tools in the forensic toolbox are, in fact, just a means to an end: deducing the cause or causes of an accident, or the reason/s why a particular product failed. Consequently, *all* evidence demands to be considered, not just specialist results from laboratory analysis. It is the way in which different kinds of evidence *support or corroborate* one another that is the clue to successful investigations. Nevertheless, however successful that path of reasoning, it must at some stage be presented in written form for consideration by others. The art of communication is thus an essential part of the entire investigative process. This point has greater significance when considering that any forensic report may well be examined by a court, with the expert being cross-examined on any results, including the way in which they were obtained. Consequently, any report prepared for litigation should always be written with an understanding that it will be examined orally and in public. Examination and cross-examination in a court will also probe the experience of the investigator, and his or her qualifications to deal with the problem at hand.

## 7.1 The Role of the 'Expert'

As a result of abuse of the system for giving expert evidence to courts in the United Kingdom, a new legal reform has been introduced in the Supreme Court over the last decade. Termed 'Access to Justice', it is fondly known as the Woolf reforms, being introduced in a bid to accelerate the process of litigation, thereby reducing costs for all parties involved. In a nutshell, the Woolf reforms ensure far-reaching powers of the courts, in allowing them to control the way a case develops. In the UK, it is Rule 35 of the Woolf (Access to Justice)[1,2] reforms that is applicable to the forensic engineer or expert witness. As should now be appreciated, the primary function of forensic engineering is the communication of knowledge from the activity of engineering to a totally different sphere of work – that of the law. When acting as an expert witness, the forensic engineer will apply his knowledge and skills to questions decided by courts of law. The role of expert evidence in any trial is to provide an independent view or opinion regarding the facts at issue in a dispute. By definition, the expert witness can offer an opinion in disputed areas, whereas such a privilege is (normally) excluded for witnesses of fact, who can*not* offer opinion on a given issue.[3,4] Opinion and/or evidence presented by experts is a duty directly to the court – emphasised in the Woolf reforms – rather than to the party who hires him or her, and pays for their time. However, the expert may feel some loyalty or responsibility to his 'paymaster' – the solicitor, individual or company paying for his or her service, for example. Nevertheless, even if the case or failure never reaches court action, expert opinion must always be completely objective, and all attempts to bias any conclusions one way or another from outside pressure (whether real or imaginary) must be resisted. Expert opinion must be (and be seen to be) a fair attempt to explain failure in a perfectly neutral way. There are several routes available to achieve this central objective, one of the most important of which is to let the real evidence speak for itself. In this way, evidence can be considered a 'silent witness' that, on closer (expert) scrutiny, will reveal detail pertinent to failure. Then again, any such detail may well not be immediately obvious to the inexperienced eye. In criminal courts, forensic evidence of fingerprints, blood type or DNA matching is often critical, particularly where there is no direct eyewitness evidence or when the witness evidence is conflicting. It must be emphasised that, in practice, conflict between witness evidence is found to be an exceptionally common occurrence.[5]

After a serious failure incident or accident, subsequent litigation may rely on expert evidence to establish the true facts of the matter. Delay in the legal process can result from a failure event that turns out to be more complex in origin than it first envisioned. Key evidence can be examined by experts appointed by all parties with interest in the dispute. This is one of the reasons why legal cases or challenges can take several years to reach fruition. Furthermore, 'delay' is built into the legal system, simply to allow time for each party to prepare its position after allegations are formalised into writs and pleadings. Further delays may accrue, by any one of the parties concerned applying to the court for discovery, or for further direction – thus pushing back any time deadlines.

## 7.2 The Expert Report

Some of the new responsibilities introduced by the Woolf reforms have in fact been used by experts over a number of years, and have subsequently become embedded in current case law. The forensic engineering expert will implicitly acknowledge that courts require impartial, objective evidence, presented in a formal manner. If the crux of the matter at hand is some specialised engineering matter, a qualified practitioner will be required to analyse and formulate subsequent questions in a manner that will permit impartial, objective conclusions to be reached. The forensic engineer may have to reconstruct a sequence of events leading to failure and attribute causal relationships. He will be required to explain any limitations imposed by nature, to interpret conformance with specifications and regulations, as well as to perform and report on any controlled experimentation or analysis. If the engineering view is pertinent and any preliminary information useful, the next stage is for the forensic engineer to prepare a written comprehensive expert report.

The expert report is a document that may be disclosed to the opposing side in the hope of reaching an out-of-court settlement. In addition, it may also be the basis for any engineering testimony presented to the court. As a formal report, it should crisply articulate the known relevant facts, any assumptions made, examinations conducted, the calculations performed, the results obtained and conclusions drawn. Although the expert report must withstand scrutiny by peer review, it must also be written for the legal practitioner, who by nature may not be versed in engineering disciplines. Any logic used should be unassailable, with the conclusions easily understood.

As stated above, in addition to standing on its own as a deliverable document, the expert report must also provide the basis for engineering testimony. Relevant photographic evidence of vital details is normally presented within the expert report, along with an explanation of any mechanical assessment (testing) undertaken on the material or structure, interpretation of analytical test results, limitations and compromise associated with each individual test, evidence from new (exemplar) products (as a direct comparator), etc. A scenario or chain of events in the accident will emerge as the evidence from the failed product is built up within the report. Finally, if possible, a reconstruction of events will allow the investigator to re-visit the facts that have been established, and so draw conclusions regarding the cause of the incident, accident or failure.

The expert should be able to say how a product had been made, and whether the material and process route had been appropriate and well-chosen. Information may well be missing at the outset of investigation. Accordingly, the expert can point out a range of additional information that may be required in order to test any failure hypothesis. If, for example, a product failed as a result of faulty manufacture, the expert should list a series of internal documents that he considers necessary for the process of discovery, a legal process initiated by the instructing solicitor.

On the other hand, if a product failed through abuse by the injured party, then an expert report will be of no benefit for any intended litigation purposes. It will be the responsibility of the instructing solicitor to assess the quality of an expert report, evaluating whether or not it is fairly based on all available evidence and whether or not it is necessary to instruct an additional expert, in areas that are outside the expertise of the initial expert for example.

Although the majority of product liability cases are settled before ever reaching a court, all expert reports must be prepared in the full knowledge they can be exposed to the full rigour of an open trial. Therefore, expert reports (and expert opinion) frequently evolve as new evidence becomes available. Therefore, the expert must be ready to modify his or her opinion to allow for any new facts.

It should be highlighted that experts may well differ as to the cause of failure or accident, particularly where critical information regarding the case is missing. Furthermore, being human, experts can also make mistakes, thereby offering an incorrect opinion. It is an accepted reality that many experts fail to delve too deeply when presented with instructions, either from within their own company (for internal reporting) or from lawyers. There are numerous examples of expert reports where the investigator/s have made a superficial inspection of the failed product – or have attempted to further their own scientific, and unproven, principles – leading them to champion a totally incorrect concluding opinion. A danger that can arise in taking a superficial approach to case-work is that the might of the legal system will be activated, on unsupported or unsubstantiated evidence. The machine often continues on unabated (sometimes for years) until another expert produces a more firmly based report, which often contradicts the original opinion. This situation creates an unwelcome build-up of costs, which may have been wasted, along with the clear potential for gross miscarriage of justice.

The Regina-v-Sally Clark case is a classic example of biased investigation and reporting. Sally Clark (August 1964–15 March 2007) was a British solicitor who was victimised by a high-profile miscarriage of justice. She was wrongly convicted of the murder of two of her sons.[6] Her first son died suddenly within a few weeks of his birth in 1996. Her second son died in a similar manner. As a result, Clark was arrested in 1998 and subsequently tried for the murder of both sons. Her prosecution was controversial, primarily because of the testimony of paediatrician Professor Sir Roy Meadow, in which he presented statistical evidence. The Royal Statistical Society would later issue a public statement expressing its concern at the 'misuse of statistics in the courts'. It arguing that there was 'no statistical basis' for Meadow's claim.[6]

Clark was convicted in November 1999 with the verdict being upheld at appeal in October 2000. However, this finding was eventually overturned in a second appeal in January 2003, after it emerged that the prosecutor's pathologist had failed to disclose microbiological reports, including a report suggesting that one of her sons had died of natural causes. Clark was released from prison having served more than three years of her sentence. The journalist Geoffrey Wansell called Clark's experience 'one of the great miscarriages of justice in modern British legal history'.[7,8] As a result of her case, the Attorney-General ordered a review of hundreds of other cases. In doing so, two other women – who had been convicted of murdering their children – also had their convictions overturned. Sadly, as a direct result of her ordeal, Clark turned to heavy drinking and would later die of acute alcohol poisoning in March 2007.

In a nutshell, the overarching duty of the expert is to the court, with the content of a forensic report including

1. The nature of the failure and its consequences
2. A statement of truth to be given with the expert report
3. Analysis of the failure itself
4. Interpretation of all evidence surrounding the failure
5. Possible causes of failure
6. The likely cause of failure
7. Disclosure of previous experience, normally accomplished by including a detailed curriculum vitae within the report body

## 7.3 Pre-Trial Meeting of Experts

The court will typically order a meeting of experts before the trial date, with the sole aim of airing both issues in dispute, and points of contention. The goal of such a meeting is for the experts to discuss points of agreement and disagreement in the matter at hand. To minimise costs in minor cases, the expert meeting may take place by phone, FaceTime, etc., rather than face-to-face at a pre-determined venue.[9] Following such a meeting, the experts are required to deliver a joint statement, stating matters on which they agree and the areas of disagreement, including the *reasons* for such disagreement.

An expert 'should note that they do not have immunity from a claim for negligence or breach of duty arising out of their preparation and presentation of evidence for the purpose of court proceedings'. However, an opposing party cannot bring a claim against an expert.[9]

## 7.4 Alternative Dispute Resolution

Commercial disputes often result in litigation, but there are other ways of resolving commercial disputes, As court costs have escalated, the alternative dispute resolution (ADR) method has emerged.[10] The role of lawyers is minimal, with experts performing a more important task than is the case in a formal court of law. Such ADR methods include arbitration before a single assessor (Single Joint Expert, Section 7.5) and mediation, where a lawyer consults the parties separately to test their evidence. It is arranged on a fixed date, with each party sitting in separate rooms together with their respective expert witnesses. The mediator moves from room to room, putting points to each party raised by the other side. In turn, the quality of the evidence raised by each expert is tested by the mediator, until one side or the other starts to yield ground. This is the turning point, as offers of settlement can then be made and modified until a deal is possible without the need for court action. The whole process concentrates minds, due to the sharp deadlines imposed by the mediator. Since decisions are reached in private, there is no public record, an advantage for some, but a disadvantage if a major design problem does not reach the public record.[11,12]

## 7.5 The Single Joint Expert (SJE)

Within the UK, the Civil Procedure Rules emphasis an advantage, 'particularly in smaller claims, for the parties to agree on and appoint a single joint expert (SJE)' rather than each party appointing its own.[11–13] So, rather than an 'expert' *per se*, the SJE role becomes that of a 'mediator', instructed to prepare a report for the court on behalf of two or more of the parties (including the claimant) to the proceedings.

In an identical manner to that of a party-appointed expert, the prime duty of the SJE is to assist the court on matters within their expertise. Clearly, a prime requirement of an SJE is to maintain independence, impartiality and transparency at all times. The aim of appointing an SJE is to reduce cost and increase the efficiency of expert testimony.

Generally, it is up to all parties involved to reach agreement on the appointment of the SJE. However, if the parties cannot reach agreement, the court may select an expert from a list prepared or identified by the parties, or can direct how the expert is to be selected. The scale of SJE fees and expenses are generally included in any terms of instruction or engagement, but a limit can be imposed by the court. When an SJE is appointed, all parties concerned can, nevertheless, instruct a different expert to act as their *advisor*, but (win or lose) they may not be able to recover the additional cost at the end of the hearing.

When compiling a report, the SJE may be required to make different assumptions of fact in the matter and may therefore require more than one set of opinions to be promulgated on any one issue. However, as in the case of a party-appointed expert, the SJE does not *determine* the facts, that being the role of the court.

## 7.6 The Role of Case Law

As highlighted in Section 7.1, the evidence given by the forensic engineering expert is a duty directly to the court, as emphasised in the Woolf reforms. In the past, problems frequently arose from conflict and confusion that some experts may have had over their duty to the court and to their 'paymaster', often raising the spectre of the 'hired gun'. Nowadays, if an expert shows bias towards their client, they can expect a rough time during cross-examination on the witness stand. There is a body of available writings explaining the verdicts in a legal case. Aptly termed 'case law', it is frequently established by judicial rulings and reasoning. Case law can be cited as a precedent in other cases, along with statutes that may have a bearing on any decision.[12,14] 'In a case from the 1980s, the Ikarian Reefer, the judge (J Cresswell) described the various responsibilities of experts. Their evidence should be: independent of the litigation; objective and unbiased. Expert witnesses should: state facts and assumptions on which the opinion is based, together with any facts that could detract from the conclusions; give evidence within their expertise; only give opinions provisionally if not all the facts are known; ensure any changes in opinion after the exchange of reports are sent to all sides in the dispute; ensure all photographs, plans, survey reports, and other documents referred to in the report, are included in the exchange before trial.[13,15,16] These rules have now been incorporated into the Supreme Court rules under the Woolf reforms.'

## 7.7 The Engineer in the Courtroom

In the courtroom, the forensic engineering expert will be required to qualify (by relating under oath) his education, training and experience in the engineering disciplines involved. The opposing barrister and legal team may well attempt to refute the expert's qualifications, perhaps by pointing to some specialised aspect of the case at hand. Consequently the engineer's expertise and standing, in his particular branch of engineering, may also be subjected to assessment before the court. However, if he/she is finally accepted by the court as an expert, that testimony will carry weight, and will be admitted. The expert can then expect to be formally questioned by the engaging barrister, and cross-examined by the opposing barrister. During any cross-examination, the forensic engineering expert will come under intense focus, where a legal team working for the opposing side will probe all matters relevant to the case. The expert must exhibit all his or her skills of effective technical presentations and aptness of expression. Despite critical commentary made in real time, the expert must maintain a level of dignity expected of the engineering profession. During the course of trial, cross-examination may also be assumed by the presiding judge who may well be a former barrister and therefore intimately familiar with critical analysis. Therefore any expert must expect a thorough and in-depth analysis of their opinion, research and report. The paramount objectives of cross-examination are to assess

- The credibility of the expert witness
- The reliability of the evidence examined by the expert witness

- The reliability of the arguments raised by the expert when offering opinion on the mode of failure
- Any ambiguities or contradictions within the expert report.
- Any conflicts between the expert evidence for the various parties in the dispute.
- In the majority of trials, the barrister will not normally be technically qualified and so will be advised by the expert acting on behalf of the opponents.
- Courts expect the expert witness to report directly to the court when preparing final submissions. The primary duty of the expert is to provide independent, unbiased opinion to the court, not to any perceived paymaster.[12,16]

The situation is found to be somewhat different in courts trying technical disputes. Here, the barrister and judge will have some degree of technical background, thus enabling them to appreciate scientific or engineering argument. Therefore, the Technology and Construction Court, the Commercial Court and the Patents Courts are usually fully equipped to deal with complicated technical issues.[17]

'Whatever knowledge a barrister can bring to a case, he will explore the logic of an opinion, so the expert must be prepared to be challenged on the arguments he or she uses to interpret the evidence. It also calls for a balanced attitude to conflicting pieces of evidence or opinions, and a fair interpretation of the facts established by the court. For instance, if cross-examination shows witness evidence, that the expert has previously relied upon, to be false' – the expert has a duty to modify his or her opinion to allow for the new facts.[16] Such a change of thinking or opinion is not unusual as a case develops at trial, actively demonstrating that cross-examination is a vital tool for revealing new and critical issues in a dispute.

## 7.8 In Summary

For the forensic engineer undertaking expert witness duties, there are numerous additional responsibilities:

- There is an overarching duty to the court.
- A statement of truth must be given with the expert report.
- There must be disclosure of previous experience. This is normally accomplished by including a detailed curriculum vitae within the report body.

Becoming an effective expert witness is an on-going process of both education and experience. The range of attributes required of the most effective forensic engineering expert witness is that they are

- True to themselves
- True to their duty
- Do their background homework
- Diligent, honest, objective, credible
- Clear, succinct, unflappable
- Strong but not arrogant
- Know their topic
- Know their own (and the case) limitations
- Know the forum

## References

1. *Access to Justice Final Report*, by The Right Honourable the Lord Woolf, Master of the Rolls, July 1996, Final Report to the Lord Chancellor on the civil justice system in England and Wales. 1995, London: The Stationery Office, ISBN-13: 978-0113800995, SBN-10: 0113800991
2. *101st Update to the Civil Procedure Rules (CPR)*. Changes come into force variously on 08 November 2018 and 30 November 2018. https://www.justice.gov.uk/courts/procedure-rules/civil
3. Reynolds, M.P., and P.S.D. King, *The Expert Witness and His Evidence*. 2nd ed. 1992, Oxford: Blackwell Scientific.
4. Smith, D., *Being an Effective Expert Witness*. 1993, Norwich: Thames Publishing.
5. Yee, K.K., Dueling experts and imperfect verification. *International Review of Law and Economics*, 2008. 28(4): pp. 246–255.
6. Surhone, L.M., M.T. Timpledon and S.F. Marseken, *Sally Clark: Solicitor, Miscarriage of Justice, Pediatrics, Sudden Infant Death Syndrome*. 2010, Devizes: Royal Statistical Society, ISBN-10: 6130485352, ISBN-13: 978-6130485351.
7. http://dbpedia.org/page/Sally_Clark.
8. https://www.reddit.com/r/MensRights/comments/26zk5r/.
9. Hadley-Piggin, J., *Expert Evidence: The Roles and Responsibilities of the Expert Witness, Keystone Law*. 2016, https://www.keystonelaw.co.uk/keynotes/expert-evidence-the-roles-and-responsibilities-of-the-expert-witness.
10. Acland, A.F., *Resolving Disputes Without Going to Court*. 1995, Evergreen: Century Business Books.
11. Lewis, P., K. Reynolds and C. Gagg, *Forensic Materials Engineering Case Studies*. 2003, CRC Press, ISBN-10: 0849311829; ISBN-13: 978-0849311826.
12. Lewis, P., K. Reynolds, and C. Gagg, *Post-Graduate Course, T839 Forensic Engineering*. 2001, The Open University, Materials Engineering Department, MMT Faculty.
13. https://www.academyofexperts.org/SJE.

14. Finch, Emily, *Law Express: English Legal System*. 6th ed. 2016, Pearson, ISBN-10: 1292086874, ISBN-13: 978-1292086873.
15. Dillow, Gavin, ed., *Halsbury's Laws of England 2016: Consolidated Tables*. 5th ed. Lexis Nexis, Reed Elsevier, ISBN-10: 140579741X, ISBN-13: 978-1405797412.
16. https://www.open.edu/openlearn/science-maths-technology/engineering-and-technology/engineering/introduction-forensic-engineering/content-section-1.8.3.
17. *Civil Procedure Rules (CPR Part 60)* – Technology and Construction Court Claims. 2018, London: Ministry of Justice,https://www.justice.gov.uk/courts/procedure-rules/civil/rules/part60

# Engineering Failure Case Study Analysis as a Teaching Tool*

**8**

I Hear, and I Forget
I See, and I Remember
I Do, and I Understand

**Attributed to the Chinese thinker and social philosopher Confucius (551–479 BC).**

The moral: tell me and I [will] forget. Show me and I [will] remember. Involve me and I [will] understand. In other words, man will learn best by 'doing'.

The premise encompassed here is that if students are engaged, and actively participating in their learning, they will remember. If students remember then they will understand and apply this knowledge to future real-life experiences.

## 8.0 Introduction

Within the UK, there exists an irresistible move away from undergraduate laboratory-based engineering teaching to that of virtual (computer-based) learning. Although not specifically emphasised in Chapter 1, it is my considered opinion that a prime driver of this trend can be found with continued pressures for cost reduction in the higher education sector. Therefore, the move away from practical laboratory exposure is unstoppable. Consequently, the churning out of 'paper' engineers having no practical experience – nor an intuitive 'feel' for the subject – could rapidly become the norm. The words 'virtual' and 'modelling' do *not* mean reality. An analogy is the airline pilot: no pilot would be given command of an aircraft based simply on time in a flight simulator. The simulator plays a *part* in the training process, along with appropriate classroom tuition and hundreds of hours of practical flying with an instructor. This practical flying element allows the trainee pilot to develop an *intuitive* feel for the response and behaviour of the aircraft. In addition to normal flying response, the practical element will include a range of unusual

---

* Certain brief passages in this chapter are incorporated, with permission from Elsevier, and used verbatim from the article – *Domestic Product Failures: Case Studies* (2005) – published in the journal *Engineering Failure Analysis*. The article was authored by the author and is cited within the References section at the end of this chapter.

situations, designed to expose the trainee to scenarios that he may (hopefully) never encounter. Similarly, the training of an engineer should be expected to include classroom elements, simulation *and* practical experience. However, with the continued demise of costly practical engineering training elements in undergraduate offerings, it is mooted that the study of current and appropriate case study analysis may offer a solution to this shortcoming. Over the last decade, researchers have proposed that one solution to this problem is the integration of research in forensic engineering into undergraduate courses.[1,2] Although not a direct replacement for practical training, case study analysis will expose the undergraduate to a vast wealth of practical knowledge and intuition. Case study exposure will offer a positive impact on students' engineering learning experience, rather than simply relying on virtual learning alone. One result of exposure to 'experience-by-proxy' is that young engineers will be armed with the (often intangible) benefit of taking this knowledge and dawning intuition with them into the workforce.

## 8.1 Historic Use of Case Study Analysis

### 8.1.1 Ancient Rome

As civil engineering is one of the oldest engineering disciplines, it is not unexpected that the first true case studies are drawn from the latter part of the 1st century AD. Vitruvius was a Roman architect, engineer and author of the celebrated treatise *De architectura*, a handbook for Roman architects. His work is divided into ten books. They address both city planning and architecture as a whole. The breadth of work is wide-ranging, incorporating building materials, temple construction, public and private buildings, clocks, hydraulics and even civil and military engines.[3] His overriding priority was to preserve a Classical Greek tradition for the design of temples and public buildings. Therefore, his treatise was in effect a handbook of case studies of classic Greek design, with his prefaces containing many pessimistic remarks about the architecture of the time. His work remained the authority on ancient Classical architecture from ancient Roman times to the Renaissance.

### 8.1.2 Late Georgian–Early Victorian Era

In somewhat more recent times, bridge designer John Roebling (1806–1869) studied bridge failures to determine each cause of the premature demise of suspension bridges. As such, he can perhaps be considered the first engineering failure analyst, and the first engineer to appreciate the power of case studies. Roebling used his knowledge to establish design criteria for his own suspension bridge designs. He tried to predict all possible ways that his bridges could fail and designed them to prevent such failure. When the rest of the engineering community had given up on suspension bridges as a reliable structure,[4] Roebling was the only engineer at that time to take this approach. As a result, his masterpiece, the Brooklyn Bridge, which opened in 1883, still stands today (Figure 8.1A). However, John Roebling never saw the opening, as he died in an accident while surveying for the bridge. After two years of working in compressed air along with his fellow workers, his son Washington became crippled, nearly blind and deaf, suffering from 'caisson's disease', now known as the bends (Figure 8.1B). Nevertheless, the bridge has stood safely for the last 126 years, reminding the engineering community of what can be accomplished by learning from past engineering mistakes. Roebling's use of failure concepts and case studies to avoid failure in his own designs shows that a case study approach has the potential for both the professional and student to learn from specific and useful examples from the past.[5] Furthermore, it is clear that the John Roebling story provides an excellent example of good engineering practice in action.

## 8.1.3 To Date

Since its inception some 40 years ago, case study analysis has been a favoured route for the teaching of engineering at the Open University in the UK. In those early days, it was clearly recognised that a prime aim of any engineering or engineering technology programme should be the dissemination of the current body of knowledge to the student.[6] Furthermore, the ethos adopted was to ensure a well-rounded exposure that would 'prime' the student and thus allow successful contribution to, and eventual mastery of, their particular field. The method chosen for delivering such engineering exposure was by current and relevant case study analysis, the foundation stones of each teaching module. The intention was that the training given to the student and the knowledge they acquire should allow them to keep abreast of their

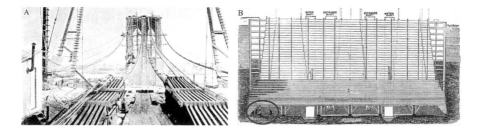

**Figure 8.1** (A) The Brooklyn Bridge under construction, with the last cable wire run on 5 October 1878. (B) East River caisson. Note the workers digging by hand at the foot of the caisson (circled in red). (Courtesy of Wikimedia Commons, public domain.)

chosen field, as well as helping them to avoid making potentially serious and fatal mistakes in their future practice. In essence, incorporation of failure case study analysis into the teaching stream is a classic example of practical 'research'-enriched teaching.

To date, case study analysis is used throughout both the undergraduate and post-graduate engineering programme.[7] When considering forensic engineering *'per se'*, an Open University postgraduate course (T839) had both its teaching and practical content based around a range of case study examples. Using real case studies, one of the central aims of the course was to provide guidance for good product design before development, thus minimising – if not eliminating – wasted effort during the product development cycle. As indicated, the approach taken was through case studies, many based on the authors' own case books,[8] with additional focus on historic catastrophes and failures.

## 8.2 Air Accidents, Mortality Statistics and Failure Case Studies

As discussed in Section 1.12.2, in the seven-month period between January and August 2018 for example, there were 64 reported air accident incidents, with a death toll in excess of 745 persons. This total can be compared with the 1,793 people who were killed in reported road traffic accidents in Great Britain in 2017.[9] Although in no way belittling each of the tragic automobile fatalities, the number of aviation fatalities certainly raises questions as to why such incidents are still occurring in what is considered a technologically advanced age. Rouse[10] has suggested a number of possible reasons for continued engineering failures:

- Practicing engineers have forgotten past mistakes that have influenced the evolution of the profession.
- Engineers try to hide embarrassing mistakes.
- Confidential settlements will always contain hidden detail; therefore there are no clear records of the failure to learn from.
- Engineers rely heavily on computer analysis rather than on 'sound engineering judgement', and an intuitive 'feel'.

Rouse also states that failure case studies can

- Imprint in the minds of students the need to check calculations repeatedly
- Encourage the student to second-guess computer analysis when it does not agree with sound engineering judgement

It is the overriding opinion of Rouse that case study analysis can be used to train students to think independently and not rely solely on the educated guesses of others, or (more importantly) of their computer.

## 8.3  What Constitutes a 'Case Study'?

The terms 'case study', 'failure case', 'failure analysis' and 'forensic engineering case' are used loosely and interchangeably in the engineering profession. However, a narrower definition for any engineering case analysis is that it is a narrative account of a situation, problem or decision usually derived from actual experience.[11] Case reports are increasingly finding use across a range of engineering curricula, whatever the terminology used. Here, they are used to enhance and enliven the learning experience of engineering principles and practices. A case report will consist of an account of an engineering activity, event or problem, and will contain the background and complexities actually encountered during investigation. Case analysis will present real-life engineering situations, reflect all sorts of engineering activity, failures and successes, old and new techniques, theoretical and empirical results. Therefore, case presentation must impart true facts, which should not be biased, or distorted to fit a pre conceived 'theory' held by the author. Engineering case analysis will differ from a technical article in which the author will make his point as directly as possible, using a logical sequence for presenting findings, conclusions and opinions.[12] This sequence seldom reflects how the investigator arrived at his conclusions. In contrast, an engineering case analysis will describe a series of events that reflect the engineering activity as it actually happened, warts and all, without fear or favour. The case writer should (attempt to) restrain his own opinion and derived conclusions, simply so the reader can assimilate the information provided and thus learn from the experience of drawing his own conclusions.

### 8.3.1  Case Study: North-Sea Ferry Engine Bolts[13]

After replacement as part of a regular maintenance schedule, a particular North-Sea ferry started to experience problems associated with its engine mounting bolts. The 30-mm diameter replacement bolts were failing at one corner of an engine mounting system. As the replacements were only lasting some three weeks 'in-service', maintenance personnel felt that replacement bolts had to be from a faulty batch. However, other views of the accelerated demise centred on corrosion-enhanced failure from operating in a sea-water environment.

The diesel engines were located mid-ship and generated some 3,000 HP at 900 RPM, through a single reduction reversible gearbox. Each of the two main engines was located near mid-ship, with a respective gearbox mid-way between the propeller and the main engines. Being some 30 m in length, alignment of the engine to the propulsion shaft-line was critical.

A marine 'engine vibrates in three different fields of motion; fore and aft (longitudinal), side to side (lateral) and up and down (vertical)'.[14] This motion is controlled by the engine mount, but not necessarily to the same degree for each type of motion. 'Most engine mounts are designed to severely restrict longitudinal, fore and aft thrust and vibration yet allow some lateral and vertical motion to dissipate vibrations.' In addition, forward and reverse shear thrust generated by the propeller must also be controlled by the engine mount. However, incorrect installation of the mounts on the engine bed can be a common issue. 'It is imperative that the mounts are perpendicular to the engine's crankshaft, so that when the mount' bolt goes through the engine bracket it is not pulled to the side, or fore and aft, preloading the mount and bolt.[14] Fine engine alignment is accomplished by chocking and shimming.

Simple observation of a number of the failed bolts (Figure 8.2A) revealed no obvious sign of corrosion on the fracture surface. Therefore the notion of environmentally enhanced failure was put to one side. Chemical analysis of the bolt material (see Table 8.1) revealed nothing untoward, with material of manufacture being of a high-tensile steel. What were clearly present on the fracture were beach markings, indicative of fatigue as the failure mechanism (Figure 8.2B). However, as the bolts failed after only three weeks in service, the chances of fatigue failure as a direct result of normal engine vibration appeared remote. The answer became evident after a site visit to the ferry. A particular corner chock and shim, being one of many used to align the engine, had worn and become loose. This in turn had allowed flexing of the engine housing, leading to a cyclic bending

**Figure 8.2** (A) The subject North Sea ferry engine mounting bolt, being 30 mm in diameter with a 3.5 mm pitch. Note: the fracture surface is shown having been sectioned for chemical analysis and fractographic observation. (B) A low-power magnification of the fracture surface, with the presence pf beach markings clearly evident.

**Table 8.1 Elemental Composition Analysis and Hardness Evaluation for the Subject Ferry Engine Mounting Bolt**

| Element | C | Si | Mn | P | S | Fe | Hardness (Hv) | UTS (MPa) |
|---|---|---|---|---|---|---|---|---|
| Weight% | 0.3 | 0.3 | 0.8 | 0.02 | 0.03 | Bal | 265 | 848 |

*Note:* UTS inferred from hardness i.e. UTS = 3.2 Hv.

load being transmitted to adjacent mounting bolts, hence providing the driver for fatigue failure in an exceptionally short working lifetime. Simple replacement of the worn chock and shim cured the problem.

This case study highlights a number of salient investigative points:

(1) Do not be led by others. Although there is a duty to consider other opinions, let the evidence speak for itself.
(2) An understanding of the operational forces at play was paramount.
(3) It required a site visit to determine the chain of events that led directly to premature demise of the ferry engine bolt.

Cases should be prepared to reflect real-world concerns, situations and issues that the engineer may encounter in practice. They are often open-ended, with no clear-cut solution, and can be presented in a variety of formats such as

- Case histories
- Case problems
- Case studies

A case history should present an account of an actual event or situation, review all variables and circumstances at issue, describe how the problem was solved and examine the consequences of decisions and the lessons learned.[15] Examples of case histories include the Challenger and Columbia space shuttle disasters, the Hyatt Regency walkway collapse, the Tacoma Narrows Bridge failure and the Three Mile Island incident. Each of these incidents can be used to illustrate both technical and ethical issues.

A case problem should be compiled to allow many possible solutions. Analysis of the problem and choice of approach are left to the student or reader. The individual must deduce a likely train of events or determine the most probable scenario at play, and be in a position to defend their deductions.

An engineering case study should be a written record of a forensic investigation or failure activity as the process actually progressed. As such, the study should document a real and appropriate engineering experience.[16] The viewpoint taken should preferably be that of the investigator and/or of the participants. A good case study should contain more than just the bare facts of the problem at issue. The reader should be exposed to as much of the supporting information as practically possible, thus establishing a framework for the problem. Furthermore, the study may present data such as

- Sketches
- Drawings
- Photographs
- The political scene
- Investigative tools/techniques
- Test data
- Calculations
- Production processes
- Reverse engineering
- Field reports
- Any additional relevant data

all of which may influence the development of probable failure scenarios and ultimate conclusions.

In short, engineering case study analysis should provide an account of a problem, technical issue, ethical dilemma or design challenge. Case studies should provide a context for the application of knowledge and experimental and analytical techniques learned by the student during the course of his or her study.[17] Clearly, the page count for any complex failure report may run into the hundreds, if not thousands. *For the sake of airing as wide a range of failures as possible, case studies presented within this book are, by (page count) necessity, all abridged versions of a full report.*

Advantages of exploiting case study analysis across the engineering curriculum include providing a broader perspective on the issues, focusing on decision-making processes, revealing the range of analytical tools required for analysis, highlighting organisational structures (and barriers) within a company, stressing uncertainty, ambiguity and risk and emphasising the role of politics (both internal and external) in the final outcomes.[18] Any case study analysis should therefore be designed to extend the learning experience beyond both the classroom and laboratory. Furthermore, as hands-on laboratory training is becoming an exception, rather than the norm, engineering case study exposure will offer the potential to bridge the ever-widening practical experience gap. Well-researched and -written case studies will impart a positive impact on students' practical engineering learning experience, rather than simply relying on virtual learning alone. One result of exposure to 'experience-by-proxy' is that young engineers will be armed with the (often intangible) benefit of taking this knowledge and dawning intuition with them into the work force.

The following, and somewhat unusual, case history will provide a short but ideal illustration of the phenomenon of hot-spot boiling erosion.

## 8.3.2 Case Study: Hot-Spot Boiling Erosion

A two-year-old (in 2017) automobile was returned to a garage with major engine power loss issues. After removal, the aluminium engine head was inverted and inspected for any sign of obvious damage. An anomaly, sited at cylinder 4 (the furthest point away from coolant entry) quickly became evident (Figure 8.3A). A closer view of the damage is seen in Figure 8.3B, where it can be said with confidence that the damage sustained was that of erosion, imparted as a direct result of hot-spot boiling. This form of erosion can be identified by intense, highly localised, areas of severe water-way erosion. Erosion from hot-spot boiling is most commonly found down the exhaust side of an alloy cylinder head water jacket, and will lack the tell-tale white oxide found with general aluminium alloy corrosion. Furthermore hot-spot failure morphology will exhibit far sharper edges than that formed by general corrosion, as clearly seen in Figure 8.4A. Hot-spot erosion failure will often occur in only one or maybe two water-ways (generally, those furthest away

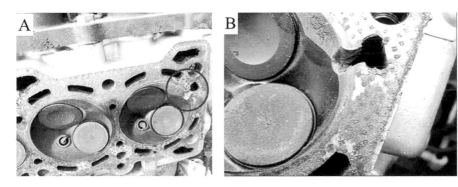

**Figure 8.3** (A) The inverted aluminium engine head in question with damage sited at cylinder 4, circled in red. (B) Closer view of the damage which is that of erosion, imparted as a direct result of hot-spot boiling.

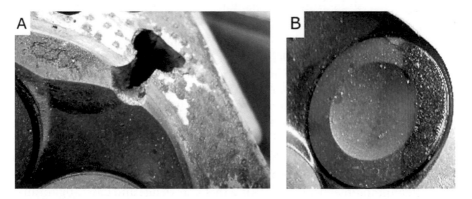

**Figure 8.4** (A) Under higher magnification, erosion damage is seen to have sharper edges and no tell-tale white oxide found with general aluminium alloy corrosion. (B) Evidence of valve head contact with the piston crown, circled in red.

from coolant entry point), leaving all the other water-ways in pristine condition. This fact alone is a clear indicator that the erosion process at play was not driven by the action of chemical corrosion. Rather, it signifies a physical manifestation of damage caused by implosion of oxygen bubbles that are formed during hot-spot boiling. The only preventative solution for such a condition is the addition of glycol at a 50% by volume to the engine coolant reservoir. This addition will raise the coolant boiling point, and thus reduce the likelihood of the occurrence of hot-spot boiling. There is an added advantage in that for cold climates, glycol will also act as an anti-freeze. However, if coolant mixtures are lower than this 50% ratio mix, the effectiveness of glycol in reducing hot-spot boiling will be compromised.

There are a number of additional situations that may exacerbate hot-spot failure, for example severe lugging. Here an engine operating at excessive loads or running the engine above the rated speed recommended by the manufacturer's specifications will equate to higher engine component temperatures. Severe lugging or lower engine revolutions will cause a decrease in fresh air volume, resulting in piston crown overheating, introducing a necessity for the cooling system to remove the additional heat load.

As a point of interest, diesel engines can also suffer a similar form of erosion on cast iron wet liners. The condition is again caused by the implosion of small oxygen bubbles on the cylinder walls. However, the difference in diesel engines is that the bubbles are formed by the violent action of diesel combustion. This form of liner will produce a multitude of small holes in a concentrated area. Even with the use of a correct heavy-duty coolant, the aggressive nature of this form of the erosion process will require liner repositioning at each scheduled service interval. If the wet liner is continually used in the same location or position, the erosion process will eventually form a hole in the cylinder bore, with the probability of hydraulic bending of the con-rod, and/or severe overheating, being the resultant outcome.

When considering engine pistons, a piston crown may have an asymmetric crown shape, or circular valve relief pockets that will allow for greater compression without the valves striking the piston heads. The crown may also have raised asperities (bumps) that improve compression by decreasing the volume of the combustion chamber. With the engine in question, it would appear that an exhaust valve head had been contacting the piston crown, as circled in red on Figure 8.4B. Evidence of imprints on the valve face pocket provided a fingerprint for piston crown contact. On balance, valve/piston contact would commonly occur as a direct result of poor engine timing. However, excessively high engine revolutions ('over-speeding') may also contribute to such valve face damage. Over-speeding can stretch the component tolerance stack-up, thus causing piston contact with the valves.

In summary, damage sustained by the aluminium engine cylinder head was that of erosion initiated as a direct result of hot-spot boiling. This mechanism is recognisable as concentrated isolated areas of severe erosion of waterways, commonly sited down the exhaust side of the cylinder head.

Until recently this problem went unrecognised as a major cause of erosion in modern alloy heads and blocks. In addition, the situation at play was exacerbated by poor timing, as witnessed by valve head imprints in a number of the piston crowns. The onset and progress of failure should most certainly have been heard by the vehicle driver, and for a considerable period of time prior to demise. Furthermore, the condition of the vehicle timing would most certainly indicate a lax or cavalier attitude to engine servicing. It eventually transpired that, in the two years of use from new, the vehicle owner had not fulfilled any mandatory scheduled servicing required for the vehicle. As such, the vehicle three-year manufacturer's warranty was deemed null and void.

## 8.4 Writing a Relevant Case Study

Providing students with case study experiences can be viewed as equipping the engineers or technologists with both the theoretical and basic 'practical' tools required to perform effectively in their chosen field. However, there are different opinions about how to develop good cases. Kardos[19] highlights the objectives and the content of engineering case studies as

Providing a medium through which learning (e.g. analysing, applying knowledge, reasoning, drawing conclusions) takes place.

He considers imparting additional specific information as relatively minor and coincidental, with a good case study

- Taken from real life (a necessity).[20]
- Consisting of one or more parts, each part usually ending with problems and/or points for discussion
- Including sufficient information for the reader to both analyse and verify problems and issues contained within the study

To make a case believable to the reader, a good case should usually include[21]

- Setting
- Personalities
- Sequence of events
- Problems
- Conflicts

An engineering case study is a written record of a forensic investigation or failure activity as the process actually progressed. As such, the study should document a real and appropriate engineering experience.[22,23] The viewpoint taken should preferably be that of the investigator and/or of the participants. A good case study should contain more than just the bare facts of the problem at issue. The reader should be exposed to as much of the supporting information as practically possible, thus establishing a framework for the problem.[24]

## 8.4.1 Case Study: Crankpin Axle from a Bicycle

'A 20 year old cyclist was pedalling up an incline when the left pedal broke without warning, causing the rider to be thrown into a ditch causing head, leg and arm injury.'[25] The nature of injury was severe, requiring three months' hospitalisation along with many more additional months of recuperation. The subject bicycle was some three years old at the time of its failure, 'had been properly maintained and had not been involved in any previous accident. The injured rider was a keen cyclist' and had used the machine regularly, covering 35 to 50 miles most weekends and 8 to 10 miles most evenings. All maintenance had been undertaken by a retailer, qualified to maintain sport cycles. 'At the time of his accident, the cyclist weighed 9 st 12 lbs and was 5'9" tall. He was absolutely certain the left pedal had come off while he was standing on it to keep his speed up the incline. Furthermore, there was no kerb at the side of the road, which might have caught against the machine whilst he was riding.'

The broken left pedal was part of a 'cotterless' crank, located onto a square section on the end of the crankpin axle and held by a screw. For comparison, an exemplar cotterless crank axle is shown in Figure 8.5A, with the subject crank axle shown in Figure 8.5B. 'Light abrasion marks were noticed near the top of the crank, but were of recent origin and consistent with the metal having contacted the road surface after it had broken off. No damage could be found on the crank or the pedal that would suggest any kind of impact of sufficient magnitude to fracture the crankpin axle. Both pedals were rigid and completely undamaged apart from light abrasions on the outside of the spindle caps, commensurate with 3 years service. Microscopic study of the ball bearing tracks on the crankpin revealed no evidence of indentation, such as would have occurred if any substantial impact force had been transmitted from the pedal crank to the frame bearings (Figure 8.5C). Any major impacting force transmitted from the pedal or lower part of the crank to the crankpin would have been expected to bend the crank out of line, but there was no evidence of any such damage. It was concluded that the crankpin' could *not* have broken instantaneously, as a result of some externally applied mechanical overload force.[26]

Two observed features were considered significant: first, there was a rim approximately 1-mm deep extending round the periphery, suggesting that the component had been case-hardened. This is a conventional method of treating cycle parts to achieve wear resistance and strength, so there was nothing unusual about this discontinuity in the fracture surface. The second feature was significant in relation to the mode of failure and is revealed by the path the

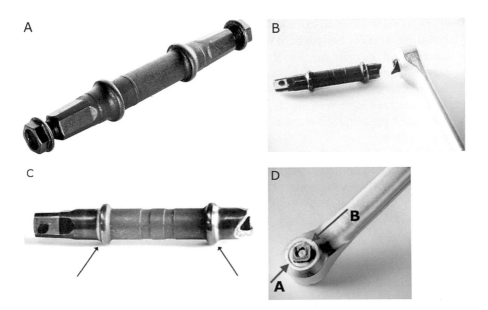

**Figure 8.5** (A) An exemplar cotterless crank axle is shown for comparative purposes. (B) Broken axle with crank arm still attached. (C) Closer view of crank axle. Note the absence of any damage to the ball tracks (arrowed, bright areas at raised shoulders), which would have been introduced by any major impact event. (D) View of fracture surface, with region of initiation marked 'A' and cusp where helical fractures terminated marked 'B'.

fracture takes across the section. In Figure 8.5D it can be seen that the fracture is at 90° to the axis of the crankpin on the sides of the square adjacent to the corner identified 'A', whereas it follows helical paths terminating in a cusp at the opposite corner marked 'B'. This is interpreted as a progressive fracture which started along the edge to the right of 'A' and progressed in a series of jumps to finish at the corner 'B'. There had been a significant change in the stressing system when the fracture reached the midway position, where the path changes from 90° to the surface to the helical route, and this was associated with the crack front bending forces which had initiated the cracking and directed its path up to this stage.[25]

There was no visual evidence to suggest this failure was related in any way to the manner in which the cycle had been maintained or that it was caused by mechanical abuse or a result of accidental or impact damage.[25] The crankpin had failed by a fatigue mechanism. A crack had initiated along one edge of the square and had been spreading across the section over a period of time until it had reached the stage when all that was required to break it completely was a slightly higher than average force on the pedal, but one at a level that had been withstood satisfactorily on many previous occasions before the cracking had spread so far. The circumstances at the time of the accident are therefore of little significance, since failure was inevitable sooner or later; all that was required as an extra hard push on the pedal. The path indicated the pedal was near the bottom of its travel when it finally broke off and this would have caused the rider to fall to the left-hand side of the machine. There would have been no warning or externally visible sign of the impending failure. The fatigue crack would have appeared as a fine, hairline mark on the surface, but this would not have been visible without removing the crank and, even then, it would have required someone with specialist knowledge to recognise its significance.

Failure of a crankpin such as described is unusual, suggesting some form of weakness stemming from an inherent material or manufacturing fault. Accordingly, the broken end was removed from the pedal crank and sectioned so that both metallographic examination and micro-hardness testing could be undertaken.

A polished longitudinal section was scanned for possible alloying elements on a scanning electron microscope using the EDAX technique. Results signified the material of manufacture to be that of a plain carbon steel, with no additional alloying elements (Table 8.2).

Longitudinal and transverse sections were prepared, the former being selected so as to include both the initiating and the terminating regions of the fracture. Etched micro-sections, illustrated in Figures 8.6 to 8.8, revealed a structural abnormality resulting from the case-hardening heat-treatment process. Case hardening is commonly applied to develop wear and fatigue resistance in engineering components and is widely used for cycle parts. It is the high hardness of the case that imparts wear resistance, with the martensitic transformation producing compressive stress in the case, which generally improves the fatigue resistance.[25] Figure 8.6A shows an etched longitudinal section of the entire cross section of the crankpin within 4–5 mm of the fracture and Figure 8.6B the same specimen at the fractured edge. Features normally observed in a correctly case-hardened component would comprise the martensitic case and a mixed structure of ferrite and low-carbon martensite in the core (which would extend to the screw threads of the centre hole in the crankpin). The depth of the case is 0.6 mm in this section. Of particular significance is the difference between the fractured edges on opposite sides of the crankpin; that in Figure 8.6B, where the fracture initiated, is fairly smooth and featureless whereas that of the opposite edge, where it terminated, is rough and irregular. These features confirm that the failure occurred in a fatigue mode.

Table 8.2   Result of Elemental Analysis of Fractured Cotterless Cycle Crank Axle

| Element | (%) | Element | (%) |
|---|---|---|---|
| Silicon | 0.2 | Chromium | 0.04 |
| Manganese | 0.73 | Nickel | 0.53 |
| Molybdenum | 0.16 | | |

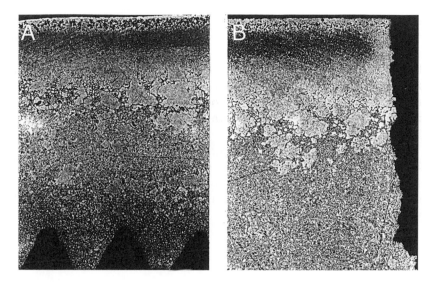

**Figure 8.6** (A) Etched cross section extending from outer surface to threaded hole, taken a short distance from fracture initiation. Magnification: 29×. (B) The right-hand micrograph is of the same section as (A) but includes the fracture surface at the right side, and at slightly higher magnification. Magnification: 36×.

Figures 8.7A and B, the region where the fracture initiated, exhibit similar microstructures to the above, except for one very significant feature of the hardened case (seen along the top of the micrographs). In Figure 8.7A a mottled band can be seen parallel with and just below the outer surface, and this continues into the fracture, becoming slightly less distinct. This same region is illustrated at higher magnifications in Figure 8.7B. The dark areas are 'pearlite', which should not be present in a correctly hardened case, and reveal that the quenching rate at this surface was too slow when the crankpin was quenched to harden the carburised case. In the lower illustration of Figure 8.7B the magnification has been increased to show the presence of cementite network outlining grain boundaries between the pearlite colonies. At the fracture surface these have opened out into crack, and it is considered that these were the initiation sites of the fatigue cracking.[25]

A transverse section taken 8 mm further away from the fracture, towards the pedal end, is illustrated in Figure 8.8. This shows the outer part of the case to be completely pearlitic, indicating the quenching rate was even slower in this region than closer to the fracture. Beneath this mottled structure at the outside is the featureless, white etching band of martensite, which merges into the duplex structure of the core towards the bottom of the photograph.

Vickers hardness tests were undertaken on the martensitic and pearlite bands. Values were in the range of 900–950 Hv and 780–820 respectively. A line survey consisting of 12 equal intercepts from the surface to 1.5 mm into the core, carried out on the section illustrated in Figure 8.8, produced a gradient which fell steeply from 950 Hv at the extreme surface to 802 Hv at 0.1 mm

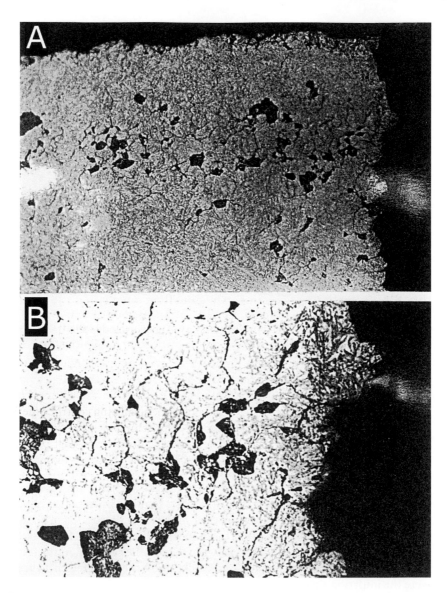

**Figure 8.7** (A) Outer surface at the top of the figure and fracture surface at right-hand side, showing detail of pearlite band and intergranular outlines of brittle cementite. Magnification: 360×. (B) High magnification of same image, with outlines of brittle cementite clearly seen. Magnification: 1080×.

into the case, thereafter rose to 940 Hv again at 0.3 mm and remained at this level to 0.6 mm, before gradually falling to 434 Hv at 1.5 mm in the core. The softer zone in the case corresponded to the position of the band of pearlite appearing in Figure 8.8.

Conclusions from metallurgical results revealed that the steel from which the crankpin was made was of suitable quality for case hardening

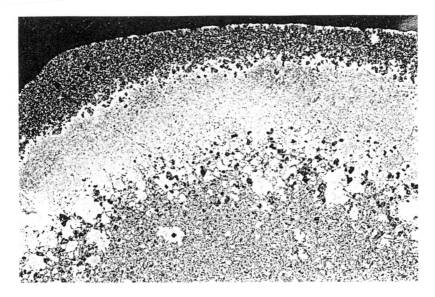

**Figure 8.8** Transverse section taken further away from the site of fracture, showing extensive pearlite region in outer part of carburised case. Magnification: 36×.

by carburising, but the structure produced by quenching was not uniform throughout the case, inasmuch that the cooling rate during the quench had not been sufficiently rapid to produce a wholly martensitic structure. The conditions described are usually the result of slack quenching control, where parts of the component do not fully enter the quenching medium or where, for some reason, heat is not extracted sufficiently quickly from that part of the component. The result is that in such regions products other than martensite appear within the carburised case, the most common being pearlite and networks of brittle cementite. The origin of such structures is a complex transformation process within the steel, and an additional factor is slight decarburisation which tends to occur during the second stage of the case-hardening process.[25]

As a consequence of the incorrect microstructure, brittle intergranular networks of cementite were produced in the case, and these eventually developed into fatigue cracks in service. An additional factor would be an uneven distribution of internal compressive stress in the surface layers. The formation of pearlite reduced the volume increase associated with the transformation, with the consequence that the case was less able to resist the service loadings applied to the crankpin. Additionally, the inter-granular networks of cementite could develop into cracks under high loadings. Both of these mechanical effects would have substantially reduced the fatigue resistance of this part of the crankpin.[25]

The metallurgical fault described above was introduced at the time of manufacture, with crack initiation occurring soon after the cycle was first ridden. Shortly before the pedal crank finally broke away, the crack had propagated (grown) to a point approximately halfway across the crankpin section. However final, mainly torsional, fracture had been rapid, with final propagation to fracture occurring during the one ride on the day of the accident. However, the fracture does not exhibit the usual appearance of a long-developing fatigue failure, and it could be that the cracks did not start to grow appreciably until the cyclist had developed the muscular strength and technique which enabled him to apply sufficient force to the pedals.

It is interesting to note that the cracking initiated on the surface of the crankpin that only came under high stress towards the bottom of the pedal's rotation. In this respect, it is significant that the cyclist stated he was standing on the pedals to gain speed up a slight incline when the pedal broke away, which is the same action that would have applied the forces necessary to initiate the cracking earlier in the crankpin's service.[25]

No blame for this accident could be attached to the cyclist, and there was no evidence that he did anything more than ride the bicycle normally and have it maintained by the retailer. The crankpin was not cracked when the bicycle was purchased, and it would have required specialist skills to have detected the developing fatigue crack shortly before the final failure and, even then, only if the pedal cranks had been removed and the exposed parts of the crankpin carefully cleaned and inspected. Furthermore, the defective structure in the crankpin was localised within the hardened case and could only have been detected by destructive metallurgical examination such as described above.[26]

At the 11th hour (basically on the courtroom steps) the crankpin axle manufacturer and his thermal treatment subcontractor accepted liability for both the failure and subsequent injury to the cyclist, with both damages and lost income being assessed over the course of the cyclist's recovery.

The case study approach is one way in which such active learning strategies can be implemented in any institution. There are a number of definitions for the term 'case study' – Fry[12] describes the case study as a complex example that will provide insight into the context of a problem, as well as illustrating the main point. A preferred philosophy regards the case study as a student-centred activity, based on topics that demonstrate theoretical concepts in an applied setting. This thinking has been further explored by the National Science Foundation (NSF) in America. The NSF initiative engaged faculty members from institutions around the nation. The goal of the initiative was 'to develop engineering ethics resource material that could be easily introduced at all levels of the engineering curriculum and in all engineering disciplines'. In response to the ASCE survey, Martin[24] developed case studies on four major engineering failures in the hopes that tutors would use those case studies as teaching aids to highlight how ethical responsibilities played a role in failures.

## 8.5 Examples of Case Study Themes

In addition to specific engineering failure analysis, case study themes can be pursued, for example, marine disasters, colliery disasters, bridge failures, etc. However, it is important that each case study portraying a theme is actually representative, and should include the widest possible range of products, materials and failure modes.[8,28,29] For example, the following brief case study illustrations focus on marine and colliery disasters as specific themes, whereas a civil or structural engineering case study theme could focus on cement and concrete as a specific material type, introducing background chemistry in addition to presenting individual failure studies. As a further example, domestic product failures that cause injury or loss of life most certainly hold potential for informative and practically oriented case analysis. Case writing skills are not difficult and can be readily learned, with the prime pre-requisite being an honest rendition of 'the truth'.

As an example, the following case study is an account of the collapse of Terminal 2E Charles de Gaulle International Airport. Its content can be readily expanded to illustrate both structural and materials issues, along with procedural (process) failures and a range of structural engineering design flaws that led to the catastrophic demise of the building.

### 8.5.1 Case Study: Collapse of Terminal 2E Charles de Gaulle International Airport

On May 23, 2004, a 110-foot section of the roof at the Charles de Gaulle International Airport in Paris collapsed, killing four travellers and injuring three others (Figure 8.9). The airport's Terminal 2E consisted of the main passenger building, the fated concourse parallel to it and an 'isthmus'

**Figure 8.9** (A) Collapse zone of the newly constructed Terminal 2E at Paris' Charles de Gaulle Airport. (B) Post-failure condition of the shell structure. (Courtesy of El Kamaria et al., Reliability study and simulation of the progressive collapse of Roissy Charles de Gaulle Airport, Case Studies in Engineering Failure Analysis, April 2015, 3, pp. 88–95, licensed under Creative Commons.)

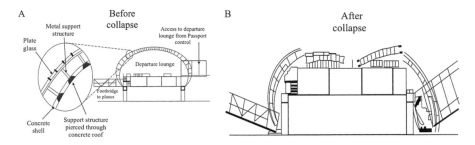

**Figure 8.10** (A) Schematic diagram of the Charles de Gaulle International Airport Terminal 2E prior to collapse. (B) Schematic diagram of the structure after collapse.

connecting the two buildings. A concrete and glass form rose from the second floor of the concourse and bulged out from its base, forming a flat arch (Figure 8.10A). With a design capacity for 25,000 people, the terminal concourse was conceived as long concrete tube having no internal supporting pillars. However, the roof collapse came less than a year after opening (Figure 8.10B). An investigative commission was instigated under the direction of Jean Berthier, engineering professor at France's École Nationale des Ponts et Chausées.

In what amounts to an indictment of the French process for executing complex public projects, the Jean Berthier technical investigation identified two likely reasons (procedural and structural) for the partial roof collapse at the 2,100-foot (650-metre) long concourse:

(a) Procedural (process) failure: a lack of detailed analysis and inadequate design checking allowed construction of a poorly engineered structure.

(b) Structural engineering failure: a number of interconnected design flaws were not caught during construction, including:

(1) Insufficient or badly positioned structural steel

(2) Lack of mechanical 'redundancy', in that the stresses were concentrated and could not be shifted to other structural components

(3) Concrete beams that offered too little resistance to stress and use

(4) The positioning of metal supports within the structural concrete

Several solutions for subsequent repair were considered, but none offered a guarantee for the security of personnel and passengers. Therefore, the entire vault of Terminal 2E was pulled down and rebuilt.

After the investigation and subsequent disassembly, the structure was rebuilt with a metal framework, erected on the existing foundation. Terminal 2E reopened in the spring of 2008.

As a post-script, on 20 October 2017, it was finally announced that the operator of Charles de Gaulle Airport would be tried for involuntary manslaughter over the 2004 collapse of the terminal. The investigating magistrate ordered that Aéroports de Paris (ADP) face trial, along with the construction, engineering and inspection companies involved in the building and certification of Terminal 2E.

## 8.5.2 Marine Disaster Case Study Examples

Recent research on parts of the RMS Titanic (1912) that were raised from the deep appear to show poor steel composition, with brittle fracture developing as a direct result of low sea temperatures.[27] Hull plate samples produced ductile/brittle transition temperatures of 32°C for longitudinal and 56°C for transverse specimens. Conversely, modern steels have ductile/brittle transition temperatures of minus 27°C. However, the steel used in constructing the Titanic was probably the best plain carbon ship plate available in the period 1909–1911. Nevertheless, it would not be acceptable at the present time for any construction purposes and most certainly not for ship construction.[27,28]

The phenomenon of brittle fracture has been acknowledged since the early part of the last century, but became recognised as a serious problem in ship construction during World War II.[29,30] As discussed in Section 4.12.6, the failure of many of the World War II Liberty ships is a dramatic and often referenced example of the brittle fracture of steel that was assumed to be ductile. A number of the older ships began to develop structural damage in the form of severe cracking of both decks and hulls. The insidious cracking grew to a critical length prior to final and rapid propagation around the hull of the ship, effectively breaking the backs of the ships. In more recent years, testing of steel quality and welding has virtually eliminated the problem, with the Charpy V-notch impact test employed to qualify steel toughness.[31,32] However, the problem of brittle fracture has been suspected as being influential in a number of off-shore oil platform and unexplained ship losses over recent years.

## 8.5.3 Case Study: Loss of the First British Off-Shore Oil Drilling Platform

The Sea Gem Disaster was a dark day for the British gas and oil industry, marking the sinking of the nation's maiden off-shore drilling platform (Figure 8.11). There were 32 crewmembers aboard the platform, but in December 1965 the legs of the rig collapsed under the wave pressure of the North Sea, triggering its total loss. As a result, 13 were killed and five more sustained serious injuries. It was determined that the disaster was due to material failure caused by corrosion, brittle fracture (ductile/brittle transition due to temperature change) and cyclic loading on the legs due to the severe nature of wave and weather conditions (a changed environment, as previously the rig had operated in the Caribbean with no problem).

**Figure 8.11** The Sea Gem, Britain's first off-shore oil drilling platform. (Courtesy of Klaus Dodds, *History Today*, February 2016, 66(2).)

An outcome of the loss was an agreement that regular inspections of off-shore platforms would be instigated, to avoid a repeat of such an incident. Regrettably, the Piper Alpha Disaster was yet to unfold and would, once again, change the face of oil-rig safety in Britain.

A report by the Transportation Safety Board of Canada into an incident involving the bulk carrier Lake Carling in the Gulf of St. Lawrence in March 2002 has demonstrated that the problem has not been eliminated.[33] The Lake Carling was only ten years old, having been correctly maintained, when it was discovered that No. 4 hold was taking on water. Further inspection revealed that a 6-metre fracture had developed on the port side shell. The vessel was properly loaded and was in calm seas when the crack went critical as a result of the steel's physical properties and the prevailing cold temperature.[33,34] At near-freezing temperatures, the relatively low fracture toughness of side plate material allowed crack initiation, and subsequent growth to failure, at a load well below the ultimate tensile strength of the plate material. The length of this crack at the time it became critical was not determined but could have been as short as 10 cm.[34] The report highlights the problem of brittle fracture as a 'live one', with bulk carriers being particularly susceptible to that mode of failure. Conclusions can be drawn that the Lake Carling incident is a classic example of lingering risk that still remains today, in spite of initiatives and advances in low-temperature applications.

Again, low temperature was found to have played a major role in the break-up of the tanker MV Kurdistan.[35] On 15 March 1979, the Kurdistan left Nova Scotia bound for Quebec. The tanker was an all-welded vessel built to construction category Ice Class I, 'carrying a heated cargo of oil for the first time. The weather conditions at the time were poor, with the ship rolling severely'. During the early afternoon, the vessel experienced a thud and a shudder during a particular downward pitch. The first sign of severe damage having been sustained by the ship was when oil started to leak from a vertical crack in the sides of a tank. Unfortunately, the vessel eventually broke in two. Almost eight hours elapsed between initial fracture of the vessel's shell and its final separation.[35,36]

The subsequent inquiry into the failure of the Kurdistan did not establish a precise sequence of failure, which showed both brittle and ductile fracture. 'The ship's shell plates were found to have Charpy transition temperatures of between 5° and 20°C; the steel in contact with the sea water was close to or below its transition; and the steel in contact with the heated cargo was above its transition. Calculations of the thermal stresses in the ship resulting from the carriage of a warm cargo in a cold sea indicated that a high tensile stress level would have been present in the shell and bilge keel. It is thought that the stresses due to the impact of a wave on the bow, superimposed on the high thermal stress and the stresses due to the moderate wave bending moments, triggered the fracture of the Kurdistan's bilge keel. The toughness of the shell plate was insufficient to arrest the propagating crack and complete failure ensued. The initiation of the fracture was due to the classic combination of poor weld metal toughness and high stresses in the presence of a defect.'[35-37]

A further example of brittle failure can be highlighted by the loss of the vessel MV Derbyshire. 'The Derbyshire was a British Oil/Bulk/Ore (OBO) carrier that was transporting ore from Canada to Japan. She was lost during the typhoon Orchid on 9/10 September 1980, sinking with all 44 onboard.'[38,39] The loss occurred rapidly, with no time for distress signal transmission. The Derbyshire was the largest British bulk carrier ever lost and has, therefore, been the subject of several investigations and discussions regarding bulk carrier safety.[39] A public enquiry was commissioned to examine the loss, in an attempt to understand more about the structural weaknesses of the vessel, in addition to increasing the safety of ships for the future. There were two theories: a fundamental flaw leading to brittle failure at rear bulkhead 65 and fragility of hatch cover in heavy seas.[39,40] The enquiry found that the hatch cover theory was the most likely explanation.

In the same context, oil-rigs can provide an interesting case study sequence of disasters and near misses. Aramid polymer ropes were selected for use in mooring the construction derrick ship Ocean Builder I in 1,045 feet of water in the Gulf of Mexico. However, on first tensioning four of these aramid lines parted, reportedly at 20% of rated break strength.[41] Subsequent investigation identified several factors contributing to the failures including rope damage and bending fatigue where one rope was too long for the water depth. An unexpected finding was a previously unknown strength degradation mechanism believed to have resulted from the interaction and response of the ropes to the ocean during their 4–6-week deployment as buoy mooring lines. Torque generated in the ropes led to rotation, causing shear and compressive strains and fibre kinking. Subsequent torsional (shear) and low-tension fatigue of these kinked fibres as the ropes thrashed about in the ocean largely unconstrained resulted in 40–50% rope strength loss, a phenomenon that probably would not have occurred if the lines had been connected to the derrick ship immediately.[41]

### 8.5.4 Case Study: Loss of the Alexander Kielland Semi-Submersible Oil Drilling Platform

Alexander Kielland (Figure 8.12) was a Norwegian semi-submersible rig in the Ekofisk oil field. In March 1980 the rig capsized, creating the worst disaster in Norwegian waters since World War II, killing 123 people.[42] 'During a storm, a particular cross-brace, one of many used to structurally stiffen the columns, broke. Loss of this brace initiated a sequence of events, culminating in the overload of a main column attachment which, in turn, fractured - separating one column and pontoon completely from the platform. Immediately the platform adopted a list of some 12 degrees. As water surged through open windows and other openings, the list increased to 20 degrees, at which point the platform became unstable and capsized.'[43,44]

Upon a detailed evaluation of the recovered platform, it was revealed that brace fracture initiation was the result of a small weld crack in a hydrophone attachment, introduced during the course of initial fabrication. As would generally be expected, this 'engineered-in' crack propagated by a fatigue mode, driven by the action of wave loading experienced over many months of normal service. It was determined that prior to the final fateful storm, the growing fatigue crack had already propagated (grown) almost completely around the brace. It can be seen that this failure again highlights the problem of fatigue. However, methods used to counteract deleterious cyclic loading from the environment led to the development of rubber–metal flex elements.[45]

**Figure 8.12** (A) The Alexander Kielland semi-submersible rig in the Ekofisk oil field. (B) The head of column D after accident. Photos: Norwegian Petroleum Museum. (Courtesy of Wikimedia Commons, licensed under Creative Commons.)

### 8.5.5 Colliery Disaster Case Study Examples

Felling mine disaster was one of the first major mine disasters in Britain, claiming 92 lives on 25 May 1812. 'As with so many prior coal mine accidents, the event was triggered by the ignition of firedamp (methane), which then detonated a coal dust explosion.' The disaster stimulated George Stephenson to design a safety lamp with air fed through narrow tubes, down which a flame could not move. It also led the eminent scientist Humphry Davy to devise another safety lamp, where the flame was surrounded by iron gauze. However, the Davy lamp was flawed owing to rusting of its wire gauze, so deaths as a result of explosions continued to rise through the 19th century. Design and material improvements led to multiple gauzes, protective bonnets and glass site tubes (Marsaut, Mueseler, Clanny).[46] However, disasters continued, reaching a climax in 1913. The Senghenydd colliery disaster occurred near Caerphilly, Glamorgan, South Wales on 14 October 1913, killing 439 miners.[47] It is the worst mining accident in the United Kingdom, and one of the most serious in terms of loss of life globally since. It was probably started by a firedamp (methane) explosion, itself possibly ignited by electric sparking from equipment, such as electric bell signalling gear. Understanding of coal dust and methane explosions was limited, even up to more recent times.[48] Sources of ignition have included

- Impact of light metals (Al, Mg alloys) against rusty tools (Silvertown, 1956). 'Light metals and their alloys have a natural affinity for oxygen and when brought into close contact with oxygen bearing material, such as iron oxide (rust) in the presence of moderate heat, then a chemical reaction occurs' – a thermite reaction.[49,50]
- Friction on rubber conveyor belts (Derbyshire, 1950). Eighty men were overcome by smoke and fumes and perished underground at Creswell Colliery,[51] with 23 bodies remaining underground for a year until it was safe to remove them.

The former led to the banning of light alloys in collieries,[52] the latter to the development of PVC conveyor belts.[53] The Markham Colliery (1973) disaster emphasises the importance of recognising, understanding and monitoring the potential for fatigue failure of lifts.[54]

## 8.6 Engineering in Medicine

Engineering itself has always been an innovative field, with the origin of ideas leading to everything from ladders to aerospace. Engineering in medicine switches the focus of attention to advances that improve human health and healthcare. Here, engineers work at the junction of engineering, life sciences and healthcare, taking principles from applied and physical science and applying them to biology and medicine. In short, engineering in medicine is the application of the principles and problem-solving techniques of engineering to biology and medicine, evident throughout healthcare, from diagnosis and analysis to treatment and recovery. Public awareness has been raised by an explosion of novel prosthetic medical devices, such as pacemakers and artificial hips, and of the more innovative technologies, such as 3-D printing of biological organs. Inherently, the field holds potential for enlightened and up-to-date case study material for student education, as highlighted by the following case study.

### 8.6.1 Case Study: The DePuy Ultamet Metal-on-Metal (MOM) Articulation Hip Replacement

On May 21, 2018 a decision by the (UK) 'High Court ruled that DePuy's Ultamet Metal-on-Metal (MOM) Articulation hip replacement was not defective' under the Consumer Protection Act 1987[55] (Figure 8.13). 'The group litigation incorporated over 300 claimants, and ended in a four-month trial during which the court received evidence from over 40 witnesses, including 21 experts ranging from orthopaedics' to engineering and behavioural psychology to statistics.[56]

The claimants had 'argued that the "defective" implants produced metal debris, which damaged patients' surrounding tissues and produced adverse

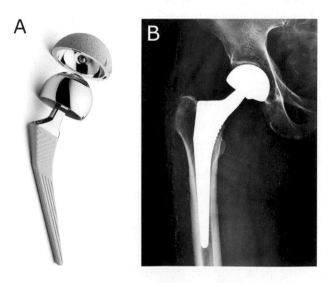

**Figure 8.13** (A) The DePuy Metal-on-Metal (MOM) Articulation hip replacement device. (Courtesy of Drug Watch, Orlando, FL.) (B) An X-ray image of the hip device after fixation within human the body.

reactions such as pain, difficulty walking, and numbness. However, the Honourable Mrs Justice Andrews DBE stated that this was a known consequence of the ordinary use of the product, and that 'material loss' was an inherent characteristic, and would occur in every type of hip implant, no matter the material.[55]

At this point, the student may be asked to give an opinion as to whether the manufacturer could be afforded protection under the 'Development Risk' defence. Any manufacturer may be allowed 'the benefit of the development risks defence only if the court concludes that harmful characteristic/s of the product is/are well known and socially accepted – as in the case of harmful products such as alcohol and tobacco. The fact that such a product has the capacity to cause injury informs the public's legitimate expectation – so it cannot be defective if it has a known characteristic that involves an acknowledged risk accepted by society. As an example, it is internationally recognised (and accepted) that a knife is sharp – therefore a knife cannot be considered defective if a user suffers a cut injury while handling it.'[57]

The court was therefore being asked a novel question '(which was the basis of the claimants' primary case on defect): whether a known and inherent characteristic of a product could in and of itself be a defect. Here, that characteristic was the '(known) tendency of the product to produce metal wear debris "which might cause an adverse reaction in some patients, giving rise to replacement revision surgery"'.[56]

'The defendant argued that a known consequence of the ordinary use of the product could not amount to a defect. Applying the Claimants' logic to this case would render all hip implants defective under the CPA, as material loss produced by the wear of the implants occurs in all types of hip implants (whatever their bearing surface: metal or polyethylene) with the potential to cause an adverse reaction.

The judge agreed with the defendant and went so far as to assess this aspect of the Claimants' case as being 'untenable'.[56]

'Claimants also argued that the Ultamet device didn't perform as well as other non-MOM implants and that it had an "abnormal" potential for causing harm, However, the judge stated that the Ultamet implant should be compared to other prostheses that were on the market *at the time the device was introduced*. The judge determined that the Ultamet implant performed as well or better than those on the market and had no increased risk of harm. This resulted in the decision that the implant' was not defective under the Consumer Protection Act 1987.[58]

'Given its decision that the safety of new products should be compared to products existing at the time they are introduced to the market, the Court's interpretation of the CPA should not discourage the development of new, potentially life enhancing, technologies in the future – as might have been the case if the Claimants' arguments had succeeded.'[56,58]

## 8.7 Concluding Remarks

With the title *Forensic Engineering: The Art and Craft of a Failure Detective*, the focus of this book is centred on both the art (forensic engineering teaching) and the principles of investigation (professional practice). Furthermore, the tome clearly champions the use of case study analysis as a teaching tool. As such, case studies discussed within the final four chapters have been assembled largely from the author's own notebook, supplemented by examples in the public domain. If failures of a similar type are to be prevented in the future, dissemination in the public domain will provide a powerful tool for educating designers and engineers of shortfalls in the behaviour of real products. At this point, background writing will be concluded, with attention being focused on a range of case studies presented within the final chapters.

## References

1. Yadav, A., G. Shaver and P. Meckl, Lessons learned: Implementing the case teaching method in a mechanical engineering course. *Journal of Engineering Education*, January 2010. 99(1): pp. 55–69.
2. Delatte, N.J., *Beyond Failure: Forensic Case Studies for Civil Engineers*. 2009, Reston: ASCE Press. ISBN: 9780784409732.
3. Pollio, Vitruvius, (Morgan. M, translator), *The Ten Books on Architecture*. 2018, CreateSpace Independent Publishing Platform, ISBN-10: 1987540816, ISBN-13: 978-1987540819.
4. Pertroski, H., *To Forgive Design: Understanding Failure*. 2012, Harvard University Press, ISBN-10: 0674065840, ISBN-13: 978-0674065840.
5. Petroski, H., Failure as source of engineering judgment: Case of John Roebling. *Journal of Performance of Constructed Facilities*, February 1993. 7(1): pp. 46–58.
6. Grolinger, K., Problem based learning in engineering education: Meeting the needs of industry. *Teaching Innovation Projects*, 2011. 1(2): 2.
7. Lewis, P.R. and C.R. Gagg, Post-graduate forensic engineering at The Open University, in *Proceedings of the International Conference on Innovation, Good Practice and Research in Engineering Education*. 24–26 July 2006, University of Liverpool, UK.
8. Lewis, P., K. Reynolds and C. Gagg, *Forensic Materials Engineering: Case Studies*. 2003, CRC Press.
9. *Aircraft Crashes Record Office*. Geneva, Switzerland, http://www.baaa-acro.com/.
10. Rouse, C.G., *Breaking the Circle: Educating Undergraduates through Failure Case Studies*. 2002, http://www.eng.uab.edu/cee/reu_nsf99/REU02/Cynthia/final.pdf].
11. Richards, L. and M.E. Gorman, Using case studies to teach engineering design and ethics, in *Proceedings of the 2004 American Society for Engineering Education Annual Conference & Exposition*. 2004, American Society for Engineering Education.

12. Fry, H., S. Ketteridge and S. Marshall, *A Handbook for Teaching and Learning in Higher Education*. 1999, Glasgow: Kogan Page. p. 408.
13. Evan, W. and M. Manion, *Minding the Machines: Preventing Technological Disasters*. 15 April 2002, Upper Saddle River, New Jersey, 07458: Prentice Hall PTR.
14. Adekoya, A. and A. Aghayere, The use of failure case studies to enhance students' understanding of structural behavior and ethics, in *9th International Conference on Engineering Education*. 23–28 July 2006, San Juan, PR.
15. Weaver, G., N.S.J. Braithwaite and C. Gagg, Lets make a strain gauge: A summer school laboratory activity in electronic materials, in *Innovative Teaching in Engineering*. 1991, Chichester, England: Ellis Horwood.
16. Raju, P. and C. Sanker, Teaching real-world issues through case studies. *Journal of Engineering Education*, 1999. 88(4): pp. 501–508.
17. Kardos, G. and K. Smith, On writing engineering cases, in *Proceedings of ASEE National Conference on Engineering Case Studies*. March 1979, http://www.civeng.carleton.ca/ECL/cwrtng.html.
18. Gagg, C., Failure of stainless steel water pump couplings, in *Failure Analysis Case Studies III*, D. Jones, ed. 2004, Oxford: Elsevier.
19. Anwar, S. and P. Ford, Use of a case study approach to teach engineering technology students. *International Journal of Electrical Engineering Education*, pp 1-10, January 2001. 38(1): pp. 1–10. https://doi.org/10.7227/IJEEE.38.1.1
20. Delatte, N.J. and K.L. Rens, Forensics and case studies in civil engineering education: State of the art. *Journal of Performance of Constructed Facilities, American Society of Civil Engineers*, August 2002. 16(3): pp. 98–109.
21. Lewis, P.R. and C.R. Gagg, Aesthetics versus function: The fall of the Dee Bridge, 1847. *Interdisciplinary Science Reviews*, June 2004. 29(2): pp. 177–191.
22. Savin-Baden, M., *Facilitating Problem-Based Learning: The Other Side of Silence*. 2003, Buckingham, UK: SRHE/Open University Press.
23. *Numerical Problems Associated with Ethics Cases for Use in Required Undergraduate Engineering Courses in National Science Foundation/Bovay Fund Sponsored Workshop*. August 1995, Texas A&M University, College Station, TX.
24. Martin, R., *Failure Case Studies in Civil Engineering Education*. 2 February 2006, http://www.eng.uab.edu/cee/reu_nsf99/rachelwork.htm.
25. Gagg, C., Domestic product failures: Case studies. *Engineering Failure Analysis*, 2005. 12(5): pp. 784–807.
26. Gagg, C., Failure of components and products by 'engineered-in' defects: Case studies. *Engineering Failure Analysis*, 2005. 12(6): pp. 1000–1026.
27. Felkins, K., J.H.P. Leighly and A. Jankovic, The royal mail ship titanic: Did a metallurgical failure cause a night to remember? *Journal of Materials (JOM)*, 1998. 50(1): pp. 12–18.
28. Deitz, D., *How Did the Titanic Sink?* August 1998, The American Society of Mechanical Engineers (ASME).
29. Hodgson, J. and G.M. Boyd, Brittle fracture in welded ships. *Transactions of the Institution of Naval Architects*, 1958. 100(3): pp. 141–180.
30. Reemsynde, H.S., *Ships that have Broken in Two Pieces—World War II US Maritime Commission Construction*. September 1990, Bethlehem Steel Internal Memo. p. 1402-1b.

31. Kent, J.S. and J.D.G. Sumpter, Probability of brittle fracture of a cracked ship, in *2nd International ASRANet Colloquium*. July 2004, Barcelona, Spain.
32. Sumpter, J.D., J. Bird and J.D. Clarke, Fracture toughness of ship steels. *Transactions of the Royal Institution of Naval Architects*, 1989. 131: pp. 169–186.
33. Drouin, P., Brittle fracture in ships - A lingering problem. *Ships and Offshore Structures*, July 2006: 229–233.
34. Canada, T.S.B.o., *Hull Fracture of Bulk Carrier Lake Carling in the Gulf of St. Lawrence, Quebec*. Marine Reports, 2002. Report Number M02L0021.
35. Garwood, S., Investigation of the MV Kurdistan casualty. *Engineering Failure Analysis*, March 1997. 4(1): pp. 3–24.
36. The Welding Institute (TWI), *M V Kurdistan Tanker*. From Report 632/1998, 1998, https://www.twi-global.com/media-and-events/insights/m-v-kurdistan-tanker.
37. Vandermeulen, J.H., Scientific studies during the Kurdistan tanker incident, in *Proceedings of a Workshop: BI-R-80-3*. 1980, Dartmouth: Bedford Istitute of Oceanography.
38. Faulkner, D., An independent assessment of the sinking of the MV derbyshire. *SNAME Transactions*, 1998. 106: pp. 59–104.
39. Ship Structure Committee, *Case Study II, Derbyshire, Loss of a Bulk Carrier*. May 2019, http://www.shipstructure.org/case_studies/derbyshire/.
40. Faulkner, D., An analytical assessment of the sinking of the M.V. Derbyshire. *The Royal Institution of Naval Architects, RINA Transactions*, 2001.
41. Riewald, P.G., Performance analysis of an aramid mooring line, in *Offshore Technology Conference*. May 1986, Houston, TX.
42. The Alexander L. Kielland accident, report of a Norwegian public commission appointed by royal decree of March 28, 1980. *Presented to the Ministry of Justice and Police*. 1981, Oslo, Norway.
43. Moan, T., Fatigue reliability of marine structures, from the Alexander Kielland accident to life cycle assessment. *J ISOPE*, March 2007. 17(1): pp. 1–21.
44. https://www.exponent.com/experience/alexander-kielland/?pageSize=NaN&pageNum=0&loadAllByPageSize=true.
45. Lei, H.H., Securing of marine platforms in rough sea. *Recent Patents on Engineering*, 2007. 1(1): pp. 103–112.
46. Hughes, H.W., *A Text-book of Coal-mining: For the Use of Colliery Managers and Others*. Digitizing sponsor; Google ed. 1901, Book contributor: Harvard University. C. Griffin and co ltd.
47. Brown, J., *The Valley of the Shadow: An account of Britain's Worst Mining Disaster, the Senghennydd Explosion*. 2009, Alun Books, ISBN 978-0-907117-06-3.
48. Du Plessis, J. and D. Bryden, Systems to limit coal dust and methane explosions in coal mines. *Safety in Mines Research Advisory Committee*, January 1997. COL 322: pp. 1–50.
49. Margerson, S., H. Robinson and H. Wilkins, Ignition hazard from sparks from magnesium-based alloys. *Safety in Mines Research Establishment (U.K.)*, 1953. Research report No. 75.
50. Lewis, P., K. Reynolds and C. Gagg, *Forensic Engineering (T839)*. 2000, Post-Graduate Course, The Open University.

51. H.M.S.O, *The Accident At Creswell Colliery, Derbyshire, 26th September, 1950.* Vol. Great Britain Ministry of Fuel and Power. 1952, The Official Accident Enquiry By Sir Andrew Bryan, D.Sc., F.R.S.E.
52. *Technical Information Sheet - Frictional Ignitions.* Vol. 2. 1980, HSE, London, UK: Health & Safety Laboratories.
53. Yardley, E.D. and L.R. Stace, *Belt Conveying of Minerals.* February 2008, Woodhead Publishing Ltd, ISBN10 1845692306, ISBN13 9781845692308.
54. Lawley, A., The Markham colliery disaster - A case study in fatigue, in *Materials Under Stress, Third Level Open University Course.* 1983, The Open University Press, UK, ISBN-10: 0335171508, ISBN-13: 978-0335171507.
55. Alexander Antelme, Q.C., David Myhill and Richard Sage, *A Case Summary and Analysis of the Pinnacle Metal-On-Metal Hip Group Litigation.* 2018, https://www.crownofficechambers.com/2018/05/21/pinnacle-metal-on-metal-hip-group-litigation/.
56. Silver, S., T. Davies, E. Brett and N. Smythe, *The DePuy Pinnacle Metal on Metal Hip Litigation.* 21 May 2018, https://www.kennedyslaw.com/thought-leadership/article/the-depuy-pinnacle-metal-on-metal-hip-litigation/.
57. Paroha, K.J. and S. Sayers, *Case Law Lessons on Defective Products and the Development Risks Defence.* 18 February 2010, https://www.internationallawoffice.com/Newsletters/Product-Regulation-Liability/United-Kingdom/Kennedys/Case-law-lessons-on-defective-products-and-the-development-risks-defence.
58. Medical Plastics News - Advancing Medical Plastics, *DePuy Wins Hip Replacement Trial.* 22 May 2018, https://www.medicalplasticsnews.com/news/medical-devices/depuy-wins-hip-replacement-case/.

# Failure Due to Manufacturing Faults

# 9

## 9.0 Introduction

The manufacturing industry has at hand a range of comprehensive systems to ensure the quality of both the incoming raw material quality and outgoing finished products. Improved quality and productivity have been achieved as the result of investment in automation and robotics, lowering the cost-per-unit, in addition to improved and automated inspection systems designed to detect issues on the first failed unit. So, with advances in modern manufacturing technology, it is unusual for components with serious faults to enter into service, principally as a result of rigorous quality control procedures used on the production line. However, from time to time components do enter service with an inherent (often latent) manufacturing shortcoming or fault. When the influence of any such fault results in a serious weakening effect, failure will often occur on the first occasion the component comes under heavy load. Case studies reviewed within this chapter will focus on the demise of engineered products as a direct result of inherent manufacturing issues.

## 9.1 Traumatic Failure

Almost every casting is a one-off, in the sense that faults stemming from the way a particular volume of metal solidifies – shrinkage, porosity, gas entrapment, blowholes, misruns, cold shuts and so on – may not be repeated in other castings made in the same foundry. Figure 9.1A shows an SEM micrograph depicting severe shrinkage in a spheroidal graphite (SG) cast iron wheel hub from a new trailer. After being loaded for the first time, the hub failed just as the trailer was turning into the roadway.[1,2] At the instant of failure, the trailer swerved across the busy carriageway, and collided with an oncoming vehicle. Figure 9.1B shows a section across an aluminium alloy brake lug from an almost new motorcycle, which broke away and rendered the brake inoperative on the first occasion the rider had to make an emergency stop.[1,2] A young child was killed. The lug was an integral part of a cast brake plate that was grossly weakened by gas porosity and shrinkage. A similar fault accounted for the failure of a robot arm casting in aluminium alloy. The robot was designed for use in a nuclear installation, but the arm

283

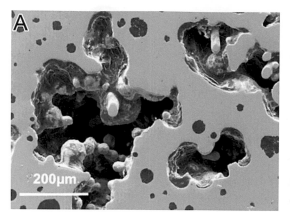
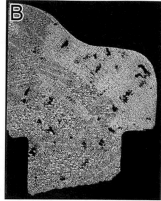

**Figure 9.1** (A) SEM micrograph of severe shrinkage in a section of a spheroidal graphite (SG) cast iron wheel hub. (B) Section across a fractured motorcycle brake lug, with fracture located at right-hand edge.

broke off the body while it was undergoing proving tests. The seriously defective condition of the flange that carried the arm was obvious from the state of this machined surface, so the casting should have been rejected out of hand and certainly not built into a robot destined to be used for handling radioactive materials. Unfortunately, the project was behind schedule, so the casting went straight into the machine shop as soon as it was delivered, and the machinist did not consider it part of his duty to report such faults. This was a difficult casting to make and a number of replacements were X-rayed and found to be similarly defective. The problem was traced to gases released from the resin binder used to form the sand moulds in the foundry coupled with inadequate feeding of the heavier sections of the body casting.

## 9.2 Femoral Stem Failure of a Total Hip Prosthesis[3]

It must be recognised that total hip prostheses do have a *finite* lifetime within the body environment and will therefore simply wear out with time. This can be aptly demonstrated by the following case where a person falling apparently caused breakage of a total hip prosthesis.

An elderly woman had a fall when she went to answer the front door of her Victorian house. Her foot went through a floorboard that had been weakened by dry rot, causing her to tumble. When help arrived, it was quickly recognised that the woman had damaged her right hip and required urgent medical treatment. At the hospital, it became apparent that the patient had been the recipient of a total hip replacement some ten years earlier and the device had broken. A surgical procedure was promptly undertaken to retrieve the broken prosthesis and insert a replacement device. After a full

 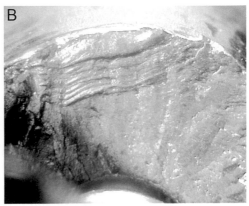

**Figure 9.2** (A) Broken femoral stem of total hip prosthesis. Note: the extraction tool protruding vertically out of fracture surface. (B) The fracture surface clearly showing fatigue striations.

recovery, the patient instigated legal proceedings against her landlord (the local council) for her injury and suffering. Although the woman stated that her fall had not been particularly heavy, she believed that her injury was a direct result of the fall. Legal action was instigated for a claim of negligence in maintaining her floor to a reasonable standard, resulting in a fall that broke her hip prosthesis.

The broken femoral stem of the device was submitted for investigation, Figure 9.2A, and is shown with the surgical extraction tool having been left in situ. Figure 9.2B shows an optical micrograph of the fracture surface, where striations associated with a fatigue failure mechanism are clearly visible. Fatigue is a mechanism that often involves many hundreds, if not thousands, of loading cycles to reach the point of failure (Sections 5.9 and 6.7) and is not an overload mechanism. If, as claimed, the hip prosthesis had suffered an overload failure resulting from a fall, the loading pattern would present a totally different fracture surface (Section 6.4).

To date, the life expectancy of a total hip prosthesis is approximately ten years. When the age of the device in question was taken into consideration and combined with evidence of failure resulting from fatigue, it was determined that the device had simply reached the end of its useful service lifetime and failure was not a direct result of the fall. The loading, experienced at the instant of impact, would in all probability have been withstood by the device if it had not been considerably weakened by the presence of a growing fatigue crack.

However, the woman had suffered a fall that was a result of poor maintenance practice by her landlord. The litigation was settled out of court by the landlord's insurers, with the woman receiving substantially reduced damages for the incident.

Within the USA, the FDA holds a large database of failures of medical devices at MEDWATCH, allowing an investigator to follow the failure history of specific hip joints, heart valves, stents and similar implants (www.fda. gov/medwatch). As there is usually a plethora of different designs, identification by tradename (or trademark) on the compilation, will generally provide the required information.

## 9.3 Failure of Freight Containers[1,2]

The next case relates to a simple manufacturing mistake that also cost an insurance company a great deal of money. It concerns freight containers made of aluminium alloy, the kind of containers universally used for transporting goods by ship, rail or road and seen all over the world in port installations and wherever freight is carried (Figure 9.3A). Contents of these containers vary enormously, from being packed throughout with light items of clothing to carrying a single item of heavy machinery. The advantage of such containers is that they may be sealed at the point of loading and remain secure until arrival at their destination. This case began[1,2,4] when a dockworker noticed a split in the end panel of a fully loaded 10-m-long container being lifted from the vessel. There was no obvious sign of external damage, and the piece of machinery was still firmly anchored inside. Shortly afterwards, a spate of similar failures was experienced at other ports. Preliminary inquiries confirmed that all of the damaged containers had been made at the same factory, and during the same two-month period.

The riveted seam between the two end panels in the side of the container had split open, from bottom to top as shown by a dashed tear line in Figure 9.3B.[1,2,4] It was subsequently determined that all the other containers had split in the same position, and all had been carrying heavy items of machinery as distinct from bulky loads distributed throughout the container length.

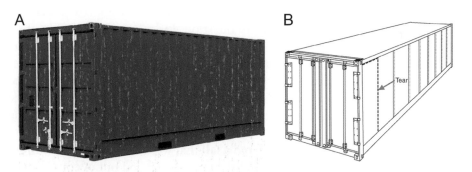

**Figure 9.3** (A) The type of freight container in question. (B) Position of seam that failed in the container wall, as indicated by the dashed line.

The costs escalated alarmingly, involving not only the transfer of the load to another container, but also finding an empty container and shipping it to the port, loss of earnings of the damaged container and cost of repair. There were at least four good reasons for requiring an urgent investigation:

1. Whether or not the owners had a case for claiming damages from any other party such as the harbour boards or the container manufacturers.
2. Identification of the fault and a satisfactory method of repair; it would be wrong simply to repair the joints according to the original design before vindicating the design.
3. Whether further failures of similar containers were to be expected and, if so, whether these might be avoided by reducing the recommended load-carrying capacity.
4. Whether partial failures of this kind might lead to complete unzipping and failure of the containers during a voyage or when offloading, with risk of serious accident and consequent damage to the cargo.

As a preliminary to this investigation, a mechanical engineer with extensive experience in riveted stressed skin structures in aircraft had concluded there was nothing wrong with the original design or method of construction of these containers. This was borne out by the continuing satisfactory performance of the vast majority of these containers, some of which had been in use for several years. This led to the view that something had gone wrong with the manufacturing processes or material quality control, which accounted for structural weakness in these particular failed containers.

The side panels were sheets of aluminium alloy that were attached to one another and the frame by riveting along vertical lap joints incorporating top-hat channel sections to act as stiffeners. All the failures involved the 'unzipping' of the vertical lap joint between the first and second sheets in from the end of the container.[1,2,4] The rivets had sheared across at the interface without causing any tearing or ovality of the rivet holes. Figure 9.4A shows a longitudinal section of a typical rivet head fracture. Failure is clearly as a result of a single shear overstress. The rivet section was then polished and etched for microstructural examination. This confirmed the shearing mode, as evident in Figure 9.4B by the way the grain structure has flowed adjacent to the fractured edge. The above interpretation explains the lack of any stretching of the holes in the sheet on either side of the rivets, so the investigation concentrated on what might be wrong with the rivets.

The first question to address was whether the rivets were the right composition. An element scan utilising pulse counts of X-rays of characteristic wavelength for the elements likely to be present was carried out in the SEM

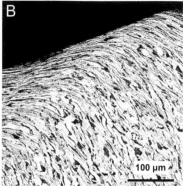

**Figure 9.4** (A) Longitudinal section macrograph of a failed rivet head. (B) Polished and etched micrograph of rivet edge, showing structural grain flow formed as a result of shear overload.

and compared with that of an unused rivet of the correct composition (HR15 alloy to BS1473). The traces were as shown in Figure 9.5. While element scans are not precisely quantitative, the fact that the two are practically identical established that the correct alloy had been used for rivets in the failed containers.

Examination of the failed rivets under a low-power microscope revealed no physical sign of any internal fault or of a progressive failure mode such as fatigue, wear or corrosion. However, if one rivet near the end of a seam had failed it would have thrown an extra load onto its neighbours, and these might then in turn become overstressed, causing the entire set to progressively unzip. Hence the next task was to examine the specification for the rivets on the engineering drawing. The relevant notes and diagram are reproduced in Figure 9.6, where it will be noticed that the rivet stock should be supplied in the H2 condition (as manufactured by the shaping and heading operations), but before use must be solution-treated and then used within 48 h.

High strength in certain aluminium alloys, particularly those used in airframes, is achieved by a precipitation-hardening mechanism. This involves a rearrangement of atoms within the crystals by soaking the alloy at a temperature not far below its melting point, so that alloying elements dissolve in solid solution, followed by quenching immediately into water. This retains the elements in the solid solution, but over a few days at room temperature, they begin to precipitate and cause a dramatic increase in yield and tensile strength accompanied by a reduction in ductility. When this process takes place at room temperature it is called natural ageing or *age hardening*; when accelerated by heating for an hour or so in the region of 200°C it is termed artificial ageing or *precipitation hardening*.

There is a problem with precipitation-hardening alloys: although they retain a high degree of ductility in the solution-treated condition, as they

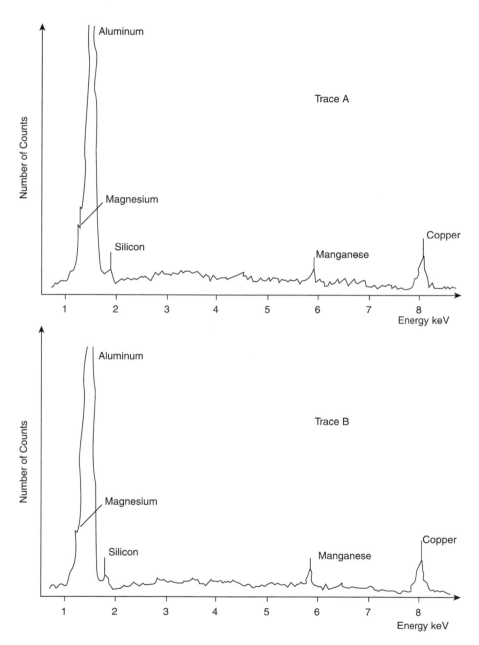

**Figure 9.5** EDAX X-ray scans from SEM analysis of broken and unused rivets.

harden they gradually lose this, and the ductility in the fully aged condition measured in the tensile test may be as low as 8% elongation, even lower in the really high-strength alloys. This means that if parts are to be deformed to any extent such as setting the head of a rivet, this operation must be carried out before the ageing gets underway, usually within an hour or so of

**A**

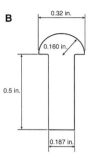

| Part | 3/16 in x ½ in Rivet for Container Sidewall (BS 614: 1951) |
|---|---|
| Material | Aluminium Alloy HR 15 (B.S. 1473 |
| Condition | TB – Stock to be supplied in H2 condition |
| | Solution treat and use within 48 hours |

**Figure 9.6** (A) Extract from workshop drawing of rivet for container seams. (B) Detail drawing of rivet dimensions.

**Table 9.1  Chemical Composition for HR15 Alloy Rivet Stock (% by Weight)**

| Element | Balance |
|---|---|
| Copper | 0.2–0.8 |
| Magnesium | 0.2–0.8 |
| Silicon | 0.5–0.9 |
| Iron | 0.7 |
| Manganese | 0.4–1.2 |
| Zinc | 0.2 |
| Chromium | 0.1 |
| Titanium* | 0.2 |

\* And/or grain-refining elements.

**Table 9.2  Mechanical Properties of HR15 Alloy**

| Condition of Supply | Heat-Treated Condition | Diameter (mm) | Tensile Strength (MN/m²) |
|---|---|---|---|
| H2 – annealed and cold drawn 20–40% reduction of area | TB – solution heat treated and naturally aged | Up to 2 | 385 |

quenching from the solution treatment. Table 9.1 and Table 9.2 are taken from BS1473 for HR15 alloy. The tensile strength given is for the TB (naturally aged) condition. In the H2 as-wrought condition in which these rivets were supplied, the tensile strength would be approximately 240 MN m$^{-2}$, but the metal would have high ductility and could be worked without any risk of cracking. In contrast, in the aged TB condition, a rivet would not possess enough ductility for it to be set without the head splitting. Thus immediately before setting them in the container seam, the stock rivets would have to be solution-treated, quenched and then set within a time window of about 1 h before the age-hardening process began.

It is quite clear from the hardness tests on the failed rivets that they had not responded to the age-hardening process as they should have and were consequently well below the strength levels that ought to have been achieved. The cause was obvious – they were of the correct composition but had not been solution heat-treated before setting. This was borne out by their microstructures and, in fact, had already been suspected by the investigating metallurgist from the microstructure of the failed rivet shown in Figure 9.4B. If the rivet had been solution-treated the grains would have been much larger, and if it were in the naturally aged or precipitation-hardened condition the amount of deformation along the sheared surface associated with the fracture would have been far less.

It was quite straightforward to establish whether the essential solution treatment had been omitted, thereby preventing the rivets from hardening to the required level. There is no way to tell from the microstructure whether a component is in the precipitation-hardened condition, but the condition may be readily established by measuring hardness. A hardness test can be carried out on a microsection as it only requires a flat surface of sufficient area for a diamond point to make an indentation less than 1 mm across. Although the rivets were too small to be tensile tested and the broken ones could not be held in grips for a shear test, there is a close relationship between hardness and tensile strength. Hence, because the measured shear strength of metals is usually about 60% of their tensile strength, hardness can be used to estimate the shear strength, even of a broken rivet, using the relationship shown in Figure 9.7. Measuring the hardness of ten rivets from the failed seam, as well as samples from the other seams in this container and from another container that had failed similarly, revealed none of them to be in the TB condition specified. The rivets used to make the seams were the correct alloy and in the as-manufactured condition specified on the order, but as they were not solution-treated before setting they could not age and consequently reach the specified level of tensile strength.

Apparently, for about a month someone at the factory making the containers had omitted the solution heat treatment. While the metallurgical investigation was in progress, a mechanical engineer was addressing the complementary questions of what level of stress the rivets were subjected to when they failed and why only containers used to transport machinery had 'unzipped'. At first sight, this may appear to be a rather academic question; the behaviour of the rivets themselves may be thought to provide a sufficient answer. Because they fractured in shear they must have experienced a set of shear stresses with little or no normal stress. Furthermore, the magnitude of this stress must have exceeded the shear strength of the rivet; the material failed by overstressing. On the other hand, fully loaded containers without the defective rivets did not fail, so the shear stress on their rivets must have been less than the shear strength of the rivets in the specified condition.

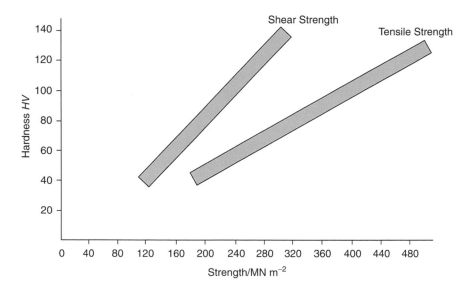

**Figure 9.7** Correlation of hardness with the tensile and shear strengths for the aluminium alloy HR155.

The shear stress is therefore known to lie between these limits, but it is not known accurately, so the magnitude of the safety margin is not known. To allow for the unknown, it is usual in design to use a static *factor of safety* (or load factor) of a component, which is defined as

$$\text{factor of safety} = \frac{\text{nominal static strength of the component}}{\text{nominal static loading of the component}}$$

'Strength' is the maximum value of load the component can bear in a given situation – for example, in tension, compression or shear – without failing in a given way (for example, by fracture). 'Static' implies that the loads are assumed to be constant and do not vary with time. The word 'nominal' is used here to mean 'according to the specification and the design calculations'. These calculations are usually based on a simplified model of the component, and may ignore some of the complications of service conditions. If these complications could be taken into account quantitatively, on the one hand, they would erode the nominal static strength and, on the other hand, they would inflate the nominal static loading. The factor of safety must be large enough to cover the uncertainties in strength and loading, which is why it is sometimes referred to as a 'factor of ignorance'.

It is essential that the static factor of safety be large enough that *at all times* within the lifetime of the component, the true strength will exceed the true loading. Whenever there is a service failure, the static safety factor should be re-examined. The investigating engineer has two options. He could choose a theoretical model of the loaded container and then analyse it to find the loads acting on the rivets, which would provide an estimate of the nominal static loading on a given rivet. Alternatively, the distortions in the sidewall of a loaded container could be measured and, using this information, the loads that are transmitted to the rivets could be inferred. This should provide a more accurate estimate of the static loading on a rivet.

It is a problem to decide how the container should be loaded and difficult to anticipate every service load that might be encountered, for example, when a container is set down heavily in transit. It should not fail under such *foreseeable* conditions, but no design can be expected to allow for *gross* misuse as, for example, being driven into the arch of a low bridge. 'Reasonable' overloads are not easy to quantify, but in general they are taken as loads that take into account foreseeable handling.

The engineer began by considering a theoretical model of a fully loaded container standing on its four bottom corner fittings, likened to a hollow box beam 2.4 m deep supported at its ends and subjected to the maximum load of 30,840 kg for this type of container, distributed uniformly along its length as indicated in Figure 9.8. The top and bottom of the box act as chords or flanges carrying the compression and tension forces due to bending, while the sides of the box act as webs, primarily carrying shear forces. The key point is that the side panels are important structural parts of the complete box. The shear force gives rise to shear stresses acting vertically on each cross section. However, these must vary in magnitude from place to place; for example, they must be zero at the top and bottom of each cross section because these surfaces are free and have no vertical forces acting on them from the outside. If the rivets were in the 'TB' condition with a shear strength of 234 MN m$^{-2}$, then the safety factor would be in the region of 3.3 whereas for the defective rivets with a shear strength of only 165 MN m$^{-2}$, the safety factor would be reduced to 2.3.

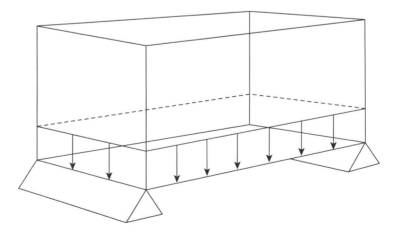

**Figure 9.8** Modelling the container as a box beam.

The simple model used for this analysis neglects two factors that could be significant. First, some shear would be carried by the frame at the floor and the roof of the container, which implies the average shear stress calculated would be high. Second, it assumes that the shear load was shared equally by all 48 rivets, whereas the distribution would in fact be parabolic, with the maximum loading being near the middle of the seam. This distribution could result in the rivets near the middle, which is where the splits appeared to have started, being exposed to a shear stress well above the average, in which case the factor of safety may have been almost eliminated. To take into account these two factors would be a daunting theoretical task. Furthermore, when the panels of a stressed skin container are loaded in shear a secondary force field develops due to buckling, the so-called diagonal cross tension field, which imposes additional shearing force on the rivets. An alternative approach to the problem would have been to measure strains in a real container loaded to its full capacity and then infer from these strains what the stresses must be. It is a technique often used to study stress distributions in complex systems. However, the investigation did not justify this costly and time-consuming analysis. Furthermore, it would only have supplied data for a static container, whereas these containers split while they were being handled, so a dynamic analysis under transient loading conditions is really required.

Whatever theoretical modelling is undertaken, what is definitely known is that the original design of these containers had been approved and tests on prototypes had met the requirements of all relevant standards, without reservation. Moreover, the vast majority of these containers, presumably not constructed with defective rivets, were still giving satisfactory service several years after the design was introduced.

Another factor that came to light during the stressing examination was the effect of load distribution inside the container. A single heavy item, such as a machine tool, places a different loading on the sidewalls than the same total load distributed evenly. If the container being handled bumps into something, then the inertia loads also add significantly to the shear stress on the rivets and are an obvious explanation as to why the splits occurred while containers were being off-loaded and why all those that failed had held heavy items of machinery. It is easy to appreciate that a container packed full of clothing, for example, would be much less vulnerable than one with a single, heavy item fastened to the floor near the middle. Additionally, the heaviest part of a machine tool might be at one end, which would increase the shear forces in the riveted seams at that end of the container.

The formal report concluded that a batch of containers had been produced with a side-wall strength below that which was intended, because the rivets had been set in the wrong condition, that is, without having been solution-treated as specified on the drawing. The factor of safety in the basic design was adequate, but in the containers that split open it was reduced to an unacceptable level because the rivets were too weak. The failures were caused by transient overstress during handling. The manufacturer of the containers was thus held liable on the grounds that they failed to solution treat the rivets before setting. There was no suggestion that there was anything wrong with the stock of rivets, which were in the condition specified and contained no fault that would have contributed to the failures. The packers, port handling authorities and shipping companies cannot be blamed. Containers inevitably get bumped from time to time, and there is no evidence that these particular containers were treated any differently from all the others.

## 9.4 Splintering Failure of a Pin Punch[5]

A pin punch was being used, by a DIY enthusiast, to drive out corroded bolts on some equipment that he was attempting to repair. The pin punch was one of a set that he had purchased from a local market. As he was hitting the striking end of the punch with a ball peen hammer, it shattered, causing splinters to embed in his cheek. Although not seriously hurt, he was shocked that the punch had failed in such a way when being (correctly) hit on its striking end.

Punches are commonly manufactured from steel of the types generally containing 0.5% carbon (BS3066:1995), quenched and tempered for high strength and toughness. More highly alloyed steels can contain manganese, silicon, nickel, chromium, molybdenum, vanadium, aluminium and boron, to enhance the properties obtainable after quenching and tempering. Alloying also helps in reducing the need for a rapid quench cooling, thereby reducing the potential for distortion and cracking. However, these alloy systems carry a cost penalty, with the products being at the top end of the market.

Chemical analysis showed that the punch had been made from a simple carbon steel, that would be quench-hardened and tempered to the level of strength desired for the application. BS3066 dictates localised hardening at the cutting/punch face and specifies a hardness zone. Microstructures (tempered martensite or bainite) produced by quenching and tempering would be characterised by a greater toughness of the 'working' face, along with enhanced capacity to deform without rupture.

Longitudinal sectioning and a Vickers hardness survey of the pin punch showed an increasing hardness as the striking face was approached, having values of 390 Hv away from the striking face, increasing to 600 Hv as the striking face was approached. This result can be clearly seen in Figure 9.9A, with hardness indents visibly getting smaller as they approach the striking face. As a result, the hardness survey was repeated using a Brinell ball indenter, producing identical results. As these results were the reverse

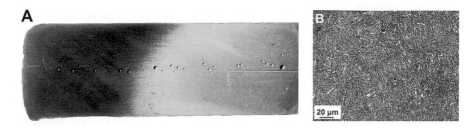

**Figure 9.9** (A) Longitudinal polished section of a pin punch. Hardness indents (both Vickers and Brinell) clearly getting smaller as they approach the striking face. (B) Microstructure of the punch striking face, showing a typical tempered martensitic structure.

of those expected, the punch was etched for microstructural examination. It was found that the punch exhibited a martensitic structure (Figure 9.9B) showing the striking face had mistakenly been hardened and tempered, rather than the punch end.

A polished transverse section through the striking end revealed large quench cracks (Figure 9.10). During the quenching process, there are two stresses involved: thermal stresses due to rapid cooling, and transformation stresses due to the increase in volume from austenite to a martensitic structure.[5] These stresses can be so severe as to cause excessive distortion or even cracks within the section.

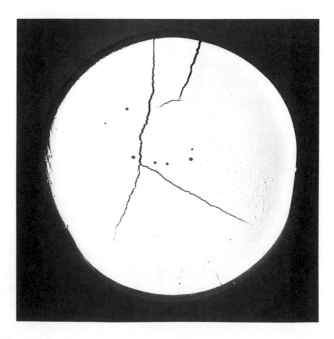

**Figure 9.10** Polished transverse section showing a number of quench cracks through the striking end of the pin punch.

For some unexplained reason, this set of pin punches had been hardened at the wrong end. The presence of quench cracks was indicative of poor thermal control, with the punches being quenched at a rate that was far too severe for the material. Furthermore, microstructural observation showed that the punches had not received an appropriate tempering treatment, having excessive hardness at the striking face and not having had 'toughness' imparted by tempering. The punch set was an imported item, with no manufacturer's name or logo, and the trader from whom it was purchased did not return to his market stall. Pin punches are considered general purpose tools for driving out pins and bushings, lining up bolt and rivet holes, removing hinge pins and the like, so there was no cause to criticise the way it was being used at the time of the incident. However, it did transpire that the user was not wearing any protective goggles; therefore he was extremely fortunate that splinters ejected from the striking face did not cause an eye injury.

## 9.5 Failure of a Vehicle Fuel Delivery Rail[3]

An estate car (station wagon) was being driven in a normal manner along a country road. Without warning the engine lost power, with a resultant loss of servo assistance to systems such as steering and braking. The outcome of this type of system failure was a rapid loss of vehicle control, which led directly to the vehicle crashing and killing the driver. Crash scene investigation immediately raised the possibility of an issue with the engine fuel delivery rail. It was therefore recovered for a more detailed laboratory analysis, and is shown as Figure 9.11A, with a closer view of the suspect fuel delivery pipe shown as Figure 9.11B. The numbering used for injector identification is not representative of engine firing sequence; it is simply a left-to-right identification notation.

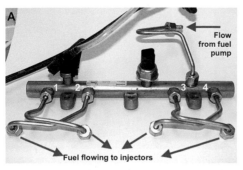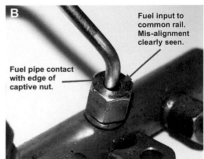

**Figure 9.11** (A) General view of fuel delivery rail, with arrows indicating pump input and outputs to injectors. (B) Close-up view of rail entry from pump. Axial misalignment of the pipe on entry to the rail is clearly seen.

Simple visual observation of the fuel pipes, at a point where captive nuts screwed onto their respective injectors, revealed axial misalignment of injector connections number 2 and 4, along with misalignment of the input from the fuel delivery pump, seen in Figure 9.11B. Now, any misalignment that brings about contact between the pipe wall and inner corner of captive nut is a recognised stress-raising situation, that will undoubtedly increase the potential for fatigue failure driven by engine vibration and hydraulically induced cycling from normal operation of the fuel pump and injectors (see stress calculations below).

Evidence of contact between captive nut and pipe was found immediately above the clamping gland. The seating footprint on both the swaged pipe ferrule and rail inlet revealed a double footprint on both the clamping gland and ferrule (Figure 9.12A), verifying that the connection had been broken and re-made post-manufacture. Mis-alignment of both fuel delivery and injector pipes would therefore have come about during a service visit, rather than at the point of manufacture.

A crack was soon spotted in the fuel delivery pipe, located at the shoulder radius of its swaged ferrule (Figure 9.12B). Crack initiation was located at a point of maximum misalignment between pipe and rail. Elemental analysis (Table 9.3) confirmed the pipe to be a seamless cold drawn single-wall stainless-steel tubing conforming to British Standard 8535-1 1996 for high-pressure fuel injection pipes. The pipe end had been swaged to provide a seal between its captive nut and injector, and the pipe had a nominal bore of 2 mm diameter. A longitudinal section of the swaged ferrule was prepared for metallurgical observation. It became clear that a single crack had initiated at a position located just before the swaged shoulder radius of the ferrule (Figure 9.13A) and had propagated almost totally through its cross section (Figure 9.13B). Clearly, the high-pressure fuel input pipe had failed

**Figure 9.12** (A) Close-up view of the ferrule showing two distinct footprints, one of which is clearly misaligned. (B) Ferrule shoulder showing crack initiation at a point immediately above the swaged internal radius.

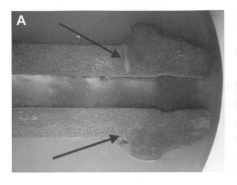

**Figure 9.13** (A) Longitudinal section of swaged ferule and section of pipe, with a clear transverse crack arrowed. (B) An etched section showing expected microstructural flow of drawn pipe and swaged ferrule, along with a clear view of the penetrating fatigue crack.

**Table 9.3   Chemical Composition, Hardness and Strength of Fuel Delivery Pipe**

| C | Mn | Ni | Cr | Fe | Hardness (Hv) | UTS (MPa) |
|---|----|----|----|----|----|----|
| 0.03 | 2 | 11 | 18 | Bal | 140 | 448 |

*Note:* Strength Inferred from Hardness/Strength Relationship UTS = 3.2 Hv

as a result of fatigue, with crack initiation being a direct result of the localised stress-raising feature (intimate contact of the pipe wall against the edge of its captive nut). However, a good microstructural flow of the pipe material, both longitudinally and at the swaged ferrule (Figure 9.13b), suggests good manufacturing process control for both the pipe drawing and swaging operations.

The initial loss of engine power was clearly the result of a major fuel leak, sited at the fuel input connection to the delivery rail. Fuel loss at this point (rather than at an individual injector connection) would result in engine starvation, with resultant (and dramatic) loss of power. In contrast, leakage from one of the injector pipes would simply increase the risk of under-bonnet fire, as well as giving poor performance, rather than the total engine power loss (and associated driver control difficulties) that would accompany any major fuel loss from the rail delivery input.

From gathered documentation, the cracked pipe formed part of the initial vehicle build and had not been changed at the time of service intervals. This was also found to be true for all high-pressure pipes feeding the injectors. Servicing was the most likely source of pipe (both fuel delivery and injector supply) misalignment, and was, therefore, responsible for sensitising the pipe to failure by fatigue. However, when considering such a short time-span for propagation and growth of a major crack in a fuel supply pipe, there must be an onus on the manufacturer to prevent or limit reoccurrence of events that overtook the vehicle in question. At a minimum, there is a moral responsibility for the manufacturer to inform service centres of the necessity to ensure axial alignment of fuel line connections on the vehicle in question, along with the requirement to inspect/replace vital fuel pipe runs when replacing components such as a fuel delivery rail.

### *Stress calculations*:

Stress induced in the walls of a tube, by pressures of fluids within that tube, is given by:

$$\text{Allowable fibre stress}\,(S) = P \times OD/2t,$$

where P = internal pressure in MPa, OD = outside diameter (mm) and t = wall thickness (mm).

   With pressure in the fuel rail being 140 MPa, stress in the tube fibres was therefore $140 \times 6/4 = 210$ MPa.

   From bending theory, strain $(\varepsilon) = y/R$, where y = distance from neutral axis and R = radius of curvature of neutral axis.

$$\text{Now, } \sigma/\varepsilon = E, \text{ therefore } \sigma = yE/R$$

   For a 316 stainless steel, E = 196 GPa. If bend radius at captive nut was 3 mm, radius at neutral axis would equal bend radius + OD/2 = 6 mm = R, y being equal to 3 mm. Therefore y/R = 0.5 and $\sigma = 96.5$ MPa at the outer fibre, still within the maximum allowable fibre stress of 210 MPa.

   However, if bend radius over capture nut was 0.3 mm, then y = 3 mm and R = 3.3 mm. Therefore stress at outer fibre = 206 MPa, remaining within but approaching the maximum allowable fibre stress of 210 MPa, thereby sensitising the region to potential crack initiation and failure by fatigue.

## 9.6  Failure of a Presentation Cake Knife[6]

A porcelain-handled cake knife was being used to cut a wedding cake. The knife was being held in the bride's right hand, with the hand of the groom over hers, a loving action that also provided additional strength to the cutting procedure (Figure 9.14A). However, the knife handle suddenly and unexpectedly failed during the cake cutting ceremony, causing a razor type injury to two fingers and the tendon of the bride's right hand (profuse blood loss rendered the bottom cake layer as unusable). The knife in question was a presentation item from the product range of a well-known and reputable porcelain manufacturer.[6] It comprised a steel blade set in a porcelain handle and was approximately three years old when the handle snapped. The knife had undergone minimal wear and tear within that period, usage being limited to four prior ceremonial occasions.

Observation of the broken parts showed there was no visible evidence of pre-existing cracks that might indicate a high degree of wear and tear.[6] Figure 9.14B shows the blade with part of the ceramic handle still attached to the tang, a porcelain sliver and the solid porcelain end of the handle. The knife tang and porcelain handle had been joined with an epoxy-based adhesive. Discolouration on the porcelain sliver (arrowed in Figure 9.15A) was as a result of blood staining from finger injuries sustained by the bride. The figure shows the end of the blade and attached porcelain fragment, with the porcelain sliver temporarily put back in place. Figure 9.15B shows the same

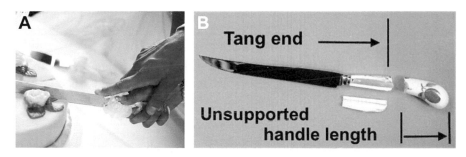

**Figure 9.14** (A) Unrelated image to demonstrate a gripping method used to cut the wedding cake. (B) The subject cake knife clearly showing the length of unsupported ceramic handle.

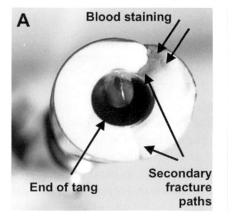
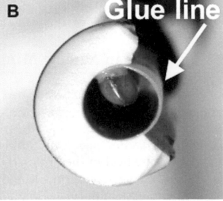

**Figure 9.15** (A) The site of ceramic handle fracture, clearly showing the end of the tang support. Note: the porcelain sliver was simply being held in place. (B) Same end view of the handle fracture, but with ceramic sliver now removed.

view, but with the porcelain sliver removed; notice the intact glue line form-ing a hollow tube.

There was no evidence of pre-existing cracking inasmuch as the entire fracture had formed at virtually the same instant, and there was no sign of intrusions or staining such as would have occurred during use or when wash-ing up.[6] The fracture surface on the rear section of the handle exhibits step and lip characteristics typical of combined bending and torsional overload failure, showing that the handle broke under heavy pressure consistent with cutting and twisting the blade, i.e. the failure of the porcelain handle was due to a combined bending and torsional overload mechanism that exceeded the strength of the porcelain at the tang end. In effect, the unsupported length of handle acted as a cantilever beam, with stresses being concentrated at the tang. Failure at this point immediately indicated that the tang was too short for the handle length. Correct design of the tang should have required it to extend the *full* length of the handle, with little or no unsupported handle overhang. It can also be seen that there was a clear issue with the glue line bonding, as illustrated by the lack of adhesion to the liberated sliver, as clearly shown in Figures 9.14B and 9.15B.

Learning outcomes from this failure highlight two design shortcomings that should have been readily foreseeable from knowledge of basic engineer-ing mechanics and materials principles:

(1) The unfortunate consequences of this breakage could have been avoided if the metal tang had extended almost the full length of the handle and if the end of the tang had not been directly in contact with the side of the hole in the porcelain.[6]
(2) Perhaps more importantly, on no circumstances should a brittle material (including brittle polymeric materials) be used for fabricat-ing a knife handle.[6]

## 9.7 Failure of an Aerobic Mini-Stepper

An aerobic step exercise machine had broken at a tension cable pulley wheel fork, causing ligament damage to the knee of the user. However, as the manu-facturer had not previously encountered product failure associated with this particular fork component, he claimed that failure had to be the result of abuse or misuse by the user. The Mini-Stepper in question is shown in Figure 9.16A, with a closer view of the fracture site shown in Figure 9.16B. General visual observation of the general fracture site revealed no evidence of physi-cal (or malicious) damage to surrounding components, chrome plating or paintwork. This single observation will rule out failure by any impact or traumatic loading.

**Figure 9.16** (A) Underside view of the Mini-Stepper in question. (B) Closer view of the fracture site. Note: no evidence, or sign, of physical abuse.

However, the tension cable pulley wheel fork had broken, with the fracture sited at a point where the tension adjustment rod connected to the fork. A crack had grown diametrically across a drilled mounting hole in the fork, thus causing the component to fracture into two halves. One fractured leg of the fork was retrieved for closer observation of the fracture surface. Under low-power magnification, river markings were clearly evident, being the fingerprint of a fast (brittle) fracture, as seen in Figure 9.17A. River marks will always point back to the crack initiation site which, in this case, appears to be more linear, rather than a point source. Fast (brittle) fracture features would also include lack of corrosion product on the surface and a generally granular surface appearance. These features were all present on the fracture surface in question, giving further confidence to the brittle nature of fracture. There is clear evidence of at least three crack arrest marks or zones as shown by red lines in Figure 9.17A. This would indicate that the fracture was not

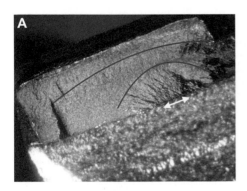
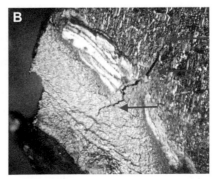

**Figure 9.17** (A) The fork fracture surface, showing the presence of river markings. The crack initiation site is arrowed in white, with at least three crack arrest marks or zones delineated by red lines. (B) The presence of additional surface cracking noted at a number of sites on the fork.

totally instantaneous, but was driven across the load-bearing cross section by at least three applications of loading. This observation was given further confidence by the presence of arrest marks evident seen around the drilled hole. Additional surface crack initiation was present on the fork, again being indicative of an enhanced mode of cracking, rather than a single traumatic overload (Figure 9.17B). The features noted pointed to an environmentally assisted failure having been at play, such as hydrogen embrittlement.

When atomic hydrogen enters a metallic material, it can result in a loss of load-carrying ability, loss of ductility or sub-microscopic cracking. As a mechanism, failure by hydrogen embrittlement requires the presence of tensile stresses, susceptible material and of course the presence of hydrogen. It is an insidious type of failure as it can occur without an externally applied load (residual stresses) or at loads significantly below yield stress. Thus, catastrophic failure can occur without significant deformation or obvious deterioration of the component. Hydrogen embrittlement is therefore considered to be the Achilles' heel of high-strength ferrous steels and alloys. Although it is high-strength steels that are the most vulnerable to hydrogen embrittlement, all materials will exhibit varying degrees of susceptibility to this form of attack.

At this point, the question any investigator should ask is 'How did the stepper tensioner fork become exposed to hydrogen?' The simple answer is that, as part of the manufacturing cycle, hydrogen can be introduced during component heat treatment, carbonising, cleaning, pickling, phosphating and electroplating. In the electroplating industry, the accepted method to alleviate hydrogen embrittlement is with a 'relief-baking' process, with the hardness and strength criteria of Table 9.4 used to determine the temperature/time for hydrogen embrittlement relief (baking).

Most specifications require relief baking within four hours of plating, while some require baking within one hour of plating. The process serves to

**Table 9.4  Specific Bake Times for Various Strength Steels**

| Tensile Strength MPa | Rockwell Hardness $H_R$ | Post-Plate Bake @190–220°C. Hour |
|---|---|---|
| 1,700–1,800 | 49–51 | 22+ |
| 1,600–1,700 | 47–49 | 20+ |
| 1,500–1,600 | 45–47 | 18+ |
| 1,400–1,500 | 43–45 | 16+ |
| 1,300–1,400 | 39–43 | 14+ |
| 1,200–1,300 | 36–39 | 12+ |
| 1,100–1,200 | 33–36 | 10+ |
| 1,000–1,100 | 31–33 | 8+ |

*Source:*  ASTM B 850-94.

remove or redistribute the hydrogen, but may further reduce the fatigue limit of high-strength steels. For the production electroplater, having to remove parts from the production line to bake – followed by a separate chromating process – is a laborious business. It is at this point of production that components can (and *do*) miss the baking process. There is no visible difference between a baked and an un-baked component. Therefore, components that have missed the baking cycle are not easily identifiable and can easily enter the production chain, ultimately entering service in a vulnerable condition.

So, from general observation of the stepper in question, and from more focused observations of the fork fracture surface, it can be said that failure was as a result of hydrogen embrittlement. Typical features of hydrogen embrittlement were clearly present on the fracture surfaces of the tension cable pulley wheel fork. In addition, it had been chrome electroplated which was, therefore, one way to introduce hydrogen into the parent metal. On the balance of probability, the fork had missed a low-temperature bake stage that is designed to drive off hydrogen that would be introduced as a result of the plating process. Failure was therefore the result of a manufacturing shortcoming, rather than the result of misuse or abuse by the claimant.

When considering the forensic aspects of hydrogen embrittlement, the crux of this failure was a missed processing (bake out) stage. This particular situation could well have been just a one-off event, i.e. the broken fork in question was the only component to miss the bake out. Equally, however, a batch of forks having missed the bake out process was also a possibility, in which case, failures of a similar nature would be expected to arise. Over the following few months, a number of identical failures did come to light; clearly, a batch of forks had indeed missed the final bake out process. The legal position, within the European Union, changed in 1997.[7] Previous to this, a prospective claimant had to prove negligence. The surface finishing industry is bound by the requirements of ISO 9001; 2008 and where applicable, records are required for seven years minimum, and the processor must be able to present a traceable record of events. This in itself does not absolve the supplier from any liability.[7] The position now is that a claimant only needs to prove the part was faulty and unfit for the use, or purpose, for which it was supplied in order to claim liability and recompense. The pulley wheel fork most certainly contained an insidious fault that had been manufactured-in at the time of fabrication.

## 9.8 Failure of Oil Pump Gearing

This is another case concerning faulty heat treatment, but this time involving three separate firms in a chain supplying a major manufacturer with oil pumps for agricultural tractors. The oil pumps were mounted inside the engine crankcase and were fairly simple devices (schematic, Figure 9.18A), consisting of two meshing spur gears driven by a shaft inside a housing. Oil was drawn in and forced around by the outside teeth of the contra-rotating gears into a delivery pipe to the crankshaft bearings and camshaft of the engine. A number of failures were experienced after a few tens of hours' work, resulting in complete engines being written off at busy times in the farming season, all of which claims had to be met under the tractor manufacturer's warranty (Figure 9.18B).

The failures were all found to result from fatigue initiating at the splines of the drive gear, so the tractor manufacturer sought to recover its costs from the oil pump suppliers. The oil pump suppliers merely assembled parts they bought as finished products, and their records established that all the faulty gears had come from one large batch supplied by the same subcontractor. This subcontractor, a specialist gear manufacturer, carried out a failure investigation on several of the gears and found every one of the fatigue fractures had initiated at a small pre-existing crack at the edge of a spline for the drive shaft (Figure 9.19B). These cracks, though small, were identified as quench cracks arising from the case-hardening heat-treatment process. They had not been detected during final inspection at the time the oil pumps were assembled. This was clearly evident from the black scale on both faces of the initial crack, which contrasted with the bright surfaces of the fatigued area propagating from it. Although the dimensions and surface hardness of the gears were satisfactory, the analysis of the steel was slightly different from what the subcontractor had ordered from its suppliers and as stated on the delivery note. The subcontractor had forged the stock, machined the gears, put them through their case-hardening heat treatment process and finish ground the teeth before dispatch to the oil pump manufacturer. Consequently, this firm accepted liability for the defective gears but sued its steel supplier, a small

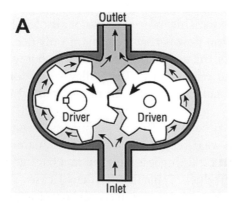

**Figure 9.18** (A) Schematic of a spur gear oil pump. (B) A failed unit showing the housing, and two spur gears within their cradle.

 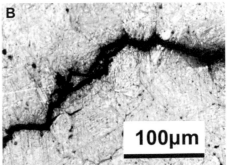

**Figure 9.19** (A) The two hydraulic pump spur gears, showing the driver gear on the left and driven gear on the right. Arrowed is a quench crack sited at the root of a spline. (B) Metallurgically prepared section through a quench crack. Black scale on crack edges provides evidence to suggest the crack was present at the time of entering service.

stockholding firm that bought steels of various compositions and rolled them into sections and specific sizes ordered by a large number of small engineering customers. Fortunately, this firm carried product liability insurance because the root cause of the fatigue failures was that the wrong grade of steel had been supplied, which was totally unsuitable for the type of heat treatment process used to case harden the gears.

Steels for general engineering purposes are alloys of iron containing up to 0.8% by weight carbon together with small percentages of other elements such as manganese, chromium, nickel and vanadium. Without a doubt, they comprise man's most versatile engineering materials, mainly because their carbon content allows a single product, say a chisel, to be heat treated so as to achieve extremely high hardness at the cutting edge, a middle section strong and tough and a softer head that is malleable. Unfortunately, when medium- and high-carbon steels are heat treated to maximum hardness they are also brittle and may sometimes crack under thermal contraction stresses during heat treatment processes. The other alloying elements modify the heat treatment response and impart specific properties in their own right, though not on such a dramatic scale as carbon.

Additional carbon may be introduced into the surface of a shaped product to a controlled depth, enabling the maximum hardness and wear resistance possible to be attained in the outside skin. This is the purpose of case hardening used to develop maximum wear resistance in parts which are in sliding contact, such as gear teeth, while retaining less hard but much tougher properties in the main body of the product. Typically, hardened cases are between 0.4 and 1 mm deep and, of course, the carburisation process can only be applied to products made from steel when they have already been manufactured to the required shape and dimensions.

**Table 9.5    Analysis of Material Ordered from the Supplier Compared to That Delivered**

|              | Carbon (%) | Manganese (%) | Nickel (%) | Chromium (%) |
|--------------|------------|---------------|------------|--------------|
| Ordered      | 0.2 max    | 0.6–1.0       | 0.6–1.0    | 0.4–0.8      |
| Failed gear  | 0.38       | 1.62          | 0.16       | 0.04         |

The oil pump gears had been forged to shape, machined to size and case hardened. Final grinding was used to correct any slight distortions resulting from the case-hardening process. The pre-existing cracks discovered in the failed gears were typical of quench cracking. So on what grounds did the gear manufacturer seek to sue its steel suppliers?

A comparison of the analyses of the steels specified to make the failed gears and the one that was delivered by the material supplier is shown in Table 9.5.

The significant figure is the carbon content. The specified steel has a low carbon content and does not undergo an appreciable volume change when it transforms during quenching; it hardens to a limited extent but does not become brittle. The small percentages of manganese, nickel and chromium increase strength but not at the expense of toughness. It is a low-alloy steel intended for case hardening.

The steel supplied was a medium-carbon heat-treatable manganese steel, widely used in engineering applications requiring strength and toughness, for which it is used in a quenched and tempered condition. If quenched too rapidly products may develop quenching cracks due to the volume increase as the metal undergoes rapid transformation from its high-temperature state to the temperature of the quenching bath. If and when such cracks form, they usually appear where there is a change in section; in this respect, the root of a spline is an ideal position. Hence, if a high-content carbon case is built up on a product made from the specified steel and is finally quenched to harden the case, the volume expansion of the underlying steel will not crack it. If the same thickness of case is built up on a product made from the medium content carbon steel, the case is very likely to crack during a severe quench.

Thus the cause of the gear failures was clearly established. Responsibility was shared by the steel supplier and the gear manufacturer. The supplier had mixed the same sized bars of two different steels and had colour coded the medium carbon–manganese steel as for the low carbon. Notwithstanding this, the gear manufacturer could have and should have discovered the quench cracks if adequate inspection and quality control procedures had been in force. Not all the gears would have been quench cracked (the tendency for this particular composition of steel depending very much on the quenching medium and the actual procedures for each batch) so they could have identified the faulty ones before they were dispatched and assembled into the oil pumps. Although the 'wrong' steel was used, gears that had not quench cracked would have performed equally well and lasted just as long as if they had been made of the specified material.

## 9.9 Replacement Gears for HGV (Heavy Truck) Gearbox[1,2]

No injury or accident was involved in this case, but it was the basis for a substantial claim against a small company that specialised in the manufacture of replacement gears for machinery and vehicles for which spares were no longer available.

The manufacturers of one particular make of heavy commercial vehicle were taken over by a larger company and, although the vehicles had long enjoyed a reputation for sturdiness and reliability for haulage work, the new owners of the business announced they would no longer supply spare parts when existing stocks were exhausted. The two largest gears in the gearbox tended to wear faster than the others, and it became impossible to replace them, even when the other parts still had a great deal of useful life remaining. The specialist firm set out to manufacture replacements and soon had orders for several dozens of sets of these two particular gears. The firm had taken measurements and characterised the hardness of the worn gears and manufactured new ones of identical dimensions using one of the strongest steels available. The specialist firm machined the gears and sent them out for heat treatment because, in addition to hardening and tempering the whole gear, the teeth had to be surface hardened to match the hardness of the originals.

All seemed to be well until the first pair of gears was returned, having failed by teeth breaking off in just over three weeks' service (Figure 9.20). Their failure was attributed to poor gear changing practice, as the gearbox had no synchromesh and double declutching was necessary when changing gear. Another two gears were fitted, but a similar failure occurred after only

**Figure 9.20** Suspect gear wheel on left, showing broken gear teeth arrowed in red. A satisfactory gear wheel made by upset forging is shown on right. Note: sections removed for metallurgical examination.

two weeks. As it was a major job of replacement the truck owners were dis-pleased to say the least because, in addition to the costs, the vehicle was not earning its keep while it was off the road. Also, the truck owners had pur-chased nine pairs of the replacement gears intending to use them for other vehicles in their fleet.

The manufacturer suspected there might be something wrong with the surface-hardening heat treatment, and discovered the hardness of the teeth that had broken was higher than they had specified. The customer returned all the unused gears, and the manufacturer sent them off to have the teeth tempered to reduce their hardness. They did so in the belief that the fractures were due to the teeth being brittle. A modified set was put in the gearbox but this set failed after four weeks, every tooth breaking off the inner ring gear. Hardness checks revealed that they had been softened to the modified hardness. The vehicle operators initiated proceedings against the manufacturer to recover their losses.

Metallurgical investigation of the failed and unused gears found nothing wrong with the composition or heat treatment of the steel. However, two of the original gears that had gradually worn out over a long period of use – distinct from teeth breaking off – were also produced. These provided the answer to the premature failures. The diameter across the outer teeth is 150 mm. These outer teeth had given no trouble; the ones that had broken off were all from the inner ring some 100 mm in diameter. In the failed gears examined, two or three had simply worn down until they no longer engaged with the mating teeth, which is why the gears had to be replaced.

At the time the original gears were made, all steel started off as ingots. In the as-cast state an ingot has no directionality but, as it is worked down to billet or bar, directionality appears as grains and inclusions are extended in the direction of working.[1,2] This is a structural effect readily demonstrated by sectioning a wrought product, smoothing or polishing the surface, fol-lowed by etching to reveal the grain structure. The directionality is revealed as flow lines, which resemble the grain in timber. For any given quality of wrought steel, toughness and fatigue resistance are much better across the flow lines than parallel with them, just like timber. The effect is equivalent to chopping a piece of wood across the grain compared with along the grain, where it splits easily.

The flow line patterns of the old gear and one of the new were studied by sectioning across a diameter. The photograph in Figure 9.21 shows how their grain flow structures were distinctly different. The original gears had been upset forged before the teeth were cut, whereas the replacements had been machined directly from round bar. The flow lines reveal how the old gears had started off as a length of round billet, about twice the length of the gear but of smaller diameter. This had been hot forged to squash it down, so that the shape finished with a much larger diameter and shorter length than when it started. This was performed specifically to develop a flow line pattern following the

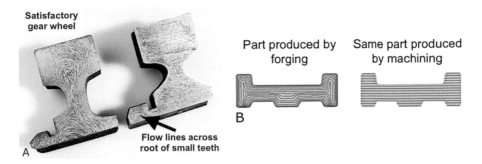

**Figure 9.21** (A) The gear wheel sections, metallurgically prepared to show material flow lines. Left: satisfactory gear from a blank made by upset forging. Right: the suspect gear wheel, having been machined directly from a slice of round billet. (B) A schematic showing the grain structure of a forged part compared with a machined part.

contour of the gear teeth so that all forces acting to bend the teeth would act broadly *across* the grain. The teeth were as tough as they could be for this type of steel, and the inclusions were oriented in directions that imparted the maximum resistance to fatigue initiation and brittle fracture.[1,2]

In contrast, the new gears had been machined from stock of the full diameter for the gear required. There was no upset forging, so the grain flow remained the same as the original billet, parallel to the gear axis. The result was that the flow lines in the teeth were in the most adverse orientation to resist fatigue and brittle fracture.[1,2] This is why they broke off so quickly under service loads, despite having the same hardness – and tensile strength – as the original gears. The schematic shown in Figure 9.21B will graphically illustrate the concept of grain 'flow' of a forged part compared with that of a machined part. Failure had little or nothing to do with drivers' needing to double de-clutch when changing gear, or with the surface-hardening heat treatment of the teeth. The manufacturer's insurers settled the claim, and all the gears sent out were recalled.

This case demonstrates the importance not only of getting the shape of the item right, but also following the necessary manufacturing route to achieve optimum properties for the application. It was considerably cheaper to machine the gears directly from round bar rather than to first upset forge individual gear blanks, especially for the small batch numbers needed for spare-part manufacture.

## 9.10 Vehicle Engine Radiator Cooling Fan

This final case is concerned with the failure of a military vehicle engine radiator cooling fan, a number of which were found to be failing in an identical way after only a few months in use. The fans in question were of a polymeric injection-moulded construction, with failure being predominantly sited at their blade root sections (Figure 9.22A). When injection-moulding a multi-blade cooling fan, complex multiple gated injection-moulding tools will be used for the job. When designing a moulding tool for this type of application, it is important to optimise flow patterns and to predict the locations and interaction of stress (load) bearing areas and knit lines (planes or interphases). Therefore, it can be said that the service (mechanical) performance of an injection-moulded thermoplastic automotive fan component will depend on both the complexity of its shape, and on the moulding tool design.

The first step taken by the forensic investigator was to interview a number of the vehicle drivers. Their descriptions of events in play at the time of fan failure were similar: each vehicle had entered deep water (which they were designed to do), with cooling fan failure immediately following. As a result, the drivers had reached a similar conclusion that their failure was the consequence of a (traumatic) blade impact with the water. When considering the number of such similar failures coming to the fore, it was clear that similar failure of identical fan units could also be expected to occur in the future. Therefore, finding the actual cause of failure was vital in order to ensure reliable operation of the military vehicle.

The cooling fans had an axial configuration designed to allow large volume airflows along with a low-pressure increase. They had the appearance of

**Figure 9.22** (A) One of the radiator/cooling fan units forwarded for inspection, clearly showing no evidence of surface impact damage. (B) A simple finite element analysis showing stress hot spots on this particular fan configuration showing the most likely mechanical failure location sited at, or adjacent to, the blade root. (Image made using the COMSOL Multiphysics® software and provided courtesy of COMSOL.)

a propeller, with the blades mostly formed as a wing profile, skewed in the radial direction for increased efficiency. The blades functioned by moving an air stream along the axis of the fan in a similar way to that of a propeller on an airplane; the fan blades generate an aerodynamic lift that pressurises the air. In normal use, the blades of any flexible cooling fans will be subjected to both combined axial and flexural loading conditions. The range of speed for all rotating under-the-hood parts typically fluctuates between 500 and 9,000 rpm. With any high rotational loading conditions, it is important to take into account the influence of inertial axial load and flexural fatigue effects initiated from airflow pressure. In addition, the resistance to creep would also be a required consideration, with long-term stress relaxation expected at the blade roots adjacent to the hub. A simple finite element analysis was undertaken to show the stress hot spots on this particular fan configuration (Figure 9.22B). This analysis would suggest that for the majority of injection-moulded fans with flexible blades, the most likely mechanical failure location would be sited at, or adjacent to, the blade root, with the expected failure mode being that of bending fatigue. An infra-red (IR) analysis was undertaken to determine the material of manufacture, which turned out to be a nylon (polyamide) plastic.

Turning attention back to the water submersion issue, if and when a mechanical cooling fan becomes submerged, it will act much like a boat propeller and pull the blades forward slightly, contacting, and possibly damaging both the radiator and fan blades in the process. However, cooling fans equipped with a clutch will typically stop turning and prevent this type of damage. Nevertheless, it is understood that the make of vehicle in question was not equipped with such a system. Therefore, if deep-water submersion was indeed responsible for the spate of failures in question, evidence of extensive fan blade tip and radiator damage will undoubtedly be clearly in evidence. However, no untoward evidence of inappropriate scuffing or traumatic impact damage (of either radiator, hub and/or blades) could be found by visual observation, thus suggesting that the cooling fan units had not been subjected to traumatic impact loading. This observation provided a first clear indicator to suggest a mechanical failure mode as having been at play.

In all cases examined, the fan blades had fractured in the vicinity of their root, where they were joined to the fan hub (Figure 9.23A). All blades had fracture propagation paths that initiated at almost identical sites, and followed an almost identical route to failure. Observation of blade stubs under low-power optical magnification revealed crack branching to be present on all units examined (Figure 9.23B). Crack branching is an indication of crack growth progressing over time. In contrast, an instantaneous overload will take the route of least resistance, forming a single crack path, *with no branching*. This feature provided the next clear evidence of the mechanical – rather than traumatic – nature of these fan failures. Although knit lines from the

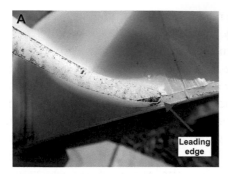
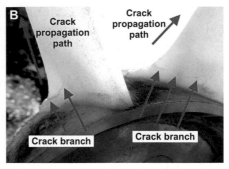

**Figure 9.23** (A) The fan hub showing blade fracture initiating at their root, this site being identical on all of the submitted units. (B) Crack branching to be present on all units examined.

original manufacturing process were clearly observable, they did not appear to have exerted influence on the fracture propagation *path*. However, crack *initiation* may well have been influenced by the presence of a (stress-raising) knit line at the leading edge of the blade roots.

Observation of the fracture surfaces clearly revealed two distinct zones: a flat smooth zone, and a rough zone, as seen in Figure 9.24A. The smooth zone was formed by a crack initiating on the upper surface in the picture. This crack had propagated (grown) over time into approximately half the blade section, by a fatigue growth mechanism. When the incrementally reducing material cross section could no longer sustain the applied service loading, the section failed by a fast overload mechanism, thus generating the rough surface seen on the bottom half of the blade thickness in Figure 9.24A. In a nutshell, the service loading spectrum and geometric root stress

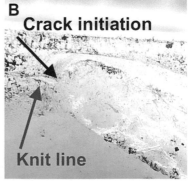

**Figure 9.24** (A) Two distinct fracture surface zones observed: a flat smooth fatigue zone (circled in red), and a rough faster fracture zone. (B) Crack initiation may well have been influenced by the presence of a knit line at the leading edge of the blade roots.

concentration (Figure 9.24B) had combined to sensitise the subject fan blades to premature failure by bending fatigue.

When considering all failure features noted on the hub blades, it was possible to formulate a failure scenario. Flexural loading would concentrate at the root of the blades (in effect, they are cantilevers acting under fluctuating load). This loading spectrum would then be combined with tensile centripetal loading and the stress concentrating influence of both the geometric section change and the knit line sited at the roots. These three features had combined to sensitise the blade to premature failure by fatigue. Once initiated, a fatigue crack would then have grown into the blade section up to a point where the remaining section could not sustain the loading conditions. A single blade would have fatigued to failure in the first instance. At that point in time, dynamic out-of-balance forces, generated by the loss of that single blade, would instantly apply additional out of balance loading forces to the remaining blades, thus driving all existing (growing) fatigue cracks to final failure in the remainder of the blades.

The advice given was to urgently discuss these failures with the fan manufacturer, who may well welcome the opportunity to (re)consider both the fan design solution and/or the material of manufacture. Clearly, points to consider would include increasing the blade root cross section, minimising or re-locating knit lines and changing the material of manufacture to one of a vibration-stabilised glass-reinforced polyamide, for example. The manufacturer may also consider it as good design practice to incorporate a cooling fan slip clutch, which would negate any future argument suggesting that accidental water submersion was responsible for failure.

# References

1. Lewis, P.R., K. Reynolds and C.R. Gagg, *T839 Forensic Engineering.* 2000, Open University Post-graduate Course, Department of Materials Engineering.
2. Lewis, P.R., K. Reynolds and C.R. Gagg, *Forensic Materials Engineering - Case Studies.* 2004, CRC Press.
3. Gagg, C.R. and P.R. Lewis, In-service fatigue failure of engineered products and structures – Case study review. *Engineering Failure Analysis*, 2009. 16(6): pp. 1775–1793.
4. https://www.asme.org/engineering-topics/articles/computer-aided-design -(cad)/forensic-reverse-engineering.
5. Gagg, C.R., Failure of components and products by 'engineered-in' defects: Case studies. *Engineering Failure Analysis*, 2005. 12(6): pp. 1000–1026.
6. Gagg, Colin, Domestic product failures: Case studies. *Engineering Failure Analysis*, 2005. 12(5): pp. 784–807.
7. https://www.fastenerdata.co.uk/hydrogen-embrittlement.

# Component Failure in Road Traffic Incidents and Accidents

# 10

## 10.0 Introduction

The car is the most common cause of sudden death and injury in most countries of the developed world and the developing world. Even if no personal injury is involved, damage to cars can be substantial and investigation of the cause or causes would generally follow. The usual reason for investigating the failure of a component found to be broken after a road traffic accident is to ascertain whether the component was broken in the trauma of collision or had suffered some kind of prior mechanical failure that caused the vehicle to go out of control. In the first instance, driver error or freak road and weather conditions might be held responsible. In the second instance, it is essential to ascertain the mode and cause of failure in order to establish where and with whom responsibility might rest. For example, failure might have resulted from a manufacturing or assembly fault, or mechanical malfunction due to wear, or some progressive weakening such as fatigue or corrosion. If the cause turns out to be mechanical, then it is essential to take into account the recent servicing history of the vehicle and what adjustments might have been made shortly before the accident occurred, for example, over-tightened wheel bearings, loose unions and a whole gamut of faults stemming from careless or unskilled maintenance.

Prior component failure accounts for only a very small minority of road traffic accidents, the vast majority resulting from collision. Human nature being what it is, a driver almost invariably seeks to blame someone else's mistake or the malfunction of some vital system like steering or brakes for loss of control, yet the experience of forensic investigators tends to support the cynical remark that the most likely cause of an accident is 'the nut who holds the steering wheel'.

The case histories that follow have been selected to represent the variety of modes of component failure. Unfortunately, these have to be limited to interpretation of the fractographic evidence in relation to the circumstances of a particular accident. Supporting details of the accident such as usually provided to the failure investigator, e.g. police 'scene of accident' photographs and road measurements, statements of witnesses, reports on the damage from vehicle inspectors and insurance engineer assessors, etc., have been omitted except where they directly relate to the findings.

317

## 10.1 Motor Vehicle Accidents as a Result of Steering Shaft Failure

The steering system of any vehicle is an integral part of its safe operation and control. When a steering system fails as a result of a broken or malfunctioning steering component, serious injury to the driver and passengers of the affected vehicle can result. Furthermore, steering failure raises the clear potential to cause injury to other road users, drivers and/or pedestrians, who may be on the road at the time of any such steering loss.[1]

### 10.1.1 Fatal Crash of a Small Family Saloon Car[2]

The driver of a popular family saloon car was killed in a road traffic accident. Although sustaining minor injuries, a passenger, who had been travelling in the rear seat of the vehicle, stated that the driver appeared to have lost steering control immediately before the vehicle crashed. Unfortunately, this type of failure was not an isolated case on the vehicle in question. Several had broken in this manner when the cars were getting old, and the majority of these vehicles were found to have tight swivel pins on the front axle due to a lack of greasing. Figure 10.1A shows the bottom of the splined steering arm where it entered the steering box. This is a complex fracture caused by cyclic torsional loads, where fatigue initiated at the outside of every one of the individual splines (Figure 10.1B). The cracks started by following a helical path across each spline, and when they entered the main cross section below the splines they continued as individual cracks toward the axis, all maintaining an orientation at 45° to the axis and producing the star-like appearance in the final fracture. However, just before they met up at the axis torsional overstress on the last remaining cross section produced a small pip of ductile fracture at 90° to the axis. The stress on a circular shaft subjected to torsion varies as the *cube* of the radius so that, when there is only a small area of sound metal near

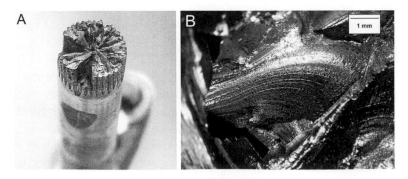

**Figure 10.1** (A) Splined steering arm fractured at the point of steering box entry. (B) Fatigue beach marks present on each individual spline.

the centre, the stress level under normal loads is much greater than when the cracking is still in the outer, splined region. Although insurers honoured all claims in this matter, it could be argued that the failure in question was the result of poor design, in that the shaft was not strong enough to withstand foreseeable loadings for the life of the vehicle. This is a particularly relevant point when considering that a number of other vehicles had broken in a similar manner. However, along with the age and uncertain history of the vehicles by the time such fractures began to be experienced, combined with a lack of evidence to show adequate maintenance, it was determined that there was not a case strong enough for a successful claim against the manufacturers.

Splined shafts with square section splines follow a similar pattern, but the width of the splines gives the final fracture a more block-like appearance and, because some of the first splines start to fatigue in two positions, the blocks are sometimes pulled out of line when the final, ductile torsional overload occurs at the axis.

## 10.1.2  Heavy Goods Vehicle (HGV) Accident[1]

An articulated truck with a flatbed trailer was travelling along a two-lane highway in the late afternoon. As it approached a slight bend, it veered off the road, drove over a steel bridge parapet and plunged onto a country road beneath, narrowly missing a small car driven by a mother with three children. However, the driver of the truck was killed. He had left his depot at 6 a.m., driven 160 miles to deliver a load and was returning with the vehicle empty. Vehicle inspectors found the front tyre to be torn and punctured and the wheel rim buckled, which they attributed to damage sustained as the wheel struck the steel parapet. They also found that the steering box splined sector shaft was broken (Figure 10.2), and pointed out this could have accounted for loss of control. The fracture surfaces, of the splined sector shaft, were smooth at the outside with a roughened area at the centre (Figure 10.3A). The break itself had occurred immediately below the bottom bush in the steering box where bending forces would have been greatest. It was noted that a 10° twist had been imparted to all splines adjacent to the site of fracture (Figure 10.3B). Otherwise the vehicle had been well-maintained and was in a fully roadworthy condition prior to the accident. To aid reader understanding of the vehicle steering gear, Figure 10.4 shows how the subject steering box had been mounted on the chassis, with the drop arm, power steering cylinder and drag link clearly seen to be projecting below the chassis.

A police traffic officer attending the accident found that steering box shaft had indeed sheared, but the rest of the steering system was properly coupled up with no excessive wear on the linkages. He also noted that the road at the time was dry and the weather clear, but tyre markings on the soft shoulder indicated the truck had left the carriageway shortly before it

**Figure 10.2** The recovered steering box with broken end of sector shaft resting in position.

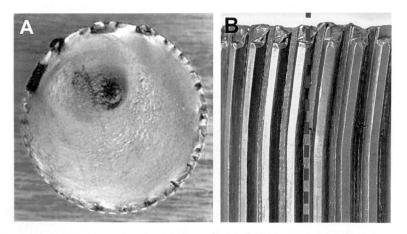

**Figure 10.3** (A) Fracture surface of splined shaft showing a predominantly smooth planar fracture, along with a rougher central area. (B) Angular distortion of every spline at edge of fracture.

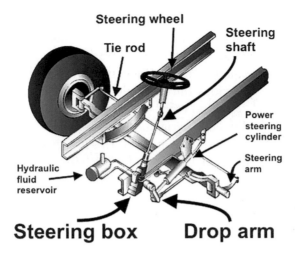

**Figure 10.4** Arrangement of linkages and drop arm when steering box is mounted on chassis. Notice the drop arm, power steering cylinder and drag link project below chassis.

mounted the crash barrier. However, there was no sign of a tyre blowout, with the damage to the front tyre consistent with the wheel striking the crash barrier. The police officer surmised that if the steering shaft had failed by fatigue, it could account for the vehicle leaving the road as the driver would have had no steering control. If the braking system had been damaged by collision with the crash barrier it would not have been possible to stop the vehicle before it reached the bridge. The officer also noted that there were no tyre marks to suggest the brakes had been applied prior to collision with the barrier.

At a subsequent inquest, a verdict of 'accidental death' was recorded, with witness statements all supporting the view that the driver lost control because the sector shaft suffered a fatigue failure at the site of its steering arm attachment.

When such parts are received for observation, the first course of action is to observe them closely, with the aid of a low-power microscope where necessary, particularly the position of the fracture and its path across the section, in addition to perceiving the nature and extent of damage present on any adjacent components.

The anatomy of the steering box, mounted on the vehicle chassis, has an internal mechanism where a line of hardened steel balls circulates in the threads between a worm screw and a nut, to give a low-friction movement. When the steering wheel is turned the worm rotates within the nut. As the nut moves up and down the thread, the forked end of the sector shaft follows it, turning the sector shaft backwards and forwards. A drop arm mounted on the splined end of the shaft transmits the motion to the steering mechanism of the vehicle (Figure 10.4). The arm is fastened onto the sector shaft over tapered splines, held in place by a nut and lock washer.

It is clear from Figure 10.3 that the fracture path is perpendicular to the axis of the shaft. It exhibits a pip at the centre and a smooth area around the outside. There is no obvious initiation site, or sites, where fatigue fracture might have originated and no crack arrests or beach marks. Furthermore, a fatigue fracture in a ductile material in torsional shear would tend to follow a helical path at 45° to the axis, not at 90° as this one does.

In contrast, a circular shaft broken by mechanical overload in torsional shear would display a degree of ductile deformation (twisting) all around the periphery before the fracture started (hence the bent splines seen in Figure 10.3B). The fracture surface would appear as a series of concentric rings, like smears, finishing as an area of tensile fracture near the centre as the two pieces finally came apart, thus leaving the characteristic 'pip' (Figure 10.3B). Hence, from these observations alone, it can be said that the fracture was caused by torsional overload, and not by bending or any other form of fatigue.

Insurers of the steering box manufacturing company subsequently appointed an independent metallurgist to review the evidence. He undertook an exhaustive examination of the failed shaft and associated steering gear components to ascertain whether they departed in any way from their material and manufacturing specifications. None did. However, when other components *inside* the steering box were examined, clear evidence was found of a massive impact against the steering arm which had forced several of the recirculating balls so hard against the worm that the case-hardened raceways had been indented.

Hence the sector shaft did not fail by fatigue, and should not have been held responsible for the loss of steering control at the inquest. The vulnerable drop arm, projecting below the chassis, was almost certainly impacted as the vehicle ran along straddling the crash barrier before it went over the bridge parapet. It was this impact force to the drop arm that had caused traumatic torsional overload failure of the splined steering box sector shaft. Sadly, this is typical of the kind of accident that results when the driver falls asleep at the wheel. The vehicle drifts off its path and there is no evidence in the tyre marks on the road of any braking or steering correction prior to the truck falling over the bridge parapet, which there surely would have been if the driver had remained alert.

### 10.1.3  Steering Shaft Failure on a Vintage Car[2]

The fracture of the steering shaft shown in Figure 10.5A is a bending fatigue failure of a long shaft that directly connected the steering wheel to the steering box on a vintage car (from the days when a frontal collision could drive the steering column like a spear into the driver's chest). When restoring the car the owner had managed to obtain a steering box and shaft from a similar vehicle that was being cannibalised for spares. He fitted the parts to his car and subsequently exhibited it at meetings and took part in road rallies. At one of these events, the steering column suffered a fatigue fracture while rounding a bend at the bottom of a hill. Unable to steer round the bend, the vehicle ploughed into the watching crowd, seriously injuring one spectator, who later died from his injuries. The shaft must have been slightly bent when fitted, with the result that every time the steering wheel was turned through more than 180° the middle of the column went through a cycle of compression and tension. Eventually the inevitable fatigue failure occurred.

### 10.1.4  The 'Boy Racer'[3]

Another case involving steering shaft failure concerned a known 'boy racer', who had a record of prior road traffic convictions for dangerous driving in his 20-year-old high-performance vehicle. The incident in question occurred late one rainy night on a left-hand bend of an unprotected (no central crash barrier) dual carriageway. The vehicle went out of control, spun sideways as it crossed over to the oncoming lanes, before impacting the off-side of an oncoming vehicle. In turn, the side impact caused a loss of control of the oncoming vehicle, which then crashed off of the carriageway, causing serious injury to the female driver and her two passengers. From the past driving record of the boy racer, it appeared that history had repeated itself, as suggested by the attending police officers. However, on subsequent examination of the wrecked vehicle, its steering shaft was found to be broken. The shaft was from the initial vehicle build, making it some 20 years of age, with the vehicle odometer having recorded some 160,000 miles. Under low-power magnification of the shaft fracture surface, an area of coarse material was clearly evident (Figure 10.5B), being consistent with the last metal to break.

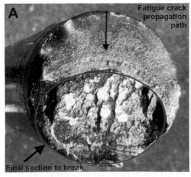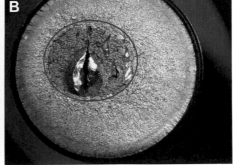

**Figure 10.5** (A) Steering shaft failed in bending fatigue. (B) Steering shaft broken by rotating bending fatigue with the final area to fail circled in red.

In addition, ratchet markings were evident all around the shaft periphery. Ratchet markings are radial steps formed where fatigue cracks, that propagate on different planes, intersect prior to coalescing into a single advancing crack front. This feature was therefore evidence of a mechanically driven failure, rather than a traumatic overload failure at the instant of impact. The lack of surface smearing (inflicted by any torsional overload or shear overload failure) provided further evidence to corroborate a rotating bending fatigue failure, rather than an instantaneous overload failure. Clearly, loss of vehicle control was as a result of mechanical failure of a critical steering component, rather than any reckless actions of the driver. No further action was taken against the driver, with his insurance company covering all costs of the injured female third-party driver and passengers.

## 10.2  Wheel Detachments

Vehicle wheel detachments may result from both fatigue failure and mechanical overload. When wheels are held on to the hub by means of nuts with conical seats tightened onto studs in the hub or brake drum, the sequence of failures can be readily deduced from examination of the individual fractures. The first ones to fail will display the greatest area of fatigue cracking and the later ones the least, often with severe erosion of the stud on the last ones to break due to the wheel moving about as it becomes increasingly loose. By identifying the first studs to fracture it is then a fairly straightforward matter to investigate the cause. The fatigue cracking invariably starts at the root of the first thread below the nut, as this is the smallest cross-sectional area and subject to the greatest stress when the nut is tightened. One of the reasons that wheel nuts are now tending to be replaced by conical seating bolts with a long, reducing taper into the threads is that these reduce the stress concentrations significantly compared with nuts that screw onto a uniform thread profile. As demonstrated by the FEA analysis of Figure 10.6, the root of the first thread in engagement is the most highly stressed and is consequently where fatigue cracking is most likely to initiate (and where a torsional overload fracture will occur if the nut is grossly over-tightened).

Wheel detachments have also been found to occur by the wheel centre itself disintegrating due to the joining up of fatigue fractures around individual bolt holes. Such detachments are usually attributed to incorrect mounting of the wheel on the hub or failing to tighten the nuts. The following two road-wheel failures are useful illustrations of the salient points made above.

A UK regional ambulance service purchased a batch of 152 ambulances, with their introduction into operation being spread out over a year. With the ambulances covering some 80,000 miles per year, the vehicles were required to have a regular maintenance and service schedule at every 9,000 miles. At

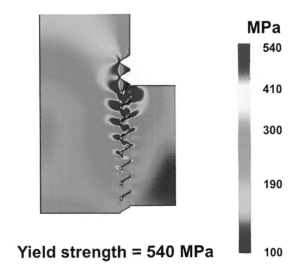

**MPa**

540

410

300

190

**Yield strength = 540 MPa**          100

**Figure 10.6** Finite element analysis (FEA) analysis of contact stress distribution in a bolted joint, with maximum stress seen to be located at the root of the first thread in engagement. (Courtesy of Jason Xu, C&C PetroGas Engineering.)

the same time, the road wheels were removed to allow inspection of braking and steering gear. Therefore, the vehicle road-wheels were removed on approximately eight occasions a year. The vehicle tyres themselves were changed at approximately 18,000 miles, or at every alternate service interval. However, two years after their introduction, the ambulance fleet started to suffer a series of serious mechanical incidents centred on the road wheels. Problems encountered included loosening of wheel nuts – which led to a far more serious wheel-off event – and (wheel) dish cracking. When considering that the fleet vehicles were of identical age and approximate use, concern was raised that the range of wheel-related failures coming to light might just be the tip of an iceberg, suggesting that many future wheel failures could well be expected.

## 10.2.1 Wheel Nut Loosening and Stud Fatigue

Initial concern, regarding the mechanical integrity of the ambulance wheels, was raised when a particular ambulance lost a wheel whilst sharply decelerating after travelling at speed. The driver became alarmed when he realised that he was being overtaken by one of his rear near-side wheels. On subsequent forensic examination, general visual observation of the wheel rims in question ruled out the possibility of mating surface corrosion being a factor with this particular wheel-off situation. The presence of corrosion products was minimal to non-existent, thereby negating the possibility of wheel nut loosening as a direct result of a corrosion action. The first and most striking

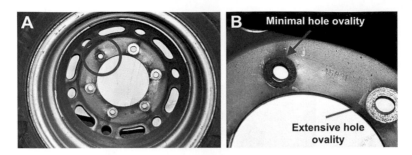

**Figure 10.7** (A) The ambulance nearside rear wheel, with a high degree of fretting at five of the six stud mounting holes. Note the lack of bright surface fretting around the sixth hole, as circled in red. (B) Minimal hole damage along with minimal surface fretting of the sixth stud mounting hole.

feature observed, on the ambulance rear nearside twin wheels, was the high degree of fretting to the wheel rim that had occurred under the nut seatings. This is clearly seen as bright damage sites on Figure 10.7A. The second striking feature noticed was that one of the sites of fretting exhibited a dull seating surface (circled in red on Figure 10.7A), the result of a surface corrosion action that could only have initiated *after* the loss of its wheel nut. On closer examination, it was also clear that the degree of surface metal loss from fretting action was minimal when compared to the remaining bolt seating sites (Figure 10.7B). These two features alone are evidence of the loss of one nut prior to continued fretting of the remaining five. However, the fact that surface fretting had initially occurred under this lost nut seating indicates that this nut/stud combination had lost some of its tensile clamping force prior to succumbing to ultimate fracture.

Rather than being circular in section, five of the wheel mounting holes had an exceptionally high degree of ovality imparted by a combined fretting and hammering action against each associated stud (arrowed on Figure 10.7B). This observation provides further confirmation of progressive loss of road wheel clamping force. At the site of the sixth stud mounting hole (where evidence had indicated initial nut loss), damage imparted to the hole was minimal when compared to the remaining five (Figure 10.7B). This feature provided additional evidence to corroborate the initial opinion that one stud had fractured prior to eventual wheel loss.

Turning attention to the wheel studs, gross material loss and surface polishing under the action of fretting and hammering was clearly evident (Figure 10.8A), with obvious 'wasting' of the parallel section, thus generating distinct shoulder. In places, metal loss had been so severe that all trace of the screw threads had been literally 'rubbed' off. Fretting results from the initial action of small amplitude movement or vibration between two mating surfaces. Even though surface stresses may initially be quite small, the cumulative effects after a large number of stress cycles may lead to fatigue. This

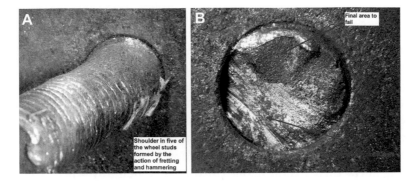

**Figure 10.8** (A) Fretting and subsequent 'hammering' had remove a large volume of metal from five of the six wheel studs. (B) The sixth stud had failed at the root of the first thread in engagement, with the presence of fatigue beach markings clearly evident on the fracture surface.

wear mechanism is frequently accompanied by corrosion, and it is this corrosive by-product that can provide a visual indicator of impending problems in the form of 'spiders' that may be seen emanating radially from around the wheel nuts. The initial fretting action would by its very nature indicate a loss of torque (and therefore, clamping pressure between rim and hub) over time. This situation would gain momentum, accelerating material loss by a 'hammering' action between the studs and their rim mounting holes. As a wear mechanism, the combined action of fretting and subsequent 'hammering' can and will remove a large volume of metal in a relatively short period of time, as evidenced by the stud shown in Figure 10.8A. The actual rate of metal removal will be dependent on the frequency of oscillation, the degree of relative movement between contacting surfaces and frictional (clamping) forces in action at that time. However, as witnessed by total lack of corrosion spiders, material loss inflicted on the wheel studs and their respective wheel rim locating holes had occurred over a very short time span, in the order of days rather than weeks.

Turning attention to the single broken wheel stud (Figure 10.8B), it had fractured at the root of the first thread in engagement, exactly where predicted by the FEA analysis shown in Figure 10.6. Beach markings were clearly evident on the surface, and indicative of crack propagation in discreet steps typical of a fatigue mode of failure. The growing crack had progressed some 60% across the section, at which point the reduced load-bearing area could not sustain the (cyclic) loading stresses generated by a normal wheel rolling action. Final failure immediately followed, with a tearing overload of the reduced sectional area (Figure 10.8B). These features are classic fingerprints of fatigue therefore it can be said with certainty that fatigue was the initiating failure mechanism responsible for the fracture of the one stud on the ambulance rear nearside axle. By definition, fatigue is a mechanical mode of failure, progressing over time with cyclically applied loading. As such, failure of the stud in question was entirely mechanical in nature, and not as a result of some traumatic overload event immediately prior to failure. Once a fatigue crack had been initiated in the stud, its final failure was inevitable.

From the above evidence, it became possible to postulate a probable scenario of events that had led to the wheel-off situation in question. Final wheel loss would have started with gradual loss of wheel-clamping force in one nut/stud combination. This situation could arise as a direct result of stud stretching (relaxing or plastic flow) or nut loosening, for example (mating surface corrosion being ruled out). When any wheel stud starts to stretch or wheel nut loosens, forces will be redistributed among the remaining studs and nuts, preferentially to those adjacent to the loose stud/nut combination. This process will accelerate with each successive stud/nut combination that loosens, as the total clamping force drops the stress concentration at the remaining nuts and studs will increase. At this point, the remaining stud/s may fracture due to overstress and/or fatigue. For the wheel-off in question, a single nut/stud remained sound, thus becoming sensitised to fatigue failure, to which it ultimately succumbed. The fractured stud was therefore the first to break (fail) as a result of being one of the last nut/stud combinations to retain its clamping force (integrity). The remaining un-broken (but loose) studs then, themselves, became susceptible to accelerated fretting and wear processes that rapidly damaged mating contact surfaces, the stud flanks and wheel rim holes. Continued loosening and unwinding of the remaining wheel nuts would follow, with nut loss and ultimate wheel loss being the final result.

Clearly, this failure was initiated as a direct result of a loss of (or an inadequate) joint clamping force. Now, the Ambulance Authority had documented evidence of adherence to published manufacturer's instructions regarding torque application and torque values. Therefore there was a clear possibility of a probable design limitation, regarding the number, strength and/or size of studs utilised for the (ambulance) service application in question. In layman's terms, there may not have been enough studs, studs of an adequate size or studs of adequate strength to ensure that the bolted wheel joint assembly complied with design engineering requirement for service loading conditions applicable to emergency response vehicles such as an ambulance.

The vehicle manufacturer remained totally non-committal regarding wheel issues on their ambulances. Rather than taking an active role in resolving wheel issues that were coming to light, their stance was that more care in maintenance would cure the problem. The implication was that as a manufacturer, they do not have design issues with wheel assemblies. However, it is appropriate to point out that during the year 2010, the manufacturer in question issued a recall, announcing a problem with a similar vehicle; the issue: their wheels may become detached:

'It has been identified that the wheel studs of Hendrickson axles fitted with 17.5 inch wheels were tightened to a torque which was higher than that specified. It is therefore possible that the wheel studs have stretched. In this condition the wheel may no longer be secure. Vehicles built from 1/1/2008 to 12/31/2009 are at risk.[4]

With the potential outcome of any wheel-loss situation being so dire, we advised that consideration should be given to immediate replacement of studs and nuts on *all* 152 vehicles that comprise the ambulance fleet. However, it was also emphasised that such action would not resolve the issue at hand; it would simply delay matters, with the nut loosening potential remaining unresolved.

## 10.2.2 Wheel Dish Cracking

The second wheel issue that was coming to light on these ambulance fleet vehicles was that of wheel dish cracking. Now wheel centre discs and dishes of automotive wheel rims are produced from sheet metal plate through a series of cold stamping operations and a final painting and low-temperature curing cycle. Sheet materials are generally characterised by a 'high ratio of surface area to thickness', and therefore the uniaxial force applied to the sheet plate during a cold stamping operation is mainly tensile in nature. This applied tensile force induces residual stresses in the component that are dependent upon the shape of the component, and vary in magnitude with position. The fact that cold-working operations enhance the tensile and fatigue crack initiation strength of a material are well-understood and documented but, in terms of fatigue performance, this may be countered by any residual stresses introduced during sheet metal forming operations. High levels of tensile forming stress act to increase the mean stress in fatigue cycling which, in turn, will effectively decrease fatigue strength.

If the fabrication process for any vehicle road wheel centre (or dish) introduced a critical weakness, a high potential for a wheel-off situation will exist, loss of a wheel arguably being one of the worst possible situations for a vehicle. In addition, catastrophic cracking or structural failure will become more likely as time and stresses accumulate. Centre or dish failure is more likely to show up in vehicles that are subjected to turning under load, even more so in those which regularly turn sharply while manoeuvring at low speed, both being typical scenarios in the every-day life of an ambulance. These are the situations that impart the most damaging stresses on wheel dishes and bolted flange regions.

The sheet stock from which the ambulance wheel dished had been fabricated was of a nominal 6-mm sheet stock thickness. When considering the general condition of all wheels under observation, the most striking feature

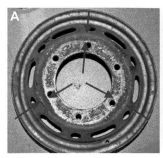 

**Figure 10.9** (A) High level of surface corrosion, with dish flatness issues clearly evidenced by the bright contact footprint. Note: some 20% of the contact footprint was not in intimate contact. (B) Wheel dish fracture path, having travelled some 100° around the wheel dish periphery.

was their high degree of corrosion (Figure 10.9A). For two-year-old wheel dishes, the degree of corrosion present was startling and unexpected. The reason for applying any paint coat is twofold: decorative and (corrosion) protective; of the two, it is the nature of corrosion protection that is paramount for structural integrity. Any protective coating applied over a steel substrate will generally fail for two main reasons: either coating adhesion was not adequate, or any straining of the steel substrate will result in cracking of the surface coating (the coating will, in effect, act as a brittle lacquer and become a 'tell-tale' sign of tensile stressing of the metal substrate; see Section 4.4.1). For any wheel dish, the predominant mode of loading is that of bending. When combined with the normal rotating (therefore, cyclic) action of a road wheel, the loading mode at play is therefore one of bending fatigue.

As long as cyclic loading stays below a fatigue threshold for the dish material, all will be well. However, as discussed above, residual stresses induced during sheet metal forming operations can achieve high tensile values that act to increase the mean stress in fatigue cycling, thus decreasing fatigue strength. By simple observation, cracking in one (nearside rear) dish had initiated at pressing/blanking marks sited around vent holes (Figure 10.9B). Once initiated, the fracture path had propagated some 100° around the dish. On the balance of probability, cyclically applied tensile stresses at this point had taken the material above its fatigue limit. The situation would be exacerbated if the material section was not of sufficient stiffness, i.e. lack of adequate stiffness would lead to increased dish flexing, thereby further sensitising the dish to failure by fatigue. A section of cracked wheel dish was removed for direct fracture surface observation under magnification. Fatigue beach markings present on the fracture provided irrefutable evidence of fatigue as the failure mechanism at play (Figure 10.10).

When considering the potentially lethal wheel-off failure, question was raised regarding the 'fitness for purpose' of the wheel for use on an emergency

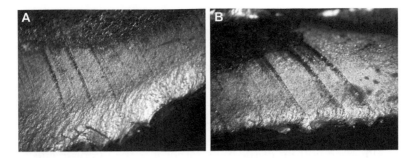

**Figure 10.10** (A and B) Macrograph of wheel dish crack surface clearly showing the presence of beach markings that are a fingerprint of fatigue as the failure mode at play.

service vehicle such as an ambulance. If matters were allowed to continue in the same vein, it was not a question of 'if' a fatal accident would occur as a result of highlighted wheel issues; rather it was a question of 'when'. There was clear evidence that both hub mounting and dish design for the ambulance fleet in question was marginal at best. Road wheels on ambulance vehicles currently in use do not appear capable of withstanding long-term operating conditions that were approaching the top end of the vehicle design specification. With adequate 'built-in' redundancy (i.e. improved factors of safety), wheel failures of this kind would become a thing of the past.

As a final point to the ambulance fleet wheel issues, an investigation undertaken by the UK Institute of Road Transport Engineers had found that wheels could come loose even though tight when checked. Stud tension and wheel-clamping load were easy to maintain when the assembly was new, but deteriorated, with sometimes disastrous results, *when wheels had been taken off many times and nuts continually re-torqued.* Their conclusion was that nut torque was often insufficient to preserve bolted joint integrity (tightness). Plastic yield of the wheel and/or studs could cause nut slackness to develop and accelerate stud fatigue. This point is of critical importance when considering the frequency of wheel removal for the ambulances in question.

## 10.2.3 Stub Axle Failure[1,2]

Another common cause of wheel detachment as a result of fatigue is when a stub axle carrying the wheel breaks away from the suspension, as illustrated in Figure 10.11A. Such failures usually occur when the vehicle is manoeuvring or moving slowly, when the bending loads on the inboard end of the axle are considerably greater than when it is travelling normally on the road. A typical fracture surface of a steel stub axle shows two fatigue crack initiation sites, a larger area radiating from the bottommost fibre and a much smaller one radiating from the uppermost fibre. The final break is a ductile fracture occupying a band across the middle.

The reason for this unequal development of the fatigue cracking is that tensile stresses acting on the bottommost fibres are caused by the weight of the vehicle tending to bend the axle upward plus cornering forces acting to push the bottom of the wheel outward. Both are of a cyclic nature when the vehicle is travelling on the road but the cornering force only becomes significant when the weight of the vehicle is transferred to the axle as it changes direction. However, the stresses on the uppermost fibres are always lower, because there is no external gravitational force bending the axle upward as there is along the bottom of the axle; rather it is the opposite effect: gravitational force produces compressive stress in the uppermost fibres and so tends to counteract tensile forces when the vehicle is cornering. The net result is that the fatigue initiates first at the bottom of the axle and propagates upward, and it is only when the cross-sectional area has been reduced and the applied loads generate higher stress levels that the cornering forces are able to initiate a fatigue crack at the top of the axle. This is also the reason why the final break almost always takes place when cornering, usually at low speed, because the upward bending forces are greatest when a heavily laden vehicle is manoeuvring.

Torsional overstress which twists the retaining nut off the end of an axle is a common cause of wheel detachment when the vehicle is travelling at high speed and usually occurs shortly after wheel bearings have been adjusted during servicing (Figure 10.11B). In the majority of motor vehicles, wheels are mounted on their axles by a pair of taper roller bearings that must be critically adjusted to allow a small amount of free play without becoming sloppy. Some front-wheel drive vehicles are fitted with a pair of ball bearings separated by a tubular spacer clamped tightly together, and the only provision for adjustment is to place shims at the ends of the spacer. The retaining nut has to be tightened to a high torque. Some wheel detachments have occurred simply because a mechanic has mistakenly tightened taper roller bearings.

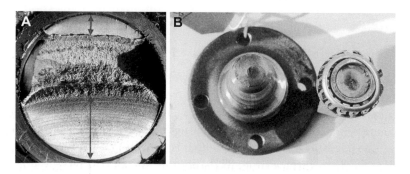

**Figure 10.11** (A) Stub axle failure by the action of reverse bending fatigue. Note: crack penetration greater at the bottom of the axle, shorter at the top. (B) Torsional shear overload failure sited at the end of the stub axle.

Taper bearings are retained on the axle by a washer and castellated nut which has to be carefully set to allow a small amount of free play and is prevented from coming unscrewed by a cotter pin or 'C' clip passing through a hole in the screwed end of the axle and lining up with opposite castellations of the nut. The method of making this adjustment is to fit the roller bearings and tighten them until they bind, then release the nut by one-sixth of a turn and check that the wheel turns freely with minimum end float. The nut is then locked in this position, or the nearest pair of castellations in the *unscrewing* direction, by inserting the cotter or spring pin through the castellations.

When disc brakes are fitted to a vehicle the friction pads and associated parts may be examined without removing the wheel bearings. However, in order to inspect the friction linings and expander of drum brakes, the drum must be removed. On small cars, this is usually integral with the hub as a single casting, which necessitates taking off the castle nut and withdrawing the outer taper roller bearing. After the brakes have been attended to, and usually the bearings re-greased, the hub/drum unit is replaced and the bearings have to be adjusted to the required amount of free play. This is where the problem is likely to arise. If the mechanic replaces the nut and turns the castellation in the tightening direction to line up the castellations with the hole in the axle (or, as has been known to happen, tightens them fully as if they were ball bearings with a spacer), then the bearings will heat up when the vehicle goes back on the road. Within about an hour or so of high-speed driving, the bearing on the right side of the vehicle seizes and twists off the end of the axle to produce the torsional shear failure illustrated in Figure 10.11B.

The reason that it is the wheel on the right-hand side of the vehicle that detaches is that the direction of rotation in forward motion tends to tighten the castle nut as the bearings expand under frictional heat. This quickly results in the castle nut binding tightly as the free play is reduced, and the friction may then cause the castellations to shear through the cotter. When this happens the bearing locks solid, the threaded end of the axle shears under torsional overload and the wheel detaches, carrying the hub with it.

Although similar frictional heating may occur if bearings on an axle at the left side of the vehicle have insufficient free play, the direction of wheel rotation as the friction increases acts to unscrew the nut, which tends to relax the frictional force compared with positive tightening on the right axle. The driver feels no immediate effect as the wheel detaches but first becomes aware of the problem when a wheel overtakes the side of the car! A loose wheel rolling into the path of another vehicle travelling at high speed often has fatal results, certainly in countries that drive on the left side of the road. The clues as to the cause of the axle shearing are readily apparent in the state of the outer bearing, the nut and cotter and the signs of frictional heating in the vicinity of the bearing.

Wheel detachment on both right and left sides of the vehicle may also occur if worn or faulty bearings fail and pieces of a broken roller or parts of the cage jam and cause seizure. This can usually be traced by careful examination of the debris to a worn out or damaged bearing. The detachment illustrated in Figure 10.11B occurred on a small car that had undergone a 20,000-mile service only the day before, and this had required removal of the rear brake drums to inspect the linings. The young mechanic, who had carried out the work, had not been properly supervised, so the garage insurers accepted full liability for the accident.

## 10.3 Brake Pipe Failures

By far the most frequent cause of accidents resulting from failures of the braking system is the loss of hydraulic fluid due to either corrosion or fatigue of the pipework. In both cases the loss begins as a slight ooze at the site of a perforation or a fatigue crack penetrating the wall and, if this passes unnoticed, all the fluid eventually escapes and the brakes become inoperative. The brake pipe failure, of Figure 10.12 shows an example of external corrosive attack in vehicle brake Bundy type pipework.[5] The failure mechanism was due to a corrosion cell being set up between the metal of manufacture, and its protective electro-deposited protective chrome coating (being metals of markedly different electrode potentials). This problem may be overcome quite simply by making the pipe from a homogeneous alloy with substantially better resistance to corrosion and thus no requirement for a protective coating.

### 10.3.1 Brake Failure by Fatigue[2,6]

Extruded homogeneous brake pipes will have swaged ends to form a union that will allow coupling to fixed components in the braking system (such as expander cylinders and the pistons of disc brakes). Fatigue of the pipework is likely to occur if there is any relative movement within the pipework, which

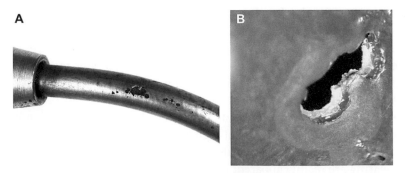

**Figure 10.12** (A) Pipe connecting brake expander inside front brake drum to hydraulic circuit. (B) Closer view of corrosion pit showing a two-layer pipe fabrication route.

**Figure 10.13** Swaged brake pipe recovered from an airport luggage 'tug' having a suspected fatigue crack sited beneath the coupling nut.

will induce cyclic bending or torsional stresses. Any such stresses will tend to concentrate at the root of the swaged pipe flange where it is held captive by the union nut.

Figure 10.13 shows a suspect high-pressure hydraulic brake pipe recovered from a 'buggy'-type vehicle used for towing strings of luggage trailers at an airport. The brakes suddenly failed, causing severe injury to one of the loaders and some damage to an aircraft. It was found that all the hydraulic fluid had been lost from the reservoir during the 72 h since the level had last been checked. The maintenance log at that service recorded the level to be slightly low, so a minor top up had been required. Cursory inspection of the underside of the vehicle at that time had revealed no signs of leakage.

After the accident it was discovered that fluid had been escaping between the pipe and the union connecting it to the brake expander mounted at the top of the brake plate of the front wheel. The escaping fluid had simply run down the pipe and dripped off the bend at the bottom, so there was practically nothing to show how much fluid had been lost. The escape had occurred through a circumferential crack in the pipe immediately below the swaged flange, as marked by the arrows in Figure 10.14. The crack had extended approximately 240° around the circumference, exactly following the ridge left by the swaging tool when the flange was formed at the end of the pipe. Its fracture face exhibited characteristic features of fatigue, showing the mark left by the swaging tool to be the initiation site. The mating (unbroken) section was then prepared for metallographic observation where it could be seen that the end of the crack had penetrated about 90% into the wall, almost penetrating the bore (Figure 10.15).

When the expander cylinder mounted on the inside of the brake plate was inspected it was discovered that the retaining nut was loose. This caused the

**Figure 10.14** Close-up of swaged end with a crack sited at the curved (swaged) bulb.

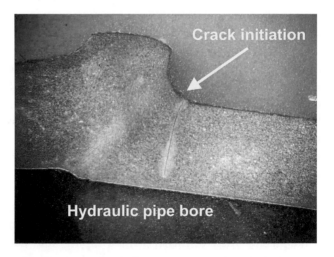

**Figure 10.15** Section through fatigue crack. It can be seen as a faint line below the bulb running through the wall thickness of the hydraulic pipe.

cylinder to be twisted sideways every time the brake shoes came into contact with the rotating brake drum, whether the wheel was rotating forwards or backwards. Cyclic stresses on the pipe were concentrated at the groove left by the swaging tool directly behind the flange. A fatigue crack was initiated and began to propagate, following the groove. Although the stress cycles would have been essentially torsional in nature at first, the sharp change in section immediately below where the flange was clamped dominated the direction of crack propagation. Later on, as the one-sided crack development continued, bending stresses from the internal hydraulic pressure would cause build-up of axial stress and accelerate the rate of crack propagation.

The fatigue cracking must have started some time before the accident, but brake fluid would only have begun to escape when the crack had penetrated the wall. Gradual reduction in the cross-sectional area of the pipe wall would result in ever-increasing stress levels, causing the crack to propagate faster, accompanied by an ever-increasing escape of brake fluid. As it was a power-assisted braking system the driver would have been unable to detect any difference in the feel or movement of the brake pedal. Hence the loss of all the hydraulic fluid from the system only three days after topping up the reservoir is entirely consistent with the way the fatigue cracking had propagated. The only prior warning would have been if fluid had been noticed underneath the vehicle, but with continual use day and night this most probably would have been spread all over the loading and baggage handling areas of the airport.

### 10.3.2 Failure of a Polymeric Air-Brake Hose

An example of brake failure not involving a metallic component is illustrated in Figure 10.16. This short length of thermoplastic pipe from an air-braking system on a commercial vehicle burst and led to a serious accident. A new vehicle had been purchased with a bare chassis and was sent to a specialist auto body worker to have a side-loading unit installed. This work had necessitated some welding above the longitudinal chassis members. The work was finished, and the vehicle was being driven out of the shop and onto a busy road. The driver applied the brakes to stop before entering the road but they failed, allowing the truck to continue forward into the path of oncoming traffic. The driver reported hearing a pop like a cork coming out of a bottle as he pressed the brake pedal.

As Figure 10.16 shows, a bubble had developed in the pipe wall and had split open at the top. Only this one short length of the pipe had been affected. The bubble was found to be on the top of the pipe a few centimetres directly below where one of the welds had been made. Clearly what had happened is that heat from the welding operation had softened the thermoplastic, and expansion of residual air inside the system had blown the bubble and thinned the wall to a fraction of its original thickness. The first time the full air pressure built up inside the pipe, the bubble had burst, which would account for the 'pop' heard by the driver.

## 10.4 Gearbox Synchroniser Ring

A two-year-old tractor was towing a trailer whilst manoeuvring down an incline, at a reported speed of 5–10 mph, when the trailer started to jack-knife. The driver panicked and put the tractor into reverse in order to stop the vehicle, allegedly imparting catastrophic damage to the gearbox. Gearbox

**Figure 10.16** A bubble in the wall of a 12-mm outside diameter thermoplastic brake pipe.

failure was therefore considered to be as a direct result of a traumatic event at play immediately prior to failure. However, a more detailed inspection of the gearbox was subsequently undertaken, where it was found that a synchronising gear ring had broken at five individual sites (Figure 10.17). Simple visual inspection confirmed that the ring had been manufactured by a powder metal (PM) route, using compacted pre-alloyed metal powder, with a final high-temperature sinter cycle. Examination of one fracture surface under

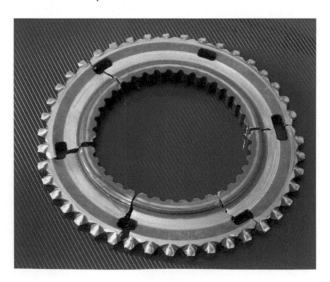

**Figure 10.17** Tractor gearbox synchroniser ring, showing fracture at five separate sites

**Figure 10.18** (A) Low-power magnification of one fracture surface, clearly showing three crack initiation points (arrowed in red) and final overload zone (arrowed in yellow). (B) Clear beach markings indicative of a mechanical (fatigue) failure mode at play.

low-power magnification clearly revealed three crack initiation points as arrowed on Figure 10.18A. In addition, beach markings were clearly evident, as can be seen in Figure 10.18B.

Fatigue cracks had initiated at sharp corners of the gear ring section, as clearly seen on Figure 10.18A. The growing cracks had propagated into the shaft cross section, effectively reducing the load-bearing section at each incremental advance of the growing crack fronts. Furthermore, the distinct zone of coarse material (arrowed on Figure 10.18A) shows how little material was actually holding the ring together at the instant of failure. This coarse material is symptomatic of a ductile fracture of the last metal to break. Under higher magnification, the presence of beach markings was clear to see (Figure 10.18B). Up to the point of final separation, the only external sign of the developing weakness would have been fine hairline cracks on the outer surface. Failure was a classic fatigue mode caused by normal 'in-service' cyclic forces that were magnified at sharp section changes.

Failure of this gear ring was the direct result of a classic fatigue mechanism, with initiation located at the site of sharp corner sections. So it could be clearly stated that failure was totally mechanical in nature, and not the result of a traumatic overload caused by the reversing incident. As a failure mechanism, it is said that fatigue is one of the ways engineered components announce that they have reached the end of their useful working lifetime. The reversing incident would have simply provided the final straw that broke the camel's back.

## 10.5 Fatal Tow Hitch Accident

Loss of a trailer being towed on a busy road is likely to have serious consequences for oncoming traffic, especially if the trailer snakes and overturns in freak weather conditions or as a result of poor driving. The critical part when seeking to find the cause of such an accident is the ball and socket of the towing hitch, or tow pin if the coupling is a simple ring-and-yoke type. If there is extensive deformation on both the vehicle side and the trailer side of the hitch, then quite clearly an overload force must have acted, and this leads inevitably to the conclusion that the hitch itself could not have been the initial cause of the trailer breaking away, especially so if there are ductile fractures in the linkages. On the other hand, if there is no evidence of mechanical overload of the linkages then the answer will usually be found in the form of a fatigue failure of some vital part in the hitch. The following case will illustrate this point.

A towing eye that had been fitted to a small family saloon car failed whilst being towed by a recovery vehicle. In a subsequent collision, the freed car sustained extensive impact damage. The car owner stated that he had purchased the towing eye from a main dealer only two or three days prior to its failure. The site of fracture was at a shoulder adjacent to the locating thread (Figure 10.19A). Observation of the fracture under low-power magnification showed that the mode of failure was not by tensile overload (no ductile dimpling) or fatigue (no beach markings or striations). The fracture surface had taken on a somewhat granular appearance, suggesting intergranular attack, thus giving confidence to an environmental issue as being a significant contributor to this failure (Figure 10.19B). To determine both the material of manufacture and manufacturing route, a sample was cut from beneath the fracture and prepared metallurgically. Structural flow, voiding and void inclusions were observed, along with clear anisotropy, as would be expected with a pressed or stamped product (Figure 10.20A). Both voiding and inclusions were considered normal and within limits for a hot-stamped product. Etched microstructural observations revealed light areas of ferrite,

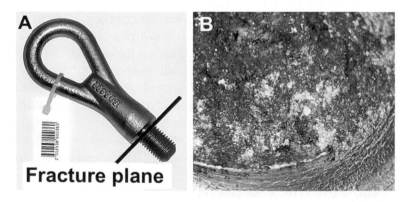

**Figure 10.19** (A) Towing bar and eye for a small family saloon car with site of fracture indicated by a red line. (B) Granular appearance of fracture, with no ductile dimpling or beach/striation markings.

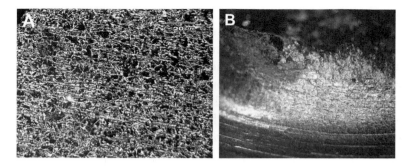

**Figure 10.20** (A) Anisotropy, structural flow, voiding and void inclusions normally expected within a stamped product. (B) Multiple surface cracking, crack jump and granular fracture surface are features characteristic of hydrogen cracking.

with darker areas of pearlite, and ferrite at grain boundaries and as plates within grains (Figure 10.20A). Microstructural observations along with chemical analysis identified the material of manufacture as low-carbon steel that had been hot worked and finished with a yellow zinc plating. In other words, the towing eye had been formed from a low-carbon steel that had been hot-pressed and finally plated for corrosion resistance, exactly as would be expected for such a component.

However, unless possessing a latent defect, it is highly unlikely that loads generated by simply towing the vehicle would be of a magnitude great enough to cause it to break. By assuming the material of manufacture to be a low-carbon steel, and assuming an average value of tensile strength to be 400 MPa, an approximate tensile breaking load can be calculated. The measured diameter of the fracture surface was 9.54 mm, equating to a cross-sectional area of 71.5 mm². Simple engineering relationships will show that it would require approximately 3 tons tensile pull to break the section in question. Now, the gross weight of the car was quoted as 1,775–1,960 lbs (0.79–0.87 tons), clearly showing that this failure was not by direct tensile overload, i.e. the threaded section of the towing eye was not under-designed.

It could be argued that the thread undercut acted as a stress raiser, thereby decreasing the tensile effort required to break the section. However, from literature, the highest stress magnifier for such an undercut is 3, reducing the tensile effort to approximately 1-ton tensile force to effect a break. It can be seen that, even taking a worst-case scenario, this failure was not affected by tensile loading induced by towing the small saloon car. This simple stress analysis clearly suggests other unforeseen mechanisms at play. It is these unknown factors that had triggered the onset of failure under service loadings well below the load-bearing capability of the fractured towing eye cross section.

Observation of fracture surfaces *per se* showed no evidence of smooth pane areas, or of beach markings or striations typical of that associated with fatigue as a failure mechanism. There was no staining from ingress of plating solutions that could point to a quench cracking mechanism of failure. However, the observed features of multiple surface cracking, crack jump and granular fracture surface are features that are characteristic of hydrogen cracking, where hydrogen evolved in the plating process is absorbed by the metal (Figure 10.20B). Normally after electroplating the hydrogen is expelled from the metal by baking at a fairly low temperature (approximately 280°C), but in this instance, if such a treatment was carried out after electroplating, it was not effective in expelling all the hydrogen. The result was that over a period of a few days minute cracks began to appear in the steel, which rendered it brittle and likely to fracture suddenly when external forces were applied. In this brittle condition the towing eye would be likely to break as soon as modest forces were applied in service.

All post-failure features observed on the towing eye were entirely consistent with that of stress-corrosion cracking, with the electroplating process providing the source of hydrogen egress. This particular failure may well have been a one-off in that, for whatever the reasons, the towing eye in question simply missed a final bake-out processing stage. Equally, however, the towing eye could be one of a *batch* that had missed the bake-out stage.

# References

1. Lewis, P.R., K. Reynolds and C.R. Gagg, *T839 Forensic Engineering.* 2000, Open University Post-graduate Course, Department of Materials Engineering.
2. Lewis, P.R., K. Reynolds and C.R. Gagg, *Forensic Materials Engineering - Case Studies.* 2004, CRC Press.
3. Gagg, C.R., Failure of components and products by 'engineered-in' defects: Case studies. *Engineering Failure Analysis*, 2005. 12(6): pp. 1000–1026.
4. https://www.vehicle-recall.co.uk/recall/R/2010/024.
5. Gagg, C.R. and P.R. Lewis, Environmentally assisted product failure – Synopsis and case study compendium. *Engineering Failure Analysis*, 2008. 15(5): pp. 505–520.
6. Gagg, C.R. and P.R. Lewis, In-service fatigue failure of engineered products and structures – Case study review. *Engineering Failure Analysis*, 2009. 16(6): pp. 1775–1793.

# Fraudulent Insurance Claims

# 11

## 11.0 Introduction

It is possible to insure against almost every kind of loss. The terms of an insurance policy are, however, carefully worded so as to specify precisely what risks are covered and the conditions under which liability can be accepted by the insurer. Where risks are foreseeable the policy clearly sets out the responsibilities of the insured to prevent such losses. Many people fail to read the 'small print' and assume that their policy will pay out for all kinds of loss or damage. Only after making a claim do they discover they did not observe their part of the contract. For example, a homeowner returned to his house after taking a long winter vacation to find the central heating boiler had failed and, as a result of a particularly severe frost, several radiators had split open. In addition, a water supply pipe had burst during the frost and had flooded the ground floor of his house when the thaw came. Unfortunately, he found that it was a requirement of his household insurance policy that he should have had his heating system serviced annually. It transpired that it had not been serviced or inspected since it was installed more than six years before the incident.

People who suffer loss usually assume that it will be covered by insurance, but when they read the terms of their policy before filling out the claim form they often find that it is not. Human nature being what it is, a few are sometimes tempted to deliberately introduce damage or describe circumstances that simply could not account for the damage or loss described. Occasionally, claims that are clearly spurious or fraudulent may be submitted in order to obtain money from an insurance company.

## 11.1 Perforated Central Heating Radiator[1,2]

A do-it-yourself homeowner claimed for the cost of redecorating two rooms in his house, allegedly necessitated by water escaping from an upstairs radiator on his central heating system. By the time the loss adjuster inspected the premises, the redecoration had been completed, new carpet laid and a new, larger radiator fitted. The homeowner produced the original radiator, which had a hole near the bottom on the side facing the wall. The hole was partway up the rear panel of the radiator, well above the bottom section where sediments collect and pitting corrosion eventually leading to a perforation tends to occur. Suspecting that this was not a genuine failure, the loss adjuster sought a metallurgist's opinion as to the origin of the hole. A piece was cut out so that the inside wall of the radiator could be examined. The wall exhibited a degree of internal corrosion consistent with several years' service, but there was no significant pitting. The damning evidence, however, was that the way the metal had deformed revealed the hole had been made using a 3-mm-diameter punch with a sharp point, almost certainly an engineer's centre punch. In addition, although the punch had penetrated the wall at an angle it would not have been possible to drive it in while the radiator still hung on the wall. Not surprisingly the claim was rejected and the insurance policy cancelled.

## 11.2 Perforated Oil Filter

A fruit farmer submitted a substantial claim for replacement of a clutch/gearbox assembly on a large tractor. He stated that he was grubbing out fruit trees in an old orchard when the clutch burned out and, upon dismantling, he found that the gearbox had seized due to complete loss of oil. There was a small hole in the bottom of the external oil filter, mounted on the side of the gearbox. The claimant stated, 'This must have been pierced by a tree pruning kicked up from the ground. Everything was satisfactory when the oil was changed only three weeks ago.' The insurance inspector was somewhat dubious, so he submitted the filter for examination.

Figure 11.1A shows the hole in the bottom of the filter. It had been formed by a penetrating object 8 mm in diameter entering at a slight angle and

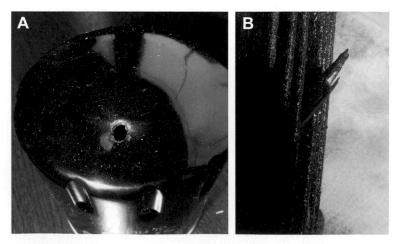

**Figure 11.1** (A) Hole in the bottom of an oil filter can. (B) Side view of piece of wood in the filter element folds, observation and analysis showing it to be the edge of a planed board, not a piece of tree pruning.

making an elliptical hole with smoothly turned-in edges. There was slight denting around the hole in the bottom of the filter but no signs of scratching of the paint or abrasion. After the bottom of the canister had been sawn away, it could be seen that the flap of metal pushed inside had an indentation at the centre, consistent with the canister having been pierced by an 8-mm-diameter conical steel punch with a nose 1 mm in diameter. The same tool had continued through the space between the canister and the filter cartridge and punched a circular hole 8 mm in diameter in the end plate. A sliver of wood was found trapped between the folds of the paper filter (Figure 11.1B), but no shreds of this could be found on the sharp edges of either hole. Instead of following the line of the two holes, the sliver of wood had entered through the hole in the canister but then changed direction and passed through the space between the inside of the canister and the top of the cartridge, to finish up trapped between the paper folds on the outside. When this sliver was examined it was found to be a strip some 5 cm long broken off the edge of a machined white wood (pine) tongue-and-grooved floorboard. It was of roughly rectangular cross section and could be passed through the hole in the canister without touching the edges.

The substantial claim was initially rejected by the insurers as fraudulent, and they threatened prosecution, but were impressed by their insured's genuine distress upon learning how the hole had been made and that he had only guessed that it was caused by a piece of tree pruning. (Laboratory tests had shown the thickness of the steel was such that all attempts to drive a pine dowel into the end of the filter case resulted in the wood splintering and only causing a slight dent in the steel.) He said the tractor had been left in the orchard overnight and he had only been running it for about 20 min when the clutch failed. He had gotten off the tractor and had seen the hole in the filter and a train of oil leading back to where he had started that morning. There were no metal punches or anything similar that could have formed the hole in the tractor toolbox. The insurance inspector made inquiries of the local police who said several instances of malicious damage to machines and tractors left out overnight had been reported in that vicinity but they had been unable to find the culprit(s). Accordingly, the insurance company met the claim in full.

## 11.3 Major Engine Damage, Allegedly Due to Loss of Oil

In direct contrast, a similar case involved a relatively new goods delivery vehicle, which suffered catastrophic engine failure that was estimated as beyond economic repair. The damage occurred after the engine unexpectedly lost its lubricating oil. The vehicle owner had been driving the vehicle at the time of the engine failure and alleged that he had driven over rough debris-strewn ground whilst making a particular delivery. He remembered hearing a 'clunk' under the vehicle, and it was his opinion that a piece of hard debris had apparently been flung upward by the front tyres. Although he had not recognised its significance at the time, the owner stated that it was this event that must have ruptured the engine oil sump. Subsequent loss of lubricant oil had gone un-noticed by him, right up to the point of catastrophic engine seizure. The owner therefore submitted a substantial insurance claim for repair costs, on the grounds of accidental damage.

However, a claim evaluator became suspicious when he noted that service records listed a requirement for a full engine service to be undertaken at 39,065 miles. With no record of any such service being undertaken, and with the vehicle digital odometer display reading showing 45,467 miles, the vehicle had clearly been driven for some 6,400 miles past its scheduled engine service. The claim evaluator therefore instructed a forensic engineer to undertake a more detailed examination of the vehicle and its engine.

In addition to extensive engine damage, the engineer promptly noted the presence of two holes in the oil pan walls. One damage site was located at the forward-facing sump pan wall (Figure 11.2A), whilst the second damage site was located to the left of the sump drain plug (Figure 11.2B). However, considering the site of forward-facing damage, it was obvious that the protective nature of the front bumper and roll-under cowling would protect the oil sump pan from a direct impact by road debris. Even objects ricocheting off of the cowling would be directed away from the site of the front-facing damage. The obvious conclusion was that it was highly unlikely that road debris could have inflicted this particular damage.

Damage location to the left of the sump drain plug was certainly in an exposed position and therefore vulnerable to accidental damage from road debris impact. However, by viewing this damage site under low-power

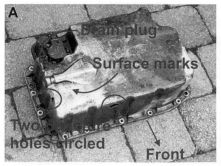 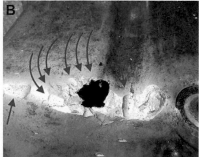

**Figure 11.2** (A) The oil sump pan showing two perforation damage sites, circled in red, under low-power magnification. (B) A number of distinct individual impact zones can be clearly seen, and indicated by the red arrows. The imprints bear a clear resemblance to those expected when impacted with a round bar.

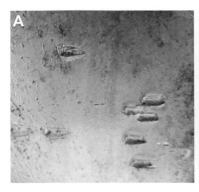

**Figure 11.3** (A) Indentation damage on the underside of the oil sump pan in question, each having a profile identical to that expected of screwdriver damage. (B) Remedial Helicoil seen to be sitting somewhat proud of the spot-faced sealing seat.

magnification, it became clear that damage had not been inflicted as a result of a single traumatic impact. Bruising was clearly present on at least eight locations around the one hole, as arrowed in red on Figure 11.3A. This bruising was the result of plastic surface deformation inflicted by a considerable number of individual impacts. From the shape of surface indentations, the object or 'tool' used to form this damage was a length of (steel) bar, having a diameter in the region of 25 mm (1 inch).

There was a third damage type noted, sited on the flat underside of the oil sump pan, consisting of a series of angular bright indentations (Figure 11.3A). Without doubt, this form of damage could not be reconciled with that of accidental road debris impact. The damage at this site is symptomatic of the oil sump pan having been attacked by a tool such as a screwdriver, in an attempt to puncture it.

On the balance of probability, the flat underside of the oil sump pan was where the first attempt at puncturing it was made. As discussed above, the numerous markings at this site clearly resemble indentations that would be made by the blade of a screwdriver. It would appear that when the perpetrator (or perpetrators) had attempted to puncture the case, they found it difficult to achieve. They had then resorted to an iron bar, battering the drain plug area with the periphery of the bar. However, an internal strengthening web had made this somewhat difficult, therefore attention was turned to the third (front-facing wall) site. This type of damage would have been inflicted by 'stabbing' at the pan. Whatever the exact scenario of damage infliction, it is clear that the pan had been overtaken by attempts to cause damage at three distinct but separate sites. Taking account of all evidence, it was possible to say that damage sustained was not as a result of a traumatic event at the instant of failure, but was a result of a considerable number of separate blows from at least two types of 'tool'.

### 11.3.1 A Probable Motive for Inflicting the Damage Observed

The presence of a Helicoil was seen in the threaded region of the sump drain (Figure 11.3B), and would provide one possible motive for the act of malicious damage to the oil sump pan in question. On the balance of probability, the drain plug thread had been damaged at some prior point, thus necessitating the remedial action of Helicoil insertion. However, the device had been fitted somewhat proud of the spot-faced sealing seat, as seen in Figure 11.3B. In turn, this had prevented plane seating of a sealing washer, which, in turn had provided an escape route for the engine oil. This oil leak/loss would not have been traumatic, but simply a steady seepage and therefore a gradual loss of engine lubricant over time. If the vehicle been serviced at the correct mileage interval, in all probability any lubricating oil loss would have been noticed, and hopefully corrective repairs instigated. As it was, the oil loss had not been noted by the vehicle owner up until the point of catastrophic engine damage. At that point in time, it is easy to envisage the idea of deliberately puncturing the oil sump pan, stating it was accidental damage, and then claiming the repair costs from the vehicle insurer. Needless to say, the insurer repudiated the claim as fraudulent, with the actions of the vehicle owner being recorded on the Motor Insurance Anti-Fraud and Theft Register (see Section 12.2), in addition to the vehicle insurance policy being annulled.

## 11.4 Engine Hairline Cracking as a Result of an Alleged Impact

A skid loader (Figure 11.4A) was supposedly involved in a rear off-side impact with a factory unit roof support girder. As a result of the 'collision', the owner/operator alleged that damage sustained by the vehicle included a broken engine mounting bracket and a hairline crack to its engine block, at a point

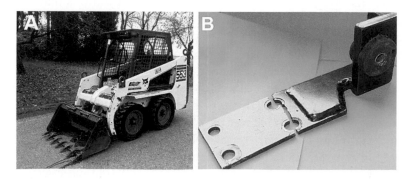

**Figure 11.4** (A) An unrelated example of a skid loader. (B) General view of broken the engine mounting bracket in question.

slightly above the mounting bracket support point. Therefore, the owner submitted an insurance claim for the cost of a replacement engine and mounting bracket. However, an insurance assessor's inspection of the vehicle found just minor rear impact damage of such minimal consequence that he just could not reconcile with the engine mounting and block damage. A forensic expert was therefore appointed to take a closer look at the damage claimed.

The broken engine mounting bracket in question is shown in Figure 11.4B, along with its rubber engine mounting bush, which was clearly in a poor mechanical condition (Figure 11.5A). Examination of the bracket revealed a 'thread' profile that had been impressed into one side of a parallel bore bolt locating hole on the bracket (arrowed on Figure 11.5B). A possible scenario that could account for this form of surface damage is that damage/wear of mounting bushes had allowed normal vibration oscillation to 'fret' between bolt thread and parallel hole bore. Under low-power magnification, two independent crack initiation points were found: at the corner of a 'thread damaged' hole, and at the edge of an undamaged bolt location hole. The actual point of crack initiation is shown arrowed in red on Figure 11.6A, being sited at the corner of an undamaged hole. Sharp drilled hole edges and surface damage from impressing of the bolt thread (Figure 11.6B) are both familiar surface stress raising features (see Chapter 5). The bracket fracture surface displayed clear beach markings, suggesting that the action of fatigue had been at play (Figure 11.6A). Beach marking features provided clear evidence of a fatigue mechanism being the culprit for failure. The distinct elliptical shape of coarse material, clearly seen on the right-hand side of the figure, shows how little of the mounting bracket was actually holding the two parts together at the instant of failure. This coarse material is symptomatic of a ductile fracture of the last metal to break. Up to the point of final separation, the only external sign of the developing weakness would have

**Figure 11.5** (A) The generally poor mechanical condition of rubber engine mounting block. (B) One of the two broken parallel sided bolting holes, with threads having been 'cut' into one side (arrowed in red) by either bolt misalignment, or by vibration resulting from the worn rubber mounting bush.

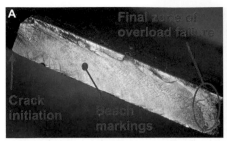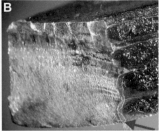

**Figure 11.6** (A) The engine mounting bracket fracture surface, with crack initiation arrowed in red along with clear beach marking, both of which are a fingerprint of fatigue as the failure mode at play. (B) Fatigue beach markings seen to initiate at the corner of the supposed parallel bore of a bolt hole. 'Threading' had inadvertently been 'introduced' on the bore, providing a stress-raiser from which a second fatigue crack had initiated.

been fine hairline cracks on the outer surface, almost certainly too fine for the layperson to see, let alone recognise their significance.

A probable train of events leading to initiation and final fatigue failure of the engine mounting bracket in question would have been instigated by an over-tightening of mounting bolts through their rubber bushes. In turn, excessive tightness of mounting led to increased transmission of vibration forces. This exaggerated dynamic loading situation was one that had not been considered or accounted for within the engine mounting design solution. In addition to initiating and growing fatigue cracks in the mounting bracket, increased levels of dynamic loading from engine vibration had also initiated a crack in the engine block at a point immediately above the mounting bracket. Furthermore, over a period of time this engine block crack had started to propagate (grow). By its very character, any fatigue failure will progress over time and is therefore mechanical in nature, not the result of any single traumatic overload event at the instant of failure.

Clearly, damage to the engine mounting bracket and cracking of the engine block was not as a result of a rear off-side impact with a shed support girder, but was as a direct result of mechanical breakdown following a period of wear and tear over its service lifetime. When presented with the evidence, the owner of the skid loader withdrew his insurance claim, but not before his details had been entered onto the Motor Insurance Anti-Fraud and Theft Register

## 11.5 Agricultural Tractor Engine Damage

An insurance company received a substantial claim on an agricultural vehicle policy. It was alleged that while working in a field, an agricultural tractor had its oil filter punctured by a stone or some similar projectile. A subsequent loss of engine oil, from the damaged oil filter, had led to catastrophic failure of the engine crankshaft. However, when in use, oil filters are under pressure, and therefore any puncture would result in the spraying of high-pressure oil that would coat just about everything in and around the vicinity of the tractor. During an initial assessment, a staff engineer of the insurance company quickly realised that the expected oil contamination was simply not apparent. He therefore recovered and took possession of both the perforated sump and the pieces of the broken crankshaft, which he then submitted for forensic examination.

The recovered oil filter is shown in Figure 11.7A, with a closer view of the puncture damage site shown in Figure 11.7B. The puncture hole had plastically deformed lips on two of its opposite edges, where the filter body material had been forced inward to form the lips. The two adjacent edges had sheared as a direct result of forces at play during lip formation. When a tractor oil filter is hit by a projectile that was picked up and thrown by an attached implement, the normally expected damage would exhibit ragged edges along with impressed dirt and grit. This was simply not the case with the damaged oil filter in question, which had a rectangular opening with sheared edges and plastically deformed lips. The most probable method of introducing this type of damage would be by stabbing the filter with an implement such as a screwdriver. This scenario is given credibility by the presence of a score mark seen in Figure 11.7B, which is the classic appearance of damage formed by the corner edge of a screwdriver skidding over a metal surface. A lack of extensive oil spray indicated that the oil filter puncture had been caused whilst the tractor engine ignition was off (the engine was not running).

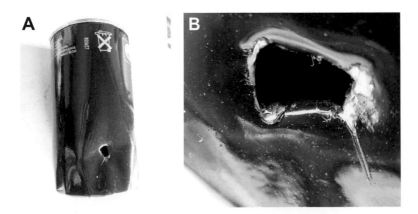

**Figure 11.7** (A) Spin-on tractor oil filter with a puncture at its lower end. (B) A closer view of the punctured hole, showing plastically deformed lips on two of its opposite edges and having undergone a shear mechanism on the adjacent two edges. Note: the clear score mark at the bottom right-hand of the puncture.

**Figure 11.8** (A) Multiple fatigue crack initiation sited along the length of the drilled oil-way hole. (B) Fatigue cracking at a tight radius between the throw bearing journal, and its adjacent web, as arrowed in red.

Turning the focus of attention to the crankshaft, its fracture surface revealed the presence of classic fatigue beach markings. Multiple fatigue cracking had initiated at two distinct locations: a tight radius between the throw bearing journal (Figure 11.8A), and around an oil-way drilling (Figure 11.8B). At the first location, the transition radius had acted as a stress raiser, sensitising the crankshaft to fatigue crack initiation at that point. Grinding chatter marks, that were visible on the ground radius and web surface, may well have accelerated the onset of fatigue crack initiation at this site. At the same time, the oil-way drilling had also begun to start to fatigue, with cracks starting at a number of places along its drilled length. On the balance of probability, this site of crack initiation would have been due to a combination of the factor-of-three stress concentration at any hole (under uniaxial loading, Figure 11.9), combined with the additional stress-raising influence of the rough surface of the drilled hole. Blueing present on the web had formed as a direct result of journal bearing surface friction, probably generated

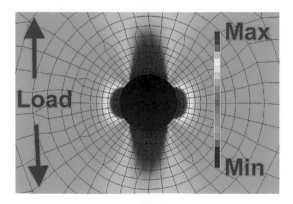

**Figure 11.9** Finite element analysis showing the uniaxial stress gradient around a hole. Note: the stress magnification factor can be as high as 4×.

immediately prior to failure, as the crankshaft progressively lost its structural integrity as a result of growing fatigue cracks.

Clearly, the engine crankshaft failure had not been caused by traumatic oil loss, but was as a direct result of mechanical breakdown. Damage sustained by the oil filter had been maliciously inflicted in an attempt to mislead and, ultimately, defraud the tractor insurer.

## 11.6 Con-Rod Failure[2]

The six-cylinder engine of a 12-year-old car was beyond economic repair due to extensive damage caused by the connecting rod of the No. 3 cylinder breaking and punching a hole in the side of the engine block. The insured driver stated:

> I was driving home from work along a dual carriageway road in a stream of traffic moving at about 40 miles per hour when we came to a section where one lane was coned off for road repairs. The road had been dug up on the inside lane and there were piles of material between the cones, which included some large pieces of what looked like reinforced concrete with steel wire sticking out. To my horror a lorry in front of me brushed one of these piles and a piece of concrete rolled into my path. I could not stop or slow down because the traffic was a continuous stream, so I just steered the car so that the lump would go between my front wheels. As I passed over I heard a loud bang and it had obviously hit the underside of the car, but I could not stop. I had to keep moving until I came out of the coned off section, which was another 1/4 mile (about 400 metres). As soon as I came clear I pulled over to the side but just as I did so I heard a loud bang from the engine and then a terrible hammering noise as the engine lost all its power. A friend was a few cars behind and he pulled over behind me, but there was nothing we could do. The engine was wrecked. There was a stream of oil along the road behind the car. I had a tow rope in my car so we hitched up and he towed me home. When I got back I could see there was a big dent with a hole in the front of the crankcase in line with where I had steered the car over the concrete. The oil must have all run out before I could pull over to stop, so the engine seized and the big end broke and knocked a hole in the side of the block.

A staff engineer of the insurance company took possession of the perforated sump and the pieces of the broken crankshaft and submitted them for forensic examination. Three different features emerged proving that the story had been concocted to justify a claim for accident damage, whereas failure was, in fact, a mechanical breakdown resulting from 'wear and tear', not covered by the insurance policy.

Figure 11.10A is a view of the fracture face at the bottom of the connecting rod, with a closer view shown by Figure 11.10B. The fracture is a fatigue

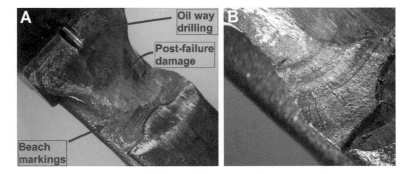

**Figure 11.10** (A) Fatigue fracture of connecting rod big end, initiating at the oil passageway hole. (B) Closer view of the fracture surface with fatigue beach markings clearly evident.

failure at the bottom of the connecting rod that had been developing over a period of time. Consideration of the load transfer path establishes that when the crack reached completion and finally broke, the big-end cap tilted sideways and the forces from normal rotation of the crankshaft bent the bolt on the side that was still intact (Figure 11.11A). The bolt at the broken side was not subject to these forces, so it had remained straight. The loose connecting rod flaying about inside the crankcase would quickly punch a hole in the side of the engine block.

There was no evidence of overheating, which would have been inevitable if the lubrication system had failed due to oil starvation. Moreover, if the engine had run out of oil the components at the bottom of the engine that are splashed by oil in the sump would be the last affected. The valve gear and combustion zones would be the first to suffer and, again, there would be blueing and galling wear over the sides of the cylinders swept by the piston. None of the signs usually associated with oil starvation could be found.

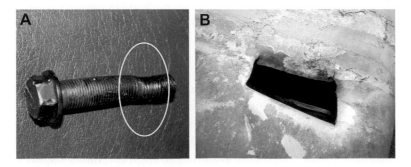

**Figure 11.11** (A) A clear bend of the big-end bolt from the unbroken side of the conrod. (B) Perforation of the crank case cover having the appearance of being inflicted by a cold chisel. Note areas of random scratching.

Turning to the crankcase cover, Figure 11.11B, the hole looked more like to have been made with the end of a cold chisel (or screwdriver blade), rather than with a piece of steel reinforcement wire sticking out of concrete. Several areas of random scratching in the paint were noted behind the hole. How could these random scratches have been produced if the car was moving forward over the piece of concrete at 40 mph? They would obviously have been deep, straight and parallel or, at most, exhibit only slight differences in direction.

The owner's handbook for the vehicle gave the engine oil capacity as 11 pints (5.75 L). The leakage rate through the hole was tested by pouring 2 L of hot engine oil into the sump and measuring the time for 1 L to run through the hole. This was found to be 22 sec. Hence assuming the sump contained 5 L at the time of the incident, the engine could have run for over 110 sec before experiencing significant distress. During that period the car running at 40 mph would have covered just over 1 mile, so the engine could hardly have been starved of oil if the sump was perforated only about ¼ mile before the car was brought to rest, as the claimant stated.

The claim was rejected. Although the insurance company could have taken legal action it cancelled the policy and advised the insured that he would need to obtain coverage in the future from a different underwriter.

## 11.7  Piston Failure on a Racing Car[3]

The owner of a high-performance saloon car used the vehicle for 'track day' racing, where both seasoned drivers and novices compete in a four-race series. The car was both raced and maintained by the owner. During one race, the (petrol) engine suffered loss of power, emission of fumes from engine breather and exhaust smoke. A strip-down of the engine found that a piston had failed, so the owner submitted a sizeable claim to his insurer on the grounds that the piston had been defective. However, very few pistons actually fail in service as a result of a manufacturing shortcoming. Piston 'failure' or damage can generally be attributed to inappropriate operating conditions such as lack of lubrication, abnormal combustion and the presence of debris within the engine. Clearance and operational issues can also play a part in early demise of a piston. It is safe to say that the life of a piston is directly related to its environment, with almost all situations resulting in damage to a piston capable of being traced to an issue unrelated to the construction and quality of the piston itself.[4,5]

The insurer handed the case to a forensic engineer, for an expert opinion on the reason(s) behind the piston failure. Simple visual observation showed that there were sections missing from the piston rings (Figure 11.12A), along with heavy erosion of the aluminium piston body/skirt (Figure 11.12B). Both

**Figure 11.12** (A) A side view of the damaged piston, with the combustion surface (crown) to the left. Note the two broken piston rings. (B) A closer view showing erosion markings flowing under the 'lost' section of piston ring.

piston rings had disintegrated under zones that had been eroded (therefore unsupported) by the passing of combustion gasses. Furthermore, there were clear signs of heavy wear on both of the remaining ring sections.

Cast pistons are usually adequate for most passenger car applications, but with high-performance engines, forged pistons are the norm. The forging process will increase the density of the metal, significantly improving its strength, ductility and thermal characteristics. Furthermore, forgings tend to conduct heat quickly and cool better than most cast pistons. However, cooling is also dependent on design of the piston, and ring contact with the cylinder. Forged aluminium pistons will be manufactured from one of two aluminium alloys: 4032 series or 2618 series. The 4032 alloy contains more silicon (11 to 13.5%) than 2618 (less than 0.25%). Silicon improves high heat strength, improves scuff resistance and reduces the coefficient of expansion, so tighter tolerances can be maintained during temperature excursions. Tighter cylinder bore clearances have the added advantage of reduced ring flex and piston rock, giving improved sealing. The 2618 alloy, by comparison, is a low-silicon alloy so it has a higher coefficient of thermal expansion and much more tendency to scuff. However, it has a tensile strength 10–15% higher than the 4032 alloy. Chemical analysis of the high-performance piston in question (Table 11.1) showed a match to that of a series 4032 aluminium alloy, with a measured average hardness of 112 points on the Brinell scale. The value of hardness would suggest a 'T-6' solution heat treatment for increased strength (of up to 30%) and durability. Simple metallography confirmed that the piston had been forged rather than cast.

**Table 11.1  Elemental Analysis of Aluminium Piston Material of Manufacture**

| Element | Si | Cu | Mg | Ni | Al |
|---------|------|------|-----|-----|---------|
| % | 12.4 | 0.98 | 1.1 | 0.9 | Balance |

In a modern high-compression petrol engine, the combustion process must proceed in a quick and controlled fashion to ensure fuel economy, performance and emissions will remain stable over the engine speed and load range. Cheaper fuels are available and generally have a lower-octane rating. It is these lower-octane fuels that can cause the combustion mixture to self-ignite ahead of the combustion flame. This end-gas self-ignition will cause a pressure wave to be set up, and its result can be heard as a ringing noise termed knocking or pinking. It is this knock pressure that will cause local erosion of the piston metal that, over time, can lead to engine failure. Blow-by is the most common manifestation of this problem. Piston rings – those that are worn past the maximum ring end gap specification – can allow combustion pressure to seep past the rings and down the piston skirt. This produces a characteristic erosion pattern, as seen on the piston in question. It is also possible that a cylinder wall cross-hatched honing pattern can also be partly to blame for blow-by. However, closer inspection of the piston discounted any possible honing issue.

Although not possible to determine beyond doubt a failure mode for the piston rings, the most probable scenario would suggest rings that were worn past the maximum end gap specification. This would allow combustion pressure to seep past the rings and down the piston skirt ('blow-by'). As erosion progressed, unsupported piston rings would fail by bending under loading from compression and ignition of the fuel. There was no physical evidence to suggest an issue with either the construction or quality of the piston itself. Clearly, the vehicle owner was not a lay-person in matters of engine operation, service and performance. On the balance of probability, he had been attempting to have the costs of his major engine repair covered by his insurer. Although 'trying-it-on' is not criminal fraud in the true sense, the insurer repudiated the claim and also issued a warning. Knowing that erosion failure can initiate from either incorrect fuel injection, injection timing or a damaged or incorrectly located injector, the advice given to the vehicle owner was to inspect and, if necessary, correct the fuel injection equipment and timing, or in simple terms, improve the quality and regularity of his engine maintenance.

## References

1. Lewis, P.R., K. Reynolds and C.R. Gagg, *T839 Forensic Engineering*. 2000, Open University Post-graduate Course, Department of Materials Engineering.
2. Lewis, P.R., K. Reynolds and C.R. Gagg, *Forensic Materials Engineering - Case Studies*. 2004, CRC Press.
3. Gagg, C.R. and P.R. Lewis, Wear as a product failure mechanism – Overview and case studies. *Engineering Failure Analysis*, 2007. 14(8): pp. 1618–1640.
4. http://kawtriple.com/mraxl/tips/piston/pistondiagnosis.PDF.
5. Heavy Duty Technology: Piston Failure Analysis, Dennis Nail, The Engine Builder, APR 2002, https://www.enginebuildermag.com/2019/12/utilizing -instagram/

# Criminal Cases 12

## 12.0 Introduction

Evidence, in any criminal case, is usually sent for analysis by government-recognised forensic science laboratories. These laboratories are usually staffed by scientists and various specialists (such as handwriting experts) who, although generally quite experienced and skilled in their respective fields, seldom have experience in manufacturing and general engineering. Occasionally, investigating police officers may instruct independent experts in relevant engineering fields. However, it is more common for the independent expert to be appointed (given instruction in the matter) in order to challenge or expand upon the evidence given to the court by sci entists. In addition, insurance companies may also have an interest in the outcome of a case. So they too may wish to have material evidence examined by an independent expert and, if necessary, to challenge the interpretation given to the court. The criminal cases that follow are selected from the authors' experience in this field and, not surprisingly, have a strong metallurgical bias.

## 12.1 Counterfeit Coins[1]

No matter what the currency and despite ongoing developments to make it 'counterfeit-proof', the counterfeiters will always somehow manage to manufacture forged notes and coins which then enter into general circulation. Government departments and forensic science laboratories try to keep one step ahead and quickly determine how this illegal currency is made and, subsequently, many of the counterfeiters are caught. However, in the meantime, the general public will suffer as the currency that they believe to be genuine is rejected when handed over as payment. Taking sterling as an example, the following three counterfeiting cases will highlight the scale and range of illegal operations undertaken over recent times within the UK.

### 12.1.1 The Use of Currency as a Feedstock Material for Counterfeiting

15 February 1971 saw the introduction of decimal currency in the UK, with the one pound (£1.00 = 100 pence) denomination being a paper note. In turn, the £1 note could be sub-divided into 'silver' coinage having a face value of 10 pence and 50 pence (Figure 12.1). This silver coinage was actually a 75/25 cupro-nickel alloy, with the coins having respective weights of just over 11 g (the 10p coin) and 13 g (the 50p coin). It did not take the criminal fraternity long to work out that, with some means of melting and casting, £1.30 worth of 10 p coins could be converted to £5.50 worth of 50p pieces.

Proof:

- the weight of a 10p coin = 11 g, and a 50p coin = 13 g,

- So, 13 10p coins (i.e. £1.30) weigh (13 × 11 g) = 143 g

- Therefore 143 g/13 g (the weight of one 50p coin) = 11 50p coins can be produced, with a value = £5.50

- Thus making £1.30 into £5.50 – a great return on investment (no pun intended!).

A simple enough calculation, but unfortunately it took the British Treasury some two years before they realised what was happening.

Now, the jewellery industry produces craftsmen skilled in making small investment castings (the 'lost-wax' process), with excellent reproduction of fine detail. Counterfeiters utilised the technique to make wax reproductions

**Figure 12.1** Reverse face of a 10 pence and 50 pence coin both of which were minted in 1971.

of 50p coins, forming investment moulds (or 'trees') containing upward of 30 or so impressions of 50p pieces. The moulds were then mounted in a small centrifugal casting machine; the right amount of 10p pieces were then melted and forced into the investment mould. It was a job that could be done in a few minutes when the boss wasn't around. It is worth mentioning that use of 10p pieces as feedstock ensured that there were no records of material purchase that could incriminate the counterfeiter. When solidified the individual castings were broken off the sprue and the tiny area where they had been connected was polished. Of course, these cast coins were not quite so sharply defined as a coin made in a coining press from solid metal, but they were practically impossible to detect visually in any normal transaction.

One physical way of detecting the counterfeit coin was by observing its microstructure. Legally minted coins start off as cast slabs that are rolled down to thin strips and the coins then blanked out and stamped in power presses with machined dies. This would then be followed by milling of the coin edges. However, the counterfeit coins were as-cast so they had a different microstructure, as illustrated in Figure 12.2, which show etched sections of the wrought and the cast cupro-nickel alloy, respectively. Figure 12.2A shows a longitudinal section of a minted coin that has been rolled from a 125-mm billet to a 3-mm-thick strip. This exhibits smaller equiaxed grains that are drawn out in the direction of working and the coring segregation is oriented in the direction of rolling. The cast structure (Figure 12.2B) exhibits large grains that are cored (that is, a compositional segregation due to the first solid to appear being higher in nickel than that formed near the end of the solidification process), which is especially pronounced in copper–nickel alloys.

A microstructural examination requires a section to be cut from the suspect coin and to be polished and etched. However, technically, it is a legal

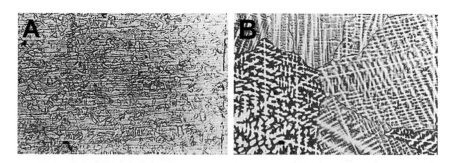

**Figure 12.2** (A) Etched microsection of legally minted coin made from wrought copper–nickel alloy showing recrystallised grains with directional coring from original cast billet from which the coinage strip was rolled (magnification: 75×). (B) Etched microsection of counterfeit coin made by casting, revealed by large irregular grains having nickel-rich cores merging into copper-rich matrix due to nonequilibrium cooling from the liquid state (magnification: 75×).

offence to deface a coin of the realm; therefore the structural differences shown in Figure 12.2 could possibly be established by non-destructive methods such as ultrasonics or electrical resistivity measurement. However, when the British Treasury finally realised what the counterfeiters were doing, a smaller 10 p version was introduced in 1992, weighing just 6.50 g, therefore effectively stopping the 'profit margin' associated with counterfeiting 50p coins from their 10p counterparts. Combined with the value of currency naturally falling over time, the risk and surreptitious effort for the counterfeiter were unjustified, so the quantity of forged 50p coins in circulation quickly fell.

## 12.1.2 The Base Metal £1 Coin

In 1983, inflation resulted in the replacement of the British one pound note with a coin (£1). To make it easy to identify, the £1 coin was thicker than other coins and 'gold' in colour. This colour difference was designed to allow the new £1 coin to stand out from the cupro-nickel 'silver' coins already in circulation. However, soon after its introduction, counterfeit £1 coins started to appear. These coins had been crudely made, by casting in a lead/tin alloy and electroplating with brass, to give a 'gold' finish (Figure 12.3A). Chemical analysis of this particular coin showed the material of manufacture to be a tin/lead binary system (Table 12.1), with microstructural observation showing a classic eutectic structure. (Figure 12.3B). These features confirmed that this £1 counterfeit coin had been cast in eutectic tin/lead. At that time,

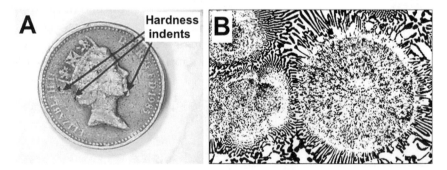

**Figure 12.3** (A) An example of a base metal counterfeit coin, having a brass electroplated coating. (B) Microstructure observation, showing bulbous primary lead dendrites within a matrix of two solid eutectic solutions (light etched: lead rich; dark etch: tin rich).

**Table 12.1** Chemical Composition of Counterfeit Coin (ppm), Showing the Material to Be a Binary Eutectic Tin/Lead System – A Conventional Solder Alloy

| Al | As | Bi | Cu | Fe | Pb | Sb | Zn | Other Metals | Sn |
|----|----|----|----|----|----|----|----|--------------|-----|
| <50 | <200 | <2,500 | <800 | <200 | 37% | <2,000 | <50 | <800 | Bal. |

eutectic lead/tin was a material more commonly used to solder printed electronic circuit boards! However, these coins had a dull ring and, as soon as the plating wore off the edges, they marked paper like a pencil. These two physical characteristics allow rapid identification, and immediate withdrawal, through standard bank checking procedures. Nevertheless, even in the first quarter of 2017, these base metal cast coins continued to make the occasional appearance amongst the change in my pocket.

### 12.1.3 The 'Struck' £1 Coin

In addition to cast counterfeit coins, from 1984 onwards, a far superior £1 counterfeit coin started to enter into circulation within the UK. This particular counterfeit coin was a 'struck' nickel–brass variant (rather than cast) that passed internal bank checking systems, with flying colours. As far as any bank was concerned, the coin was genuine. To determine the quality of these struck counterfeits, a suspect £1 coin was checked for its visual, physical and structural characteristics (Figure 12.4). Simple visual comparison between the

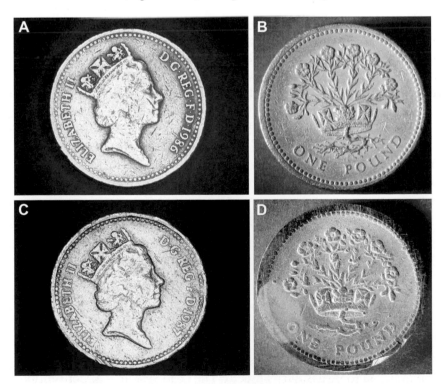

**Figure 12.4** Obverse and reverse faces of £1 coins. (A and C) Suspect coin on the left. (B and D) Known genuine coin on right. Both coins have a flax pattern in a diadem (coronet) on the reverse face, but with different year dates on the obverse face.

suspect coin and a genuine article showed the two coins as having the same reverse face pattern. However, it was blindingly obvious that the two coins had different year dates (Figure 12.4A and D). Furthermore, information gathered from the Royal Mint established that the reverse design for coins issued in 1986 was flax in a royal diadem (coronet) representing Northern Ireland, while the reverse design for coins issued in 1987 was an oak tree in a royal diadem representing England. Now, the British Royal Mint just does *not* produce coins with the wrong obverse and reverse faces for the year date. This one observation instantly confirmed that the suspect coin was indeed a forgery. However, simple curiosity called for further in-depth examination, simply to establish the quality and possible manufacturing route of the forgery.

Observation under low-power magnification revealed a range of feature differences between the suspect and a genuine coin. Detail of lettering and design on both obverse and reverse faces of the suspect coin was not of a crisp sharp quality. Although bearing the same edge inscription as the genuine one ('DECUS ET TUTAMEN'), the suspect coin had lettering of a different font (that was probably hand-stamped), along with uneven reeding pitch (Figure 12.5). In total, six feature differences were found on the suspect coin.

Chemical data were gathered by EDAX analysis (Figure 12.6 and Table 12.2) and compared to a genuine coin as detailed by the Royal Mint (Table 12.3). It was found that elements of both nickel and zinc were 'lean' in the suspect composition, being 36.4% less for nickel and 6.5% less for the zinc constituent, copper being the balance. At the time of writing, prices on the London Metal Exchange show nickel to be more than five times the cost of copper and zinc. Therefore reduction of the nickel content would dramatically reduce material cost to the counterfeiter. Moreover, such compositional differences would account for both the weight and hardness variation that was found between the suspect and a genuine £1 coin seen in Table 12.2.

**Figure 12.5** Edge detail differences, suspect coin at top, genuine coin at bottom.

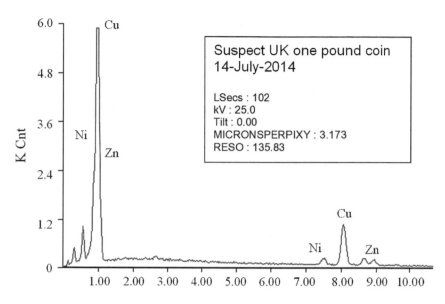

**Figure 12.6** Energy dispersive X-ray analysis (EDAX) of the suspect coin.

**Table 12.2  Weight, Hardness and Compositional Data for Genuine and Counterfeit £1 Coin**

| Coin Type | Physical Dimensions Diameter:Thickness | Weight (g) | Hardness (Hv) | Composition (Weight %) | | |
|---|---|---|---|---|---|---|
| Genuine | 22.5 mm:3.15 mm | 9.5 | 184–210 | Ni 5.5 | Cu 70 | Zn 24.5 |
| Suspect | 22.5 mm:3.15 mm | 9.35 | 141–165 | Ni 3.5 | Cu 73.6 | Zn 22.9 |

**Table 12.3  Composition of the Suspect Coin along with the Expected Composition, as Supplied by the Royal Mint**

| Metal | Suspect Coin (Weight %) | Expected Composition (Weight %) | Relative Deviation (%) |
|---|---|---|---|
| Ni | 3.5 | 5.5 | −36.36 |
| Cu | 73.6 | 70 | +5.14 |
| Zn | 22.9 | 24.5 | −6.53 |

Metallographic examination was undertaken to determine the most probable manufacturing route. A polished and un-etched section of the suspect coin is shown in Figure 12.7A, followed by an etched section to reveal the bulk microstructure (Figure 12.7B). The obvious feature of the polished and un-etched section is the amount of air voids and inclusions contained in the suspect coin material. As nickel is completely soluble in copper, copper–nickel alloys are single-phase alpha structures. Subsequent etching to show the microstructure revealed equiaxed twinned grains of alpha copper,

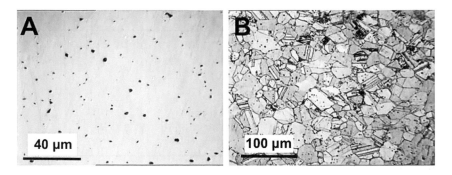

**Figure 12.7** (A) Polished and un-etched section of the suspect coin. (B) Etched to reveal the bulk microstructure. Note: prolific microstructural voiding and inclusions are clearly evident.

as would be expected in a wrought product (Figure 12.7B). Again, a high degree of voiding and contamination, seen as 'dust' particles (on both micrographs) was found to present within the suspect material. Gas inclusions and contamination, at the level found in the suspect coin, would not be found in a commercially produced stock. As the suspect coin exhibited a wrought microstructure, it is logical to suggest that the suspect coin had been stamped or 'coined'.

However, the level of contamination and gas voiding found in the suspect coin microstructure would *not* be present in a coin produced by the Mint. The presence of such contamination suggests the material had been alloyed by the counterfeiter and used as parent stock for his activities. This scenario provides an explanation for the chemical composition of the material being off, and also explains why there were both hardness and weight differences between suspect and genuine coins. An additional reason for producing 'in-house' feedstock material is the availability of individual material elements, bulk purchase of which is unlikely to arouse suspicion. This point is more credible when considering the illegal nature of the product.

When considering a possible route for making dies to strike a counterfeit coin, the spark erosion technique comes to the fore. Spark-erosion requires the use of a coin as a sacrificial electrode that will be submersed in an electrolytic bath, facing the counterfeiter's die steel. An electrical current is applied to the coin electrode, producing a spark that jumps across the gap between coin electrode and the die, thus burning or etching the coin's design onto the steel die. It would be normal practice to erode both obverse and reverse faces simultaneously, thereby requiring two coins, one for each face. As two separate coins are required, this is where it becomes possible to sink incorrect reverse patterns for the year date, i.e. using two coins of different year dates. After both the obverse and reverse have undergone the spark erosion process, the dies would be highly polished. This is a necessity as the sparking process will produce somewhat pitted surfaces. At this stage, care with polishing will make the dies 'sharp'; these counterfeits appear to be extremely well-struck, with knifelike edges and rims. On the balance of probability, it was this technique that was used to manufacture dies to strike the counterfeit coin in question.

### 12.1.4 Footnote to Counterfeiting

By the year 2014, conservative estimates suggested that there were some 48 million counterfeit British £1 coins in general circulation. It was only after the scale of counterfeiting became a major news story that the British Treasury felt the need to take action; in short, British £1 coin counterfeiting had reached a 'level of political significance' (whatever such a phrase was supposed to mean). In March 2017, a new £1 coin was introduced, having a number of features that make it much more difficult to counterfeit (Figure 12.8). Features of the new £1 coin include

- Twelve-sided – its distinctive shape makes it instantly recognisable, even by touch.
- Bimetallic – it is made of two metals. The outer ring is gold-coloured (nickel–brass) and the inner ring is silver-coloured (nickel-plated alloy).
- Latent image – it has an image like a hologram that changes from a '£' symbol to the number '1' when the coin is seen from different angles.
- Micro-lettering – it has very small lettering on the lower inside rim on both sides of the coin: 'One Pound' on the obverse 'heads' side

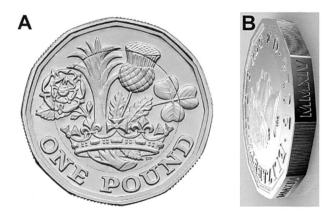

**Figure 12.8** (A) The new 12-sided British £1 coin, introduced in March 2017. (B) Milled edges ribbed or 'grooved' on alternate facets.

and the year of production on the reverse 'tails' side, for example 2017.

- Milled edges – it has grooves on alternate sides.
- Hidden high-security feature – an undisclosed high-security feature is built into the coin to protect it from counterfeiting in the future.

Only time will tell as to how effective the above features are in preventing illegal manufacture of the new £1 coins.

However, counterfeiting of coinage is not limited to the British pound. Struck counterfeit euro coins started appearing in April 2002 with the detection, in Germany, of a few two-euro counterfeits having Belgian and German national faces. The highest denomination euro coin, the €2, is bi-metallic (two different metals fused together), also showing two different colours (Figure 12.9A) to hinder counterfeiting. The inner part (gold colour) is made of three layers: nickel brass, brass and nickel brass. The outer silver-coloured part is made of cupro-nickel. The coin has a diameter of 25.75 mm, a 2.20 mm thickness and a mass of 8.5 grams. The edges of the coin also vary between national issues, but most of them are finely ribbed with edge lettering (Figure 12.9B). When introduced, the coins were considered counterfeit-proof. However, in 2015, a European Commission report suggested that some 365,000 (out of an estimated total of over 10 million fake euro coins in circulation) had been identified and withdrawn by the authorities since the introduction of the common European currency in 2002. As an idea of the probable scale of euro counterfeiting, in December 2014 it was reported that Italian police had uncovered the biggest haul of counterfeit coins in the history of the euro; the haul was estimated to contain €500,000 worth of fake euro coins, probably forged in China.

**Figure 12.9** (A) The two-euro bi-metallic coin. (B) Finely ribbed edge along with edge lettering.

## 12.2 Staged or Fraudulent Road Traffic Accidents Within the UK[2]

Criminal fraud is calculated to cost insurers in the UK around £1 billion per annum, with car insurance fraud costing a staggering £200 million a year. Although it is clear that motor insurance swindles are becoming big business, all costs are directly passed on to motorists through higher premiums and excesses. Insurers are now attempting to clamp down on fraudulent claims with the opening of a Motor Insurance Anti-Fraud and Theft Register (MIAFTR2) introduced in the spring of 2005. Here, UK insurers log details of all motor insurance claims on the database, thus enabling insurers to identify potentially fraudulent claims where vehicles are falsely reported as stolen, or where multiple claims are fraudulently made for a single loss-event.

The aim of such a register is to dissuade individuals – and criminal groups – from seeing insurers as a 'fair game' for fraud (Section 12.2.1). However, there is a growing trend where fraudsters are staging car crashes that involve innocent drivers and businesses (Section 12.2.2) and acquiring large sums of money from false compensation claims. Investigators have found that highly organised gangs are 'staging' accidents across the country, submitting claims for damage to the car, loss of earnings, legal costs and compensation for injuries, including whiplash – to the driver and non-existent passengers – all backed up by 'witnesses' (Section 12.2.3).

Collaborating with motor engineering assessors, it is increasingly common for the failure or forensic engineer to be instructed to offer opinion on suspect claims. The engineer will endeavour to evaluate the genuine from that of the 'staged' accident based on available mechanical evidence, documentation and statements.

### 12.2.1 Deliberate 'Damage'

A Mercedes 300D had allegedly hit a white people carrier when losing control on a greasy road surface (Figure 12.10A). The third-party vehicle had suffered only minor damage, whereas the Mercedes had the appearance of having collided with a solid upstanding such as a bollard or fire hydrant. The vehicle had been recovered and transported to a storage facility where a local independent engineer inspected it. His report concluded that the extent of damage sustained had made repair of the vehicle an uneconomic proposition – the vehicle should be written off, with the owner being compensated for its full market value. However, when inspected by an assessor, the general interior condition of the vehicle – steering wheel shiny, seats worn, peddle rubbers worn, seat belts frayed, leather stitching coming undone, etc. – along with only 46,000 miles recorded on its odometer, aroused his suspicion. An examination by a forensic engineer then followed, where regular vertical

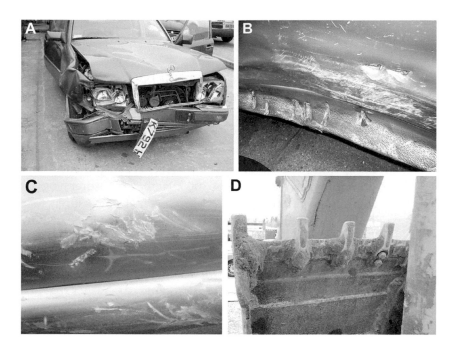

**Figure 12.10** (A) Frontal damage to Mercedes. (B) Vertical score marks at 30-cm pitch. (C) Evidence of yellow paint at damage sites. (D) JCB (back-hoe) bucket with a 30-cm tooth pitch.

score marks at a 30-cm pitch (Figure 12.10B), and yellow paint (no sign of white paint) on damaged areas (Figure 12.10C), was found. On investigation, a JCB (back hoe) was noticed at the rear of the compound. Measurement of its bucket tooth pitch (Figure 12.10D) was found to be 30 cm, and furthermore, the JCB was yellow! It transpired that both parties were working with the recovery storage company and deliberately inflicting damage to old vehicles in order to fraudulently recover an inflated insurance value. Furthermore, there were at least four other old vehicles in the facility, perhaps waiting their turn to 'sustain damage'.

## 12.2.2 Car Hire Insurance Fraud

A claimant had hired a car for the weekend, opting for the hire company's fully comprehensive insurance policy. During the course of the weekend, the claimant had an accident where he had run into the rear of another car as a result of slipping on snow. Damage included buckling the chassis of the third-party vehicle, effectively writing it off. The third-party driver was also seeking compensation for whiplash from the hire company's insurers. Subsequent investigation showed extensive damage to the third-party vehicle (Figure 12.11A and B), with little or no damage to the front of the

**Figure 12.11** (A) Third-party vehicle, showing damage from rear impact. (B) Closer view of damage with rear bumper removed. (C) Car hire vehicle, with minor damage to badge area and chrome trim. (D) No distortion to the engine compartment, radiator or fan.

hired vehicle (Figure 12.11C and D). With suspicions raised, it only took a few enquiries to find that the third-party vehicle had been declared a total loss some months prior. For a few pounds' outlay on vehicle rental, insurance cover and a scrap vehicle, the two conspirators were attempting to defraud the hire company's insurers of vehicle replacement cost for the third-party scrap vehicle, in the order of some £4,000. In the UK, it is becoming a common occurrence for hire car policies to be abused in this manner.

### 12.2.3 Staged 'Accidents'

A car had driven out of a private driveway on a side road in Derbyshire, UK. When the owner got out of his vehicle to close his gate, a white minibus hit the side of his car with a glancing blow (sketch, Figure 12.12A). The minibus driver then complained of back pains and accused him of careless driving. Shortly afterwards three passengers emerged and said that they, too, had been hurt in the accident. A few days later the car driver's insurer received 12 personal injury claims. The claims included ones from the driver and passengers of a second minibus that had apparently run into the back of the first.

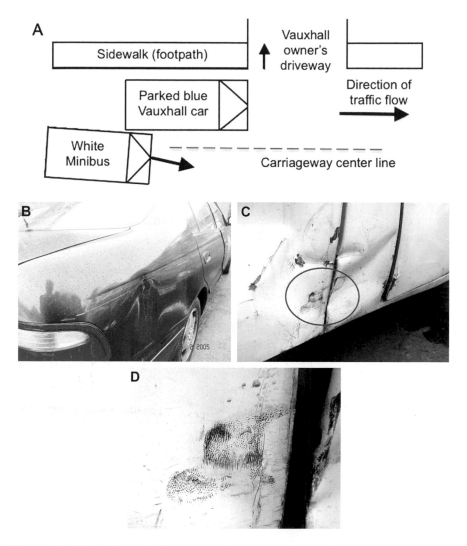

**Figure 12.12** (A) Sketch plan of 'accident'. (B) Minimal damage to rear off-side of Vauxhall. (C) Damage to front wing, pillar and door edge of minibus. Note: damage height, total lack of feathering expected in a moving incident with such extensive damage, indents and folds with no feathering. (D) Close view of damage with black markings that may well have been inflicted with a rubber mallet.

Straightforward observation revealed absolute minimal damage to the Vauxhall rear off-side (Figure 12.12B). There was a total lack of feathering expected in a moving incident with such extensive damage to the front wing, pillar and door edge of minibus (Figure 12.12C). A closer inspection of the minibus damage revealed black markings consistent with having been inflicted with a rubber mallet (Figure 12.12D).

From all observed evidence, it could be stated that this particular incident was a classic example of a cleverly staged accident. The gang had been monitoring the car driver's daily movements for weeks and knew exactly when to strike, before putting in a collective insurance claim of £45,000.

## 12.3 Shotgun Pellets[3]

A drug dealer was murdered, shot with a double-barrelled shotgun. Two cartridges, similar to that shown in Figure 12.13A, had been discharged at close range, the pellets from the first shot hitting him in the chest and those from the second in the lower abdomen. A week or so before the murder he had received a threatening letter about defaulting on payments to which had been taped ten shotgun pellets, found to be the same size as those of the 48 recovered from the chest wound (Figure 12.13B). Pellets recovered from the

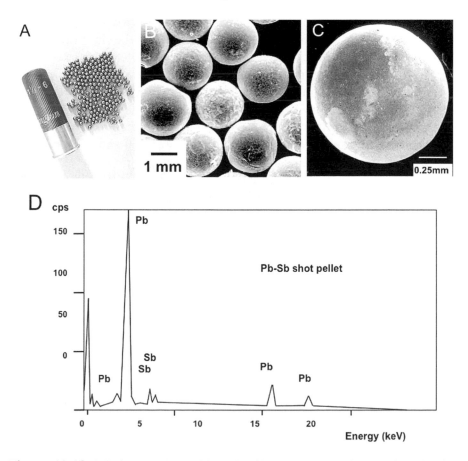

**Figure 12.13** (A) Shotgun cartridge and pellets similar to that used in the discharged weapon. (B) Shotgun pellets similar to the 48 recovered from the victim's chest wound. (C) One of ten pellets individually photographed. (D) Chemical analysis of pellet.

lower wound were larger. The prime suspect arrested was found to own a double-barrelled shotgun and there was evidence to suggest this person was the sender of the letter, although this was vehemently denied.

The ten pellets on the letter were individually photographed (Figure 12.13C), weighed and accurately analysed for 12 elements, including impurities usually present at low levels in lead shot as well as the principal alloy elements, antimony and arsenic (Figure 12.13D). Both harden the shot, but the arsenic also increases surface tension of the molten metal so that it will form spherical droplets in the shot-making process. These results were compared statistically with similar analyses of 20 pellets from the chest wound. Mean weights of these two batches were found to be 0.1107 and 0.1088 g with standard deviations of 0.0077 and 0.0083 g, respectively. The forensic science laboratory that conducted this investigation had built up a database of over 600 analyses of shotgun pellets covering the full range of commercially produced shot sizes. The laboratory reported there was a closer match between the pellets from the letter and the ones recovered from the chest wound than could be found when matching either group of these pellets with its database. On this basis, the laboratory concluded that the murder pellets were from the same cartridge as those on the letter. At trial, this conclusion was challenged, by three different experts.

A metallurgist pointed out that the various alloys are prepared and their compositions adjusted in a furnace holding between 50 and 100 t of metal. This is cast into 25 kg ingots and sent to various shot-makers, who re-melt the ingots and pour the molten metal through a sieve at the top of a tower. As it falls, the molten streams break up into droplets and solidify. Although considered individually, quantities of impurities were close in the two groups of pellets analysed; when certain critical ratios were compared there were significant differences. For example, while the lead alloy is held in the molten state, copper, being completely insoluble and less dense than lead, tends to rise to the surface and will be higher in the first ingots to be cast than the last, whereas bismuth, being soluble in the liquid lead and having no tendency to segregate, will tend to oxidise near the surface and will thus be higher in the last ingots to be cast than the first. Individual ingots from the same cast will thus vary slightly, particularly when their analyses are reported in parts per million. The average copper/bismuth ratio in the ten pellets was slightly more than twice that in the pellets recovered from the wound. This suggests that even though these two batches of pellets might have originated from the same cast in the refining furnace they had been poured at different times during the shot-making process and could well have been made by different shot-makers who had purchased their ingots from the same refiner.

A single kettle of liquid alloy at the shot-maker's would hold typically 30 to 50 t of metal and would be continually replenished with additional ingots as liquid was drawn off. A total of 50 t of alloy would be enough to make 500 million pellets if each pellet weighed 0.1 g. Under cross-examination the analyst agreed that his data showed that the two samples of pellets most probably came from the same source but not necessarily from the same cartridge.

The firearms expert explained that although there would be no difficulty in un-crimping the end of a cartridge, removing the wadding, taking out ten pellets and then re-crimping, he did not believe the chemical analysis did, or could, prove the ten pellets were taken from the same cartridge as the other 20. He pointed out that the statistical mean weight of the two batches differed slightly despite exhibiting almost identical standard deviations and that there appeared to be two distinct groups even within the batch of 20.

The retired shot manufacturing manager explained how the shot is made by melting 50 t or so of ingots in a kettle (so-called because the metal poured is drawn from the bottom), and is poured through a screen at the top of a tower. The streams of liquid metal break up into droplets and solidify as they fall down and finish up in a water bath that quenches them. They are then polished and rolled over a series of sloping plates that reject the pellets that are not truly spherical, which are returned to the kettle to be re-melted. Ingots of different alloys are segregated in their stockyard, but there is no certainty that all the ingots charged to the kettle on the same day will have come from the same supplier or have been delivered at the same time. Finished pellets are sent as small orders to various cartridge makers, and they sell boxes of cartridges to a wide variety of outlets. Thus sportsmen thousands of miles apart could be using shotgun cartridges containing pellets made from the same kettle of metal in the same shot manufacturing plant on the same day.

As the Crown's case against the accused depended almost entirely on the analyst's conclusion that the pellets on the letter were from the same cartridge as those in the fatal wound, the jury not surprisingly reached a verdict of not guilty.

## 12.4 Metal Theft[3]

Nonferrous metals are of considerable value, and many small enterprises make a living by collecting scrapped items at source and selling them to scrap merchants who sort and re-grade them before sending them to refineries. At the very bottom of this pyramid are individuals who ask no questions about the source of their scrap and who make their living by selling it to the highest bidder. This case concerns one such individual who made a serious mistake.

An electroplating firm ordered 9 tons of copper cathode from a copper refinery (Figure 12.14A). This is a pure form of copper that is produced only by electrolysis; the cathodes are somewhat irregular, 5 to 10 mm thick, with a smooth surface on one side and numerous tiny nodules standing proud on the other, as illustrated in Figure 12.14B. This sample is a piece of cathode of similar size to those stolen and is much larger than the guillotined pieces that it was to have been cut into for the electroplating process. It is included here to show the characteristic nodules on the surface as well as along the edge at the right-hand side. A freshly produced cathode is a bright salmon pink colour when it leaves the refinery but oxidises to a dull brown after a few weeks. For use as anode in barrel electroplating the cathode has to be reduced to small pieces about 25 mm$^2$, so the electroplating firm needed to send its metal to another firm that had a powerful guillotine. The electroplating firm arranged to have this done on a Monday but, because the consignment arrived late on Friday afternoon, the truck was driven inside the plating firm's yard and the gates locked. It was not visible from the outside. On Monday morning when the gates were opened up the truck was gone, and with it 9 tons of copper worth more than £42,500 ($56,000) in terms of today's money.

The nonferrous metal trade has a system of notifying all dealers and metal merchants of a theft and the nature of the material stolen. Some four to five days after the theft a small van drove to a scrap dealer 250 miles away from where the truck was taken. The driver offered to sell a quarter ton of scrap copper, saying he had more if the price was good enough. The scrap merchant realised that in the load of mixed copper – a few old hot water cylinders, pipes, wire and so on – were pieces of cathode, so he purchased that load and offered to buy all that the driver could deliver. He also notified the police. Within an hour the driver was back with another quarter ton, most of which was cathode. He was arrested, and the police accompanied him back to a small yard where they found the stolen truck containing just over 8 t of copper cathode.

The man arrested claimed that he was just a small-scale scrap dealer, who had bought all the metal in small lots from local housing estates and small industrial units. It hardly needed a metallurgist to identify it as ex-refinery cathode, which is totally different from swarf or clippings arising from manufacturing processes, or discarded items from plumbing systems

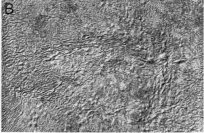

**Figure 12.14** (A) A stack of copper cathodes similar to those stolen. Notice the side lug hooks. (B) Surface of the copper cathode. Notice nodules and surface and wrinkling.

and electrical wiring. The thief was obviously unaware that the only uses for cathode copper are in electroplating and for re-melting and alloying with other metals. In no way could the metal in his possession have been collected from housing estates and light industry. It was particularly incriminating that the total amount was 9 t and was just starting to develop the brown oxide coating. It could only have emanated from the one refinery still operating in the UK, and its colour was consistent with it having been produced some three to four weeks before it was recovered. Further inquiries revealed that the man arrested had been living with his family in a caravan not far from the refinery at the time of the theft. The suspect was charged with theft and subsequently served a custodial sentence.

## 12.4.1 An Additional Metal Theft Case[3,4]

Another example of metal theft concerned a factory manufacturing extruded copper alloy tubing for steam condensers. One evening, the factory suffered a break-in and a large quantity of metal was stolen. The metal in question was a rather unusual brass of 76Cu-22Zn-2Al, and the factory was the only one in that area using aluminium bearing brasses in its manufacturing process. At that time, commercial brasses used for steam condensers and general manufacture of tubing were 70Cu-30Zn and 70Cu-29Zn-1Sn. A suspect who worked in a nearby iron foundry was apprehended in connection with the robbery. He claimed that he had spent the evening in question drinking with his four buddies, and had never been near the brass tube factory. However, swarf was dug out of the suspect's boot, and compared with a piece of swarf from the floor around a large lathe used to skim billets prior to extrusion. The SEM comparison is shown as a montage in Figure 12.15, where the top half

**Figure 12.15** An SEM montage of two separate pieces of swarf. The top half was recovered from the boot of a suspect, and the bottom half was recovered from the crime scene. Note the perfect tool wear pattern match between the two pieces.

is swarf dug out of the suspect's boot and the bottom half is swarf recovered from the crime scene. As a lathe tool wears, it develops a serrated cutting edge that will produce characteristic score marks rather like the marks on a bullet fired down the barrel of a particular rifle. It can be seen that there is a perfect wear pattern match between the two pieces of swarf that can be considered comparable to a 'fingerprint' in conventional detective work. Elemental X-ray analysis on the SEM confirmed that the sample removed from the suspect's boot was an aluminium bearing brass, having a composition of 76Cu-22Zn-2Al. Therefore the combination of wear fingerprint and alloy composition was more than sufficient evidence to place the suspect at the scene of the crime. It later transpired that the suspect *and* his drinking buddies were perpetrators of the robbery.

## 12.5 Chain from Murder[5]

A woman abruptly left her group of friends in a bar, allegedly because she was being pestered by a man. She walked home and locked her door, securing the chain, but the following day she was found raped and brutally murdered. The chain was hanging broken from the door handle. The man who had been pestering her was known to her friends and lived in the same district. He was questioned by police and claimed that he had walked home soon after the woman left and had watched a late-night movie on television. The police searched his premises and found a small crowbar under sacking at the side of his house with four links of brass chain twisted tightly round the hooked end, shown close up in Figure 12.16A. Figure 12.16B shows part of the broken chain that the woman had used to secure her door alongside the four links twisted round the end of the crowbar.

Forensic examination revealed that the chain was composed of welded links made from brass wire. The diameter of the wire and the pitch of undeformed links away from the fractures of both exhibits were practically identical. The chemical compositions of one link from the chain on the door and one from the crowbar were not identical but were both within the specified range for 70/30 copper–zinc alloy.

The man was arrested and charged with murder. DNA tests later confirmed that he had raped the woman. At trial, the defence sought to claim that the scientific tests had not established beyond all doubt that the links on the crowbar were indeed from the same length of chain used to secure the door. At this point an independent engineer was called. He found that the links had been electrically welded on the same type of machine, but the damning evidence was that the torsional fracture surfaces where the wires had been twisted to destruction were perfect matches to the crowbar and the chain still hanging from the door.

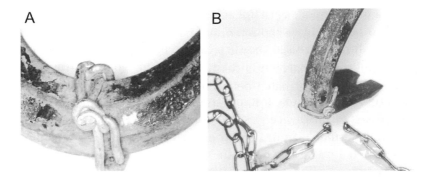

**Figure 12.16** (A) Links of brass chain wrapped around a crowbar. (B) Twisted links of ornamental brass chain used to secure door.

## 12.6 Coin Box Busters[5]

For security reasons, it is not possible to give a full description or include photographs of the items referred to, but this case is a good example to show the role of the engineer with experience in manufacture in the criminal forensic field. Essentially they were three ingenious, though crudely constructed, devices for breaking open coin boxes in gaming machines, public telephones and similar situations.

A police detective produced the first device, found in the trunk of a car seen driving away from a broken telephone box that had been kept under observation. The driver was found to be in possession of a large quantity of small coins and a device of the type described below, so he was arrested and charged with criminal damage and theft. Suspecting that he was one of a gang systematically breaking open coin boxes in different areas of outer London, the police searched the vehicles of other suspects and found two other similar devices concealed in the spare wheel compartments. Both suspects claimed the devices were used in car repair work and denied any knowledge of each other or of the person arrested.

The detective in charge of the investigation wanted to know whether there was anything in common in the design of the three devices and in the way they had been constructed. There certainly were similarities. All three devices were of similar design, based on a hydraulic car jack widely obtainable from automotive accessory stores. In this instance, all three were of the same make and had the same colour paint on the body of the hydraulic cylinders. Parts of the jack had been removed and replaced by two steel bars of square section, slightly bent in the middle and with sawn-off ends to which were crudely welded the cutting ends from two cold chisels. These formed a pair of claws capable of piercing the sides of the coin boxes. Surface score marks from the manufacturing operation on the bars of all three devices were an exact match, and the microstructures and hardnesses were the same.

The bars had obviously all been cut to the same length from the same bar stock and had been bent to the same angle while heated with a gas welding torch, which had oxidised the surface to a blue-grey colour. The chisel ends were slightly different, but all had been cut to the same length from 1-in. (25-mm)-wide cold chisels. As each chisel has only one cutting end, six would have been required to construct three pairs of claws.

Another part of the device was a length of M18 screwed steel rod. Such rods are readily available from DIY stores in 1-m lengths. On each device, these rods were 0.5 m long and had an M18 nut welded on one end so that they could be turned with a spanner to open and close the jaws. The hydraulic ram had been converted into a punch by welding on a stub cut from the top of a high-speed steel twist drill. All three were of slightly different diameters, increasing consecutively by 1/32 in. apparently taken from a sized set of HSS twist drills, but they were cut to similar lengths and rather poorly welded to the rams.

It was clear that all three devices had been constructed in the same workshop using the same design concept – it is unlikely there would have been a drawing or a materials specification – and with parts obtained from the same sources. On each device, one pair of welds had been made with a gas torch and the others with electric arc welding equipment, using flux-coated mild steel electrodes. All the welds exhibited similar imperfections and looked as if they had been executed by the same person, who was by no means a skilled welder.

The police had established that a considerable number of coin boxes in different localities had been broken into with these particular devices, mainly because the holes in the sides matched the claws formed by the cold chisels and the locks had been pushed in by a circular object of similar diameters to the twist drills. The engineering evidence proved complicity in their design and manufacture and led to the conviction of all three suspects found with the devices in their cars.

## 12.7  Re-Melting Beer Barrels[1,3]

A number of public houses in a large city had experienced a spate of thefts of aluminium beer barrels. Whole batches of empty barrels placed in their yards awaiting collection by the brewery had disappeared in the early hours of the morning. In a suburb of the same city, a large warehouse and yard were occupied by small firms in the metal recycling trade, among which were vehicle dismantlers, one of whom had a small metal melting unit used for separating assemblies that contained aluminium alloy parts. As a vehicle was taken apart components containing alloy were placed onto a sloping hearth under oil burners so that the aluminium alloy melted and ran down into a bath while the iron and steel were left on the hearth to be raked out from time to time. The molten alloy was cast into ingots and sold to a metal refiner.

The police suspected this warehouse yard might have been the destination of the stolen beer barrels, so they kept watch in a number of ways, including helicopter surveillance, but were never able to observe any barrels on vehicles entering the site. (It was later learned that the reason for this was that the barrels were carried in closed vans and the sheet metal gates of the yard were always kept locked 'to prevent entry by nosey parkers and casual passers-by'.) Although most of the firms on the site were car dismantlers, one of them ran a one-man business stripping old electrical cable to recover lead and copper. He owned the lease of the site and lived a prosperous lifestyle, despite the small throughput of his business. Early one morning the police raided the site and found the beer barrel melting process in full swing, with a closed van still partly filled with barrels backed up to the door and five men, including the cable stripper, apparently engaged in operating the furnace and casting 50-kg ingots. All five were arrested, ordered to strip and change their boots, and the burners were shut down by the cable stripper. However, the cable stripper claimed he had nothing to do with the melting process but had merely attended to open the yard and carry on with some cable stripping in a different part of the building while the others were working. He admitted that he did this from time to time but had no idea where the barrels came from and said this job was usually done at night because the furnace was in use by the other firms for car parts during normal daytime hours.

His association with the beer barrel melting became clear after forensic examination of metal particles found on the soles of the boots and on the clothes he was wearing when arrested.

All the suspects' clothes were brushed, and a number of tiny particles of metal were collected. In addition, there were numerous particles in the heels and soles of their boots, such as illustrated in Figure 12.17. Examination under the SEM revealed that all were rounded particles characteristic of splashes and droplets such as those emitted by molten metal running into open moulds. Every one of 20 samples taken from boots and clothing of all five men were found to be aluminium alloy containing low percentages of magnesium and silicon. This composition spread would include the alloy used for beer barrels, but not the ones commonly used for automobile castings, which usually contain greater amounts of silicon. Most significantly, no particle recovered from the cable stripper's clothing or boots was copper or lead but, even if there had been any from this source, they would have taken the form of mechanical slivers or clippings, not solidified droplets.

The clothing exhibited numerous holes formed by hot particles landing on them, particularly below the trouser knees. There were two holes joined together in the side pocket of a nylon jacket worn by the cable stripper. These had been formed by one, or possibly two, splashes of hot metal striking the

**Figure 12.17** (A) The heel of a suspect's boots. (B) A closer view showing beads of aluminium alloy embedded in the heel, picked up by walking on a floor close to the metal casting operation.

fabric and melting through the nylon. The lower leg of a pair of corduroy trousers had a series of holes, formed where small hot particles had landed. Although most of the above observations are essentially scientific, the engineering dimension becomes important in identifying the particles and clothing damage as typical of that expected when workers are standing close to foundry operations. Few scientists have ever observed the splashing that occurs and the distance droplets may carry when a stream of molten metal poured from a furnace first strikes a launder and then runs into an open ingot mould, or the fine spray of droplets that are thrown up as it solidifies when the metal has picked up hydrogen due to being melted under reducing conditions. In this instance, the damage to the clothing worn by the cable stripper clearly established that he must have been standing very close to the stream of molten metal as the ingots were being cast.

## References

1. Gagg, C.R. and P.R. Lewis, Counterfeit coin of the realm - Review and case study analysis. *Engineering Failure Analysis*, 2007. 4(6): pp. 1144–1152.
2. Gagg, C. and D. Harris, Current status of road traffic accidents within the UK - Case study analysis. *International Conference on Engineering Failure Analysis (ICEFA II)*. 2006, Toronto, Canada: Elsevier.
3. Lewis, P.R., K. Reynolds and C.R. Gagg, *Forensic Materials Engineering - Case Studies*. 2004, CRC Press.
4. Gagg, C.R. and P.R. Lewis, Wear as a product failure mechanism – Overview and case studies. *Engineering Failure Analysis*, 2007. 14(8): pp. 1618–1640.
5. Lewis, P.R., K. Reynolds and C.R. Gagg, *T839 Forensic Engineering*. 2000, Open University Post-Graduate Course, Department of Materials Engineering.

# Author's Closing Remarks

As emphasised throughout this book, the work of a failure detective (forensic analyst) demands a wide range of expertise and experience (both practical and academic), along with constant exposure to specialists from many other walks of life. Additional pre-requisites demanded of the practitioner will include firm professional ethics, an open mind and a commitment to continued professional development. As a tool, forensic analysis will provide an iterative feedback loop to further inform the processes of design, manufacture, operation and maintenance. As such, forensic engineering (and failure) analysis can be considered an essential design and management tool – one that will help minimise, or prevent, future disasters, accidents and failures. There is, however, immense scope for improvement in the arena, particularly when considering the increased transparency of publicising poor product performance. Probably one of the best ways of preventing future accidents or disasters is to promulgate details of previous incidents or casework. In the last 20 years, there has been a rapid growth in both failure analysis and forensic science literature, with more volumes of case studies, journal papers, dedicated conferences and, increasingly, information posted on the Internet. There is no doubt that journals, such as *Engineering Failure Analysis* (EFA) in the UK, have made great strides in publishing both revealing and incisive case studies from investigators. This expansion in recent years is a reflection of the growing desire to publish detailed case studies in an attempt to limit further re-occurrence of identical failures (if not eliminate them altogether).

As a penultimate point, I would suggest from experience that there seems to be an irresistible trend toward producing paper-based or computational engineers who have little or no feeling for the real (engineering) world. Time and again the cry is heard, 'it (the answer) came from a computer, so it must be correct'. However, 'We learn best by doing' is an old adage that, I would suggest, any trainee failure or forensic engineer would do well to encompass, along with another old favourite, 'We learn best from our mistakes' (something that I most certainly have experience of, with embarrassment alone being my greatest teacher). Therefore, to peers and colleagues heavily engaged in both forensic engineering practice and/or teaching, I would say, 'get in there, roll up your sleeves, get down, get dirty, and find out how things *work*'. Only then will the seasoned engineer and/or 'apprentice' be in a position to *intuitively* understand the most likely point (and reasons) for any *failure*. Experience will then dictate that any deviation from the expected norm would be as a result of a cause or causes unknown – thus requiring further in-depth investigation.

Finally, when considering the theme of this manuscript, I now think it fitting to modify my opening quotation:

This book has a beginning but not an ending; it is, like both the writer and subject arena, 'a work in progress'.

# Index